AF326498

Quantum
Algorithms and Applications

A Scaffolding Approach

VOL. 1 (PARTS I + II)

Peter Y. Lee

Ran Cheng

Huiwen Ji

ISBN 978-1-961880-11-5 (vol 1+2, ebook, color)

ISBN 978-1-961880-12-2 (vol 1, paperback, b/w)

ISBN 978-1-961880-13-9 (vol 2, paperback, b/w)

ISBN 978-1-961880-14-6 (vol 1, hardcover, b/w)

ISBN 978-1-961880-15-3 (vol 2, hardcover, b/w)

Library of Congress Control Number (LCCN) 2026938313

First edition, March 2026

This document is typeset using LaTeX.

Quantum circuit drawings are created using the yquant package from
https://github.com/projekter/yquant.

Publisher website: https://polarisqci.com

About This Book

Quantum computing is emerging as a new paradigm for computation, one that can exploit uniquely quantum resources to solve certain problems more efficiently than is known to be possible classically. Recent progress in hardware, error mitigation and correction, and experimental demonstrations has sharpened a practical question: where can quantum devices deliver reliable advantage and sustained utility?

At the heart of this question lies quantum algorithms. Since Shor's discovery that factoring can be solved in polynomial time on a quantum computer, the field has grown into a toolkit of recurring primitives: Fourier-based methods, amplitude amplification and estimation, Hamiltonian simulation, and modern polynomial-approximation frameworks such as block encoding, quantum signal processing, and quantum singular value transformation.

This book develops these algorithmic ideas with an explicit balance between *algorithms* and *applications*. You will learn how core primitives are assembled into workflows for major application areas, including physics and chemistry simulation, optimization, quantum machine learning, and quantum methods for linear systems and differential equations. Throughout, we emphasize what is being computed, what must be assumed about data access and outputs, and how resources and verification shape real performance.

Following the *Scaffolding* approach, topics are introduced in a carefully staged progression, reinforced through in-text exercises and end-of-chapter problem sets. Highlight boxes, reference tables, and recurring design patterns help you keep moving through dense material while building a coherent mental model.

A guiding theme is conceptual durability. Programming tools and platforms will evolve, but principles travel well. The aim is not only to present algorithms, but to cultivate understanding that lets you translate a problem statement into a quantum workflow, identify dominant costs, and evaluate advantage claims with clarity.

This book is written for senior undergraduates, beginning graduate students, and practitioners who want both conceptual grounding and technical competence. Whether you are encountering quantum algorithms for the first time or organizing existing knowledge into a unified structure, the goal is to support confident and long-lasting mastery of quantum computation.

About the Authors

Dr. Peter Y. Lee: Holds a Ph.D. in Electrical Engineering from Princeton University. His research at Princeton focused on quantum nanostructures, the fractional quantum Hall effect, and Wigner crystals. Following his academic tenure, he joined Bell Labs, making significant contributions to the fields of photonics and optical communications and securing over 20 patents. Dr. Lee's multifaceted expertise extends to educational settings; he has a rich history of teaching, academic program oversight, and computer programming.

Dr. Ran Cheng: Earned his Ph.D. in Physics from the University of Texas at Austin, with a specialization in condensed matter theory, particularly in spintronics and magnetism. Following a postdoctoral position at Carnegie Mellon University, he joined the faculty at the University of California, Riverside, where he was honored with the NSF CAREER and DoD MURI awards.

Dr. Huiwen Ji: Holds a Ph.D. in Chemistry from Princeton University. She is a materials chemist whose research spans solid-state functional materials, quantum materials, and energy-related materials. After appointments at the University of California, Berkeley and Lawrence Berkeley National Laboratory, she joined the University of Utah as a faculty member in Materials Science and Engineering. Her honors include an NSF CAREER Award.

Contents

Level Indicators

Unmarked content provides the main pathway through the book and includes the core algorithmic ideas, standard tools, and representative applications.

✳: More technically demanding material that often requires stronger mathematical preparation or more background in physics and algorithm analysis. Assumes familiarity with the unmarked core material.

✳: Optional exploration beyond the core narrative, including deeper case studies and specialized topics.

I Foundations of Quantum Algorithms

IV   Other Applications

<table><tr><td>V</td><td>Supporting Materials</td></tr></table>

Preface

The *Scaffolding* Series

This book is part of the *Scaffolding* Series, a set of textbooks designed to provide a systematic and pedagogically grounded route into quantum computing. The series currently includes:

- *Mathematical Foundations of Quantum Computing: A Scaffolding Approach*
- *Quantum Computing and Information: A Scaffolding Approach*
- *Quantum Algorithms and Applications: A Scaffolding Approach* (this book)

Each book can be read on its own, but together they form a coherent progression: mathematical foundations, core quantum computing concepts, and then algorithmic primitives and applications.

This book is written for senior undergraduates and beginning graduate students. To support readers with different backgrounds and goals, the text includes level indicators and navigation aids described below.

Parts of the Book

This book is organized into four parts:

- Part I: Foundations of Quantum Algorithms. We establish the computational viewpoint used throughout the book, including algorithmic efficiency, the circuit and query models, and the complexity language needed to reason about scaling, precision, and success probability.

- Part II: Core Quantum Algorithms. We develop the core algorithmic primitives that reappear across quantum speedups: Fourier-based methods (quantum Fourier transform and quantum phase estimation), period finding and Shor's algorithm, amplitude amplification and estimation, and modern polynomial-approximation frameworks centered on block encodings (including linear combination of unitaries, quantum signal processing, and quantum singular value transformation).

- Part III: Quantum Simulation in Physics and Chemistry. We show how the toolkit is used for simulation tasks motivated by physics and chemistry. The emphasis is on how physical structure becomes computational structure through Hamiltonians, encodings, and measurement strategies, and on how accuracy, cost, and feasibility trade off in practice.

- Part IV: Other Applications. We survey additional areas where the same primitives recur, including optimization, quantum machine learning, and quantum methods for linear systems and differential equations, with attention to input/output models, verification, and resource accounting.

The appendices provide supporting reference material, including brief refreshers on quantum computing fundamentals and curated overviews of problem landscapes with quantum potential.

For print production, this single book is issued as two physical volumes due to length limits: Volume 1 contains Parts I–II, and Volume 2 contains Parts III–IV and the Appendices.

Recommended Use

Readers are expected to be comfortable with linear algebra and basic quantum computing concepts. For a refresher, we recommend the first two books in the series, *Mathematical Foundations of Quantum Computing* and *Quantum Computing and Information*.

For a two-semester course, a natural path is to cover most chapters in sequence and assign a substantial portion of the exercises and problem sets. For a one-semester course, instructors can select modules based on emphasis (for example, algorithmic primitives, simulation, or applications). Self-learners can also treat the book as a reference: the highlight boxes and end-of-section summaries are designed to support selective reading.

The Scaffolding Approach

Quantum computing and information spans mathematics, quantum physics, and algorithmic design. The scaffolding approach used in this series aims to reduce barriers by building understanding in stages and revisiting core ideas as the reader's toolkit grows. The main strategies in this book are:

- Progressive Learning: new ideas are introduced through concrete examples before abstraction and scaling arguments.

- Spiral Reinforcement: key primitives reappear across chapters in multiple contexts, strengthening retention and transfer.

- Active Engagement: exercises and problem sets are integrated to reinforce skills through application.

- Cognitive Load Management: diagrams, tables, and highlight boxes summarize structure at conceptual choke points.

A guiding theme is conceptual durability. Programming tools and platforms will evolve, but principles travel well. The goal is to help readers translate a problem statement into an algorithmic workflow, identify dominant costs, and evaluate assumptions and claims with clarity.

Navigation Aids

To assist readers, the book incorporates the following features:

Highlight Box

Highlight boxes summarize key takeaways at conceptual choke points and near the ends of sections. They may record a frequently used identity or pattern, clarify what to remember before moving on, or provide a big-picture interpretation of a construction.

Exercise 0.1 Exercises are interspersed throughout the text to sharpen your skills and reinforce your understanding of the material. Each chapter concludes with a Problem Set, providing more challenging problems that encourage deeper engagement.

R This box outlines prerequisites for the current chapter or section, which may include topics from other books in the *Scaffolding* Series.

i This indicator is used for tips, alerts, connections between concepts, and pieces of advice.

Info Box
This box provides supplementary context, associated concepts, and additional information.

Level Indicators

To support readers with different backgrounds and goals, some sections are marked by level indicators.

Unmarked content provides the main pathway through the book and includes the core algorithmic ideas, standard tools, and representative applications.

Items marked with $*$ contain material that is more technically demanding, often requiring greater mathematical maturity or more background in quantum mechanics, Hamiltonian methods, or algorithm analysis.

Items marked with $*$ are intended for further exploration. They include optional extensions, specialized topics, deeper case studies, and broader perspectives beyond the core narrative.

Acknowledgements

We extend our sincere thanks to Ms. Liang Zhou for her creative work in designing the book cover and for her thoughtful visual interpretation of the subject matter.

We are also deeply grateful to Ms. Nathalie Chiao, whose editorial insight greatly improved the manuscript's readability, coherence, and clarity.

To our colleagues—Dr. James M. Yu, Dr. Jason Wang, and Dr. Joseph Zhao— your thoughtful discussions, encouragement, and feedback have been invaluable in shaping this book into a more practical and effective educational resource.

We also offer our sincere thanks to our expert reviewers, whose comments and suggestions helped strengthen both the rigor and accessibility of this text (listed in alphabetical order by last name):

- Omar Alnaseri, Adjunct Professor at DHBW, Germany; Researcher in Quantum Communication Systems and Quantum ML/AI; Senior Member of IEEE

- Jacob Biamonte, Professor, Principal Ministry Research Chair in Quantum Computing (MEIE), ÉTS Montréal, Université du Québec

- Daan Camps, Researcher in Advanced Technologies Group, NERSC, Lawrence Berkeley National Laboratory

- Steven Frankel, Rosenblatt Professor, Faculty of Mechanical Engineering, Technion – Israel Institute of Technology

- Aram Harrow, Professor of Physics, Massachusetts Institute of Technology

- Jaewan Kim, National Distinguished Research Fellow, Korea Research Institute of Standards and Science (KRISS); Professor Emeritus, Yonsei University and Korea Institute for Advanced Study (KIAS)

- Keith King, former communications engineering leader with experience supporting the US government

- Zlatko Minev, Google Quantum AI; CIFAR; formerly at IBM Quantum, Yale, and UC Berkeley

- Zheng Qu, Associate Professor in Data Science, Fei Tian College–Northern, New York

- Fujio Yamamoto, Professor Emeritus, Information and Computer Sciences Department, Kanagawa Institute of Technology, Japan

- Naoki Yamamoto, Professor, Department of Applied Physics and Physico-Informatics, Keio University, Japan

Finally, we express our appreciation to the broader quantum computing community, whose continuing advances inspire and guide this work. Our goal is to distill and share that knowledge in a way that makes quantum computing both accessible and exciting for students.

AI Use Disclosure

This work includes material prepared with assistance from AI-based tools, including large language models, to support tasks such as outlining, language refinement, and literature-oriented drafting. All such material was reviewed, edited, and validated by the authors, who take full responsibility for the final content.

Dedication

This book is dedicated to all inquisitive minds and steadfast spirits who believe in the power of learning and in the boundless potential of human intellect.

Reviews

_{{_ A carefully structured guide to the core ideas of quantum algorithms, connecting foundational primitives to real application domains and helping readers build lasting intuition for quantum computational design."

— Zlatko Minev
Google Quantum AI; CIFAR; Formerly: IBM Quantum, Yale, and UC Berkeley

_{{_ The one-stop resource for everything quantum computing. Whether you are developing new algorithms or exploring practical applications, this book has it all. True to the clear, signature style of the author's earlier titles, this latest installment brings complex concepts into sharp focus through masterful presentation."

— Steven Frankel, Rosenblatt Professor
Faculty of Mechanical Engineering, Technion - Israel Institute of Technology

_{{_ This is an excellent resource for a student or professional coming from a classical STEM background. It manages to be technically rigorous without being impenetrable. If you find standard texts like Nielsen & Chuang too dense for a first pass, this "Scaffolding Approach" provides the necessary rungs to climb that ladder of complexity."

— Omar Alnaseri, Adjunct Professor at DHBW, Germany
Researcher in Quantum Communication Systems and Quantum ML/AI; SMIEEE

_{{_ Professor Peter Y. Lee and his coauthors, who have been building the Quantum Information Science series through a carefully scaffolded approach, have now published the long-awaited third volume, _Quantum Algorithms and Applications_, following _Quantum Computing and Information (Vol. 1)_ and _Mathematical Foundations of Quantum Computing (Vol. 2)_. I have used the first volume in teaching quantum information science to a broad range of undergraduate students and have seen an overwhelmingly positive response. This new volume is exceptionally well designed, enabling students to acquire a broad and up-to-date understanding of quantum algorithms and their applications in a clear, systematic, and accessible manner. I expect that many future quantum computer programmers will learn the foundations of using quantum computers from this book."

— Jaewan Kim, Professor Emeritus, Yonsei University and KIAS
National Distinguished Research Fellow,
Korea Research Institute of Standards and Science (KRISS)

" The field of quantum algorithms is advancing at a very rapid pace, and it is not easy to learn enough to reach the current research frontier. However, with this textbook, readers can efficiently study a wide range of topics, from the fundamentals to state-of-the-art algorithms. I would recommend it as an excellent first introduction for anyone who wishes to pursue research in this field."

— Naoki Yamamoto, Professor
Department of Applied Physics and Physico-Informatics, Keio University, Japan

" This book begins with a review of the fundamental concepts of quantum algorithms, followed by detailed explanations of key techniques such as the Quantum Fourier Transform (QFT) and Quantum Phase Estimation (QPE). It then bridges these foundations to Shor's factoring algorithm. After demonstrating Shor's algorithm through concrete examples, the discussion expands into the more general framework of Hidden Subgroup Problems.

The book also highlights the importance of Hamiltonian simulation, explaining time evolution as governed by the Schrödinger equation. And variational algorithms based on Ansatz are treated with a rigor and depth that is particularly commendable.

Unlike many CS-oriented books, this book devotes substantial space to simulations in physics and chemistry. The Hamiltonian introduced earlier plays a central role here as well. In doing so, the book provides a concrete and efficient approach to simulating nature, staying true to the vision originally envisioned by Feynman.

In addition, readers can explore modern applications such as quantum optimization and quantum machine learning. Together with the other two volumes in the Scaffolding series, this book is likely to become a definitive reference in quantum computing for researchers, engineers, and students alike."

— Fujio Yamamoto, Professor Emeritus
Information and Computer Sciences, Kanagawa Institute of Technology, Japan

Foundations of Quantum Algorithms

Part I establishes the computational viewpoint used throughout this book. We begin with algorithmic thinking and the basic language for analyzing efficiency and scaling, since "quantum advantage" is ultimately a statement about resources as a function of problem size. We then introduce the two standard models of quantum computation used in algorithm design: the circuit model and the query (oracle) model. These models clarify what an algorithm is allowed to do and what counts as cost.

We also review the complexity-theoretic perspective needed to interpret speedups responsibly, including classical and quantum complexity classes. Finally, we discuss the practical split between the NISQ and FTQC eras and why this distinction affects which algorithmic ideas are realistic today. The goal of Part I is to give every reader a common baseline before we turn to the main algorithmic toolkit in Part II.

1. Algorithmic Thinking

Contents

Every scientific discipline develops its own way of thinking. In mathematics, we cultivate proof techniques; in physics, we rely on models and approximations. In computer science, the central way of thinking is *algorithmic*: the art of designing precise procedures for solving problems.

In this chapter, we aim to clarify what algorithms are, why they are fundamental, and how we analyze their efficiency. These ideas form the backbone of computer science and provide the essential language for understanding quantum algorithms later on. We will ask questions such as:

- What exactly is an algorithm?

- What roles do algorithms serve in computation?

- How do we measure the efficiency of an algorithm in terms of time, space, and other resources?

By answering these questions, we lay the foundation for studying quantum algorithms, which promise dramatic speedups over classical methods in certain problems.

 This chapter builds on the following materials from:

- QC Math[1], Chapters 1–4.

1.1   Algorithms and Their Roles in Computation

Suppose you need to find a specific contact in a phonebook with thousands of names listed in alphabetical order. The goal is to locate the contact as quickly as possible.

One approach is to start at the first page and check each name sequentially until the desired contact is found—a method resembling *linear search*. In the worst case, with 10,000 names, this requires checking all 10,000 entries, making it time-consuming for large phonebooks.

An alternative is to use the phonebook's alphabetical order by opening to the middle page, checking if the target name comes before or after the middle entry, and repeating this process on the relevant half—a method known as *binary search*. For 10,000 names, binary search requires at most 14 checks, as each step halves the search space.

This example underscores how algorithms leverage problem structure (here, alphabetical order) to achieve efficiency. Binary search is exponentially faster than linear search for large, sorted datasets, illustrating the power of algorithmic design in reducing computational effort.

Formally, an algorithm is a well-defined sequence of instructions for carrying out a computation. Algorithms dictate not only *what gets computed* but also *how efficiently* it is accomplished. They are the bridge between abstract problems and practical computation, determining the time, memory, and energy required to solve tasks—especially as input sizes grow large.

A Basic Workflow for Algorithm Design

When analyzing an algorithm, it helps to separate five steps:

1. Problem: What is the input, and what output is required?

2. Model: What operations are allowed, and what counts as "cost"?

3. Algorithm: What procedure transforms input to output?

4. Cost: How do time and space scale with the input size n?

5. Correctness: Why does the procedure produce the correct output?

Classical algorithms form the foundation of digital computation. They span a wide range: from elementary arithmetic to sophisticated recursive procedures, from simple search routines to the transformative *Fast Fourier Transform*.

In the remainder of this section, we will introduce algorithms through a series of illustrative examples, and define algorithms more formally and explain their role in computational theory.

1.1.1   Introduction Through Examples

To make the concept of algorithms concrete, we will begin with examples that highlight efficiency, creativity, and impact.

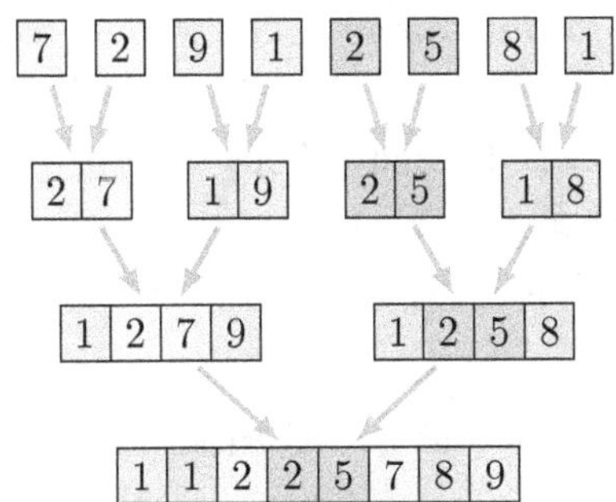

Figure 1.1: Merge Sort as a Divide-and-Conquer Procedure

1   Merge Sort vs. Bubble Sort

Merge sort exemplifies a powerful algorithmic design paradigm known as divide-and-conquer. It operates by recursively dividing the input array into smaller subarrays until each subarray contains only one element (which is inherently sorted). These sorted subarrays are then merged to produce the final sorted array. Let n be the number of elements in the initial array:

- Divide: At each level of recursion, the array is split into two roughly equal halves. This continues until we reach subarrays of size one, resulting in approximately $\log_2 n$ levels of recursion.

- Merge: At each level, the merge step combines pairs of sorted subarrays. Merging two arrays of total size k requires at most $O(k)$ elementary operations (comparisons and copies). Since each level processes all n elements, the work per level is $O(n)$.

Thus, the total number of operations across all $\log_2 n$ levels is $O(n \log_2 n)$.

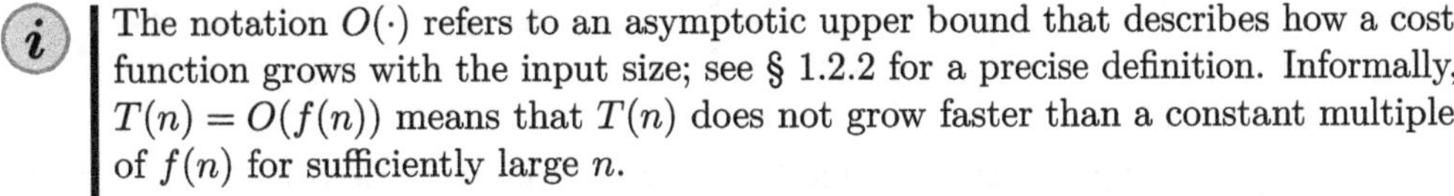

The notation $O(\cdot)$ refers to an asymptotic upper bound that describes how a cost function grows with the input size; see § 1.2.2 for a precise definition. Informally, $T(n) = O(f(n))$ means that $T(n)$ does not grow faster than a constant multiple of $f(n)$ for sufficiently large n.

Throughout this book, we write $\log n$ without specifying the base, since all logarithms differ only by constant factors in asymptotic analysis. When intuition matters, you may think of $\log n$ as $\log_2 n$, reflecting binary computation.

In contrast, a simpler (but less efficient) algorithm like bubble sort repeatedly iterates through the array, comparing adjacent elements and swapping them if they are out of order. In the worst-case scenario (e.g., a reverse-sorted array):

- The first pass requires $n - 1$ comparisons (and possibly $n - 1$ swaps),

- The second pass requires $n - 2$, and so on,

- The $(n - 1)$-th pass requires 1 comparison.

The total number of comparisons is $1 + 2 + \cdots + (n - 1) = \frac{(n-1)n}{2}$, which is $O(n^2)$.

To appreciate the impact of this difference in efficiency, consider a large input of $n = 10^9$:

- Merge sort requires approximately $n \log_2 n \approx 10^9 \times 30 = 3 \times 10^{10}$ operations. On modern hardware, this might take seconds or minutes.

- Bubble sort would require about $n^2 = 10^{18}$ operations—equivalent to decades of processing time.

The choice of algorithm can thus determine whether a problem is solvable in practice or not.

> **Exercise 1.1** Consider four hypothetical algorithms whose running times scale as follows:
>
>   (a) $T(n) = n$
>
>   (b) $T(n) = n \log_2 n$
>
>   (c) $T(n) = n^2$
>
>   (d) $T(n) = 2^n$
>
> Assume that one step corresponds to a single basic operation, and the computer can perform 10^9 operations per second. For each case, estimate the largest input size n that can be solved within 1 second.

2   Fast Fourier Transform

The Fourier Transform is a fundamental tool in science and engineering, used to decompose a signal into its constituent frequencies. The *Discrete Fourier Transform* (DFT, see QC Math [1], Chapter 11: *Functions of Vectors and Matrices*) applies this concept to a discrete sequence.

Given a sequence of n complex numbers $x_0, x_1, \ldots, x_{n-1}$, the DFT produces another sequence $\widetilde{x}_0, \widetilde{x}_1, \ldots, \widetilde{x}_{n-1}$:

$$\widetilde{x}_k = \sum_{j=0}^{n-1} x_j \, e^{-\frac{2\pi i \, kj}{n}}.$$

Computing a single $\widetilde{x}_k$ directly requires n multiplications and $n-1$ additions, so computing all n outputs takes $O(n^2)$ operations. This estimate assumes that the exponential factors $e^{-\frac{2\pi i \, kj}{n}}$ (the so-called *roots of unity*) are either precomputed or can be evaluated in constant time, and thus their computation is not included in the count.

The *Fast Fourier Transform* (FFT) dramatically improves upon this by employing a divide-and-conquer strategy, much like merge sort. The Cooley–Tukey algorithm, for instance, splits a DFT of size n into two DFTs of size $n/2$ (assuming n is even), reuses intermediate results, and then combines the outputs. This recursive process continues until reaching trivial size-1 DFTs.

As a result, the FFT achieves a running time of $O(n \log n)$—a striking improvement over $O(n^2)$. For example, with $n = 10^6$, a direct DFT would require about 10^{12} operations, while an FFT reduces this to about 2×10^7. This efficiency makes possible many modern technologies, from real-time audio processing to image compression.

3   Generating All Subsets of a Set: Exponential Scaling

A common technique in algorithm design is to represent subsets of a finite set using binary numbers.

Let $S = \{s_0, s_1, \ldots, s_{n-1}\}$ be a set with n elements. Each subset of S can be uniquely represented by an n-bit binary number. Specifically, the j-th bit of the integer i (where $0 \le i < 2^n$) indicates whether the element s_j is included in the corresponding subset.

For instance, let $S = \{a, b, c\}$ and consider $i = 5$. The binary representation of 5 using 3 bits is 101. This corresponds to the subset $\{a, c\}$, since the 0th and 2nd bits are set to 1 (indicating inclusion), and the 1st bit is 0 (indicating exclusion).

This method enables systematic enumeration of all 2^n subsets (i.e., the power set) by iterating over i from 0 to $2^n - 1$ and checking which bits are set in each i.

Since there are 2^n possible subsets, any algorithm that explicitly lists all of them must produce an exponentially large output. Therefore, the overall runtime must be at least exponential in n. The binary-encoding method above matches this scale up to constant factors, so it is essentially optimal for the task of generating the full power set.

This example is useful because it clarifies an important point about quantum algorithms. A quantum computer can prepare a superposition involving all 2^n basis states using Hadamard gates, but this is not the same as explicitly outputting all 2^n subsets. Measurement returns only one outcome per run. The strength of quantum algorithms is therefore not that they can list exponentially many outputs at once, but that they can sometimes use superposition, interference, and amplitude amplification to extract a specific global property more efficiently than classical algorithms. For the task of writing down the entire power set, however, the exponential output size itself is already an unavoidable bottleneck.

Exercise 1.2 Consider the set of 26 letters in the English alphabet. Implement the subset-generation algorithm referenced in § 1.1.1.3 to generate all possible subsets of this set, where each letter appears at most once and order does not matter. Assume all subsets are saved to disk as compact 26-bit bitmaps. For estimation, assume a computer clock speed of 1 GHz, and treat generating or writing one bitmap as one basic step. Estimate the total number of subsets, the approximate running time in seconds, and the disk space needed to store the output.

Next, consider generating all possible *strings with unique letters* (i.e., sequences where order matters and no letter is repeated). Calculate the total number of such strings. Then estimate whether it is feasible to write all of them to disk under the same assumptions.

4   Boolean Satisfiability (SAT)

Suppose you have a rule like: "The light turns on if the switch x_1 is up (represented as 1) OR both switches x_2 and x_3 are down (represented as 0)." Whether the light is on or off depends on the truth values of the variables x_1, x_2, x_3. This rule is an example of a Boolean formula, $x_1 \lor (\neg x_2 \land \neg x_3)$, where $\land$, $\lor$, and $\neg$ represent the logical operations AND, OR, and NOT, respectively.

The Boolean satisfiability problem (SAT) asks: is there some assignment of truth values to the variables that makes the entire formula evaluate to true? For the above example with light switches, possible satisfying assignments include $(x_1 = 1,\ x_2, x_3 \in \{0,1\})$ and $(x_1 \in \{0,1\},\ x_2 = x_3 = 0)$. But for more complicated Boolean formulas with many variables, the search for satisfying assignments quickly becomes much harder.

More generally, let $x_1, x_2, \ldots, x_n$ be Boolean variables, each taking a value in $\{0,1\}$, where 0 denotes false and 1 denotes true. A Boolean formula $\varphi(x_1, x_2, \ldots, x_n)$ is constructed using logical connectives such as $\wedge$, $\vee$, and $\neg$. The SAT problem asks: does there exist an assignment $(x_1, x_2, \ldots, x_n) \in \{0,1\}^n$ such that $\varphi(x_1, x_2, \ldots, x_n) = 1$?

Since there are 2^n possible truth assignments, a naïve algorithm must evaluate the formula for each one to determine if any satisfy the condition. In the worst case, this requires checking all 2^n possibilities, resulting in exponential time complexity. This combinatorial explosion makes SAT a prototypical example of problems where exponential scaling $(O(2^n))$ arises from exhaustive search, similar to subset generation.

SAT is also a canonical NP-complete problem (see § 4.1.4), meaning that many other computational problems can be reduced to it in polynomial time, and it typifies the challenges associated with exponential scaling in algorithm design.

5 Shor's Algorithm: Exponential Speedup in Factoring

Integer factorization—the task of finding the prime factors of a composite number—is a foundational problem in number theory and cryptography.

A simple algorithm like "trial division"—which attempts to divide N by every prime up to $\lfloor \sqrt{N} \rfloor$—has a running time that is exponential in the bit length $n \equiv \log N$. More precisely, since $\sqrt{N} \approx e^{n/2}$, the number of potential divisors (and the corresponding operations) grows exponentially with the bit length n.

The bit length of N is $\log_2 N$, while the number of decimal digits is approximately $\log_{10} N$. These differ only by a constant factor, since

$$\log_2 N = (\log_2 10)\, \log_{10} N \approx 3.32\, \log_{10} N.$$

More generally, logarithms in different bases differ only by constant factors, so in algorithm analysis we often write simply $\log N$ when the base is unimportant.

Clever algorithms such as the general number field sieve have a running time given by:

$$O\left(e^{c \cdot n^{1/3} (\log n)^{2/3}}\right).$$

This function grows faster than any polynomial in n (hence it is superpolynomial), yet slower than any fixed exponential of n (hence it is sub-exponential).

In 1994, Peter Shor discovered a quantum algorithm that solves the factoring problem in polynomial time, more precisely $O(n^3)$. Thus, unlike merge sort and the Fast Fourier Transform which offer a polynomial speedup, Shor's algorithm offers an exponential speedup. (We will delve into Shor's algorithm in Chapter 8.)

To illustrate this speedup, suppose N is a 2048-bit number (typical in RSA encryption). Classical methods might require more than 10^{20} operations to factor it, while a quantum computer using Shor's algorithm could, in principle, factor it using only millions of operations—a difference of more than 10 orders of magnitude.

This exponential separation between classical and quantum runtimes is a prime example of quantum speedup. However, it hinges on the assumption that large-scale fault-tolerant quantum computers can be built—a challenge that remains open.

6 Dimensions of Algorithmic Thinking

The above examples each illustrate a different dimension of algorithmic thinking:

- Merge Sort vs. Bubble Sort: efficiency gained through divide-and-conquer.

- Fast Fourier Transform (FFT): transforming a quadratic-time computation into near-linear time, enabling practical applications like signal processing and data compression.

- Subset Generation: exponential scaling as a fundamental barrier when the output itself is exponentially large.

- Boolean Satisfiability (SAT): computational hardness exemplified by NP-complete problems, where exhaustive search dominates.

- Shor's Algorithm: exponential quantum speedup over the best known classical algorithms.

These examples will serve as stepping stones toward a formal definition of algorithms and a discussion of their roles in computation.

1.1.2 Definition and Roles in Computation

The preceding examples illustrate the power of algorithmic design in concrete settings. To understand these ideas more systematically, we now step back to ask: what exactly is an algorithm, and what role does it play in computation?

1 Definition

Formally, an algorithm is a finite, well-defined sequence of instructions designed to perform a task or solve a class of problems. In theoretical computer science, algorithms are treated as abstract procedures—such as functions or transition systems—that operate over inputs drawn from a finite alphabet.

Consider the example in § 1.1.1.1. Sorting algorithms like merge sort or bubble sort consist of basic instructions (e.g., comparing or swapping elements) arranged into a structured procedure. Mathematically, such algorithms can be modeled as *functions* that map inputs (unsorted arrays) to outputs (sorted arrays), or as *transition systems* that describe step-by-step changes in the computation's internal state. The term *finite alphabet* refers to the predefined set from which input symbols are drawn—such as binary digits, decimal numbers, or characters.

Algorithms are evaluated based on three principal criteria:

- Correctness: Produces the correct output for all valid inputs.

- Termination: Halts after a finite number of steps.

- Efficiency: Uses reasonable computational resources, typically measured in time and space.

It is equally important to note what is *not* an algorithm. For example, the instruction "Keep flipping a fair coin until it lands heads, then stop" is not an algorithm in the strict sense, because it does not guarantee termination in a bounded number of steps, although it will terminate with probability 1. Similarly, vague instructions like "Sort the array in some way" are not algorithms, because they do not specify precise, executable steps.

 Algorithmic thinking predates modern computing. The term "algorithm" derives from the 9th-century Persian scholar Al-Khwarizmi, whose treatises introduced systematic methods for solving mathematical problems.

2   Roles in Computation

An algorithm is more than a mere recipe—it is the core formal object that defines what it means to compute. Algorithms are fundamental to computer science for several reasons:

- Definition of Computation: Algorithms provide a precise model for computation, specifying how inputs are transformed into outputs via finite, discrete steps.

- Generality and Reusability: Algorithms solve entire classes of problems, not just specific instances. They can be abstracted, optimized, and reused across diverse domains.

- Machine Independence: The same algorithm can be executed on different physical platforms or programming languages—classical, quantum, or otherwise—making algorithms the universal substrate of computing.

- Resource Analysis: Algorithms allow us to measure efficiency by analyzing time and space complexity, leading to rigorous comparisons of solution strategies.

Thus, algorithms are not just tools within computation—they are the language through which we understand, describe, and analyze all computational processes.

Algorithms also permeate nearly every aspect of modern life: they power search engines, route airplanes, compress images and videos, detect spam, train machine learning models, and secure communications. Their influence extends well beyond computer science into physics, biology, economics, and the social sciences, where algorithmic thinking provides new methods for analyzing complex systems.

1.2 Measuring Algorithmic Efficiency

As illustrated in § 1.1.1, an essential aspect of algorithm design and analysis is evaluating how resource demands scale with input size (typically the number of bits if the input is encoded as binary strings). Two fundamental resources are considered: *time*, corresponding to the number of basic computational steps, and *space*, corresponding to the amount of memory required. These notions are formalized through complexity measures that allow us to classify and compare algorithms independently of hardware or implementation details.

Throughout this chapter, we use n to denote the input size (for example, the length of an array to be sorted). When a separate symbol such as N appears, it denotes a problem-specific quantity (for example, an integer to be factored), and we state explicitly how n depends on N.

1.2.1 Time and Space Complexity

The *time complexity* of an algorithm quantifies the number of elementary computational steps as a function of the input size n, while the *space complexity* measures the maximum memory usage during execution. Both are usually expressed asymptotically, with emphasis on the dominant growth behavior as $n \to \infty$.

For instance, the merge sort algorithm has a time complexity of $O(n \log n)$ and a space complexity of $O(n)$, meaning that both runtime and memory usage grow nearly linearly with the input size.

This asymptotic viewpoint is essential because hardware- and software-dependent factors, such as processor speed or compiler optimizations, affect runtime only by constant factors. Such effects are negligible compared to differences in scaling behavior: an algorithm with $O(n^2)$ complexity will eventually be outperformed by one with $O(n \log n)$ on sufficiently large inputs, regardless of implementation details. Asymptotic analysis therefore captures the intrinsic efficiency of algorithms rather than their performance on any particular machine.

In quantum algorithms, resource analysis refines classical notions of time and space by distinguishing gate types, ancillary qubits, and additional scaling overheads (see § 3.2.2). Beyond time and space, quantum algorithms may incur an additional overhead that reflects factors not captured by simple gate or qubit counts. Nonetheless, asymptotic scaling behavior remains a key consideration.

1.2.2 Asymptotic Notations

Asymptotic notations describe how the resource usage (time or space) of an algorithm grows with input size. The most common notations are summarized in Table 1.1: Big-O gives an upper bound, Big-Ω gives a lower bound, Big-Θ gives a tight (two-sided) bound, and little-o denotes strictly smaller growth order. These bounds are up to constant factors and hold for sufficiently large inputs.

Asymptotic notation deliberately ignores constant factors and lower-order terms, because they become negligible for sufficiently large n. For example, $3n^2 + 10n + 7$ is $\Theta(n^2)$. Also, the notation itself does not specify whether we are analyzing worst-case, average-case, or best-case cost; unless stated otherwise, we use worst-case cost as the default because it provides a uniform guarantee.

More precisely, for functions $T(n)$ and $f(n)$ with $T, f \geq 0$ for large n:

$$T(n) = O(f(n)) \iff \exists\, c > 0,\, n_0 \text{ such that } T(n) \leq c\, f(n) \text{ for all } n \geq n_0,$$
$$T(n) = \Omega(f(n)) \iff \exists\, c > 0,\, n_0 \text{ such that } T(n) \geq c\, f(n) \text{ for all } n \geq n_0,$$
$$T(n) = \Theta(f(n)) \iff \exists\, c_1, c_2 > 0,\, n_0 \text{ such that }$$
$$c_1 f(n) \leq T(n) \leq c_2 f(n) \text{ for all } n \geq n_0,$$
$$T(n) = o(f(n)) \iff \text{for every } c > 0,\, \exists\, n_0 \text{ such that } T(n) < c\, f(n) \text{ for all } n \geq n_0.$$

Notation	Informal description
$O(f(n))$	$T(n)$ grows *no faster than* a constant multiple of $f(n)$ (asymptotic upper bound).
$\Omega(f(n))$	$T(n)$ grows *at least as fast as* a constant multiple of $f(n)$ (asymptotic lower bound).
$\Theta(f(n))$	$T(n)$ is sandwiched between two constant multiples of $f(n)$ (tight asymptotic bound).
$o(f(n))$	$T(n)$ grows *strictly slower than* $f(n)$: it is negligible compared with $f(n)$ for large n.

Table 1.1: Asymptotic Notations in Complexity Analysis

1   Tight vs. Loose Bounds

When writing asymptotic bounds, it is technically correct to use a looser upper bound than necessary. For example, if an algorithm takes

$$T(n) = 3n^2 + 10n,$$

then it is still true that

$$T(n) = O(n^3),$$

even though this is not the tightest upper bound, because n^2 grows no faster than n^3 up to constant factors. However, this bound is not very informative: it may suggest cubic growth, when in fact the algorithm is quadratic.

A *tight* bound captures the dominant growth rate of $T(n)$ as closely as possible. In the above example, the tight asymptotic expressions are

$$T(n) = O(n^2), \qquad T(n) = \Omega(n^2), \qquad T(n) = \Theta(n^2).$$

Thus, when asked to provide $O(\cdot)$, $\Omega(\cdot)$, or $\Theta(\cdot)$ for a given function, we usually mean the *tightest natural bound*: the simplest function that correctly describes the dominant term.

Exercise 1.3 Suppose an algorithm takes $T(n) = 100n + 0.01n^{1.5}$ steps to execute. Find suitable functions $f(n)$, $g(n)$, and $h(n)$ such that

$$T(n) = O(f(n)), \qquad T(n) = \Omega(g(n)), \qquad T(n) = \Theta(h(n)).$$

(Choose the tightest natural bounds.)

2   Best-Case vs. Worst-Case Complexity

Big-O is often used to describe the worst-case behavior of an algorithm, and Big-Ω the best-case behavior. However, the notations themselves do not imply worst-case, best-case, or any other type of analysis. They simply describe bounds on whichever cost function is being considered. It is therefore necessary to state explicitly which cost function (worst-case, best-case, or average-case) is being bounded.

■ **Example 1.1** Consider an algorithm whose running time depends on the input configuration. For a problem of size n, suppose the algorithm requires at least

$5n + 10$ steps in the best case (e.g., when the input is ideally structured) and at most $9n^2 + 5n + 20$ steps in the worst case (e.g., when the input is adversarially structured). Then:

$$T_{\text{worst}}(n) = O(n^2), \qquad T_{\text{best}}(n) = \Omega(n).$$

Here T_{worst} and T_{best} are different cost functions, and the asymptotic notation specifies bounds on each. ∎

1.2.3 Common Scaling Behaviors

We classify algorithmic efficiency by growth rates, i.e., how resource demands scale with input size n. Common scaling types include:

- Constant: $O(1)$. The number of operations does not depend on the input size. Example: accessing a specific element in an array by index.

- Logarithmic: $O(\log n)$. Extremely efficient—better than linear. Example: binary search in a sorted list.

- Linear: $O(n)$. Each input element is processed a constant number of times. Example: summing all elements in a list, or linear search in unsorted data.

- Linearithmic: $O(n \log n)$. Slightly worse than linear. Examples: merge sort (§ 1.1.1.1), and the Fast Fourier Transform (FFT, § 1.1.1.2).

- Polynomial: $O(n^k)$, $k > 1$. Higher powers grow more rapidly. Example: bubble sort has $O(n^2)$ complexity (see § 1.1.1.1); regular matrix multiplication has $O(n^3)$ complexity.

- Polynomial with logarithmic factors: $O(n^k (\log n)^c)$. Polynomial-time scaling with an additional logarithmic overhead. Example: many divide-and-conquer algorithms (including merge sort) have $O(n \log n)$ time complexity.

- Polylogarithmic: $O((\log n)^k)$. Very slow growth, common in data-structure queries and some randomized or algebraic subroutines.

- Exponential: $O(b^n)$, $b > 1$. Grows rapidly and is often intractable. Example: generating all subsets of a set (§ 1.1.1.3) and boolean satisfiability (§ 1.1.1.4); the Held-Karp dynamic programming algorithm for the Traveling Salesman Problem runs in $O(n^2 2^n)$ time.

- Subexponential: A narrower class between polynomial and exponential. Typically written as $2^{o(n)}$, these runtimes grow faster than polynomial time but slower than 2^{cn} for every constant $c > 0$. Example: the general number field sieve for integer factorization runs in $O\left(e^{c\, n^{1/3} (\log n)^{2/3}}\right)$, which is subexponential. Some lattice reduction algorithms and cryptographic attacks also fall here.

- Factorial: $O(n!)$. Extremely inefficient growth, worse than exponential. Example: brute-force enumeration of all tours in the Traveling Salesman Problem or exhaustive scheduling.

- Superpolynomial: Growth that eventually outpaces every polynomial n^k but may be as fast as exponential or even factorial. Example: $n^{\log n}$ is superpolynomial, as is 2^n. This class is very broad and not considered "efficient."

Such scaling functions are graphed in Fig. 1.2 for visual comparison.

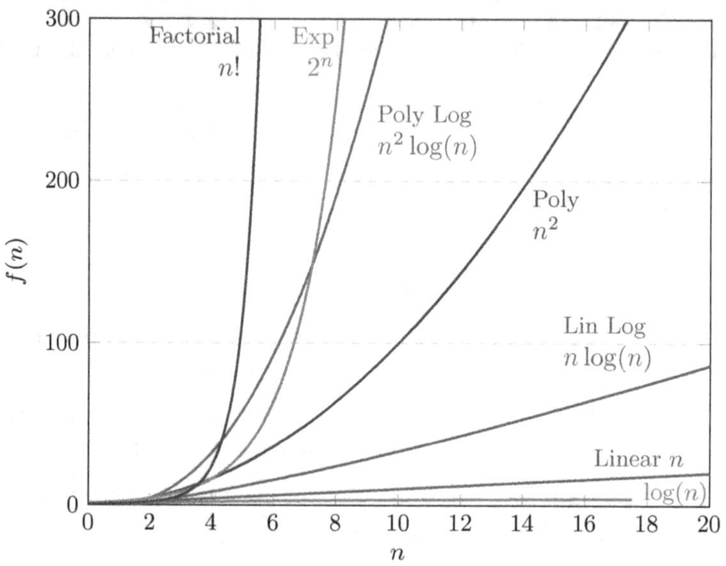

Figure 1.2: Growth Rates of Common Complexity Classes

Key Takeaway

$$O(1) < O(\log n) < O(n) < O(n \log n) < O(n^2) < O(2^n) < O(n!)$$

Exercise 1.4 For each of the following functions, determine the dominant term as $n \to \infty$ and express the growth rate using Big-O notation.

(a) $0.3n + 5n^{2.5} + 500n^{1.8}$

(b) $10n + n \ln n + 0.1n \ln(\ln n)$

(c) $2^{(\log_2 n)^2 + 0.001n}$

(d) $n^{100} + 2^{0.01n}$

(e) $n! + 2^{9n}$

1.2.4 Exponential vs. Polynomial Scaling

The distinction between polynomial and exponential scaling is fundamental to understanding algorithmic efficiency. Polynomial scaling refers to problems whose resource requirements grow as a power of the input size, such as n^2, which remain tractable even for relatively large inputs. In contrast, exponential scaling involves functions such as 2^n, where resource demands increase so rapidly that they quickly become infeasible as the input grows.

As illustrated in Fig. 1.3, we compare the polynomial function $P(n) = n^2$ with the exponential function $E(n) = 0.001 \cdot 2^n$. At small input sizes ($n < 18$), the polynomial curve exceeds the exponential one. However, once $n > 20$, even the small constant factor (0.001) cannot prevent the exponential function from surpassing the polynomial by a wide margin.

If we set a practical threshold for feasible computation at 10,000, the polynomially scaling problem remains manageable up to about $n = 100$, whereas the exponentially scaling one becomes intractable beyond $n = 20$. This example clearly illustrates how exponential growth ultimately dominates computational cost at scale.

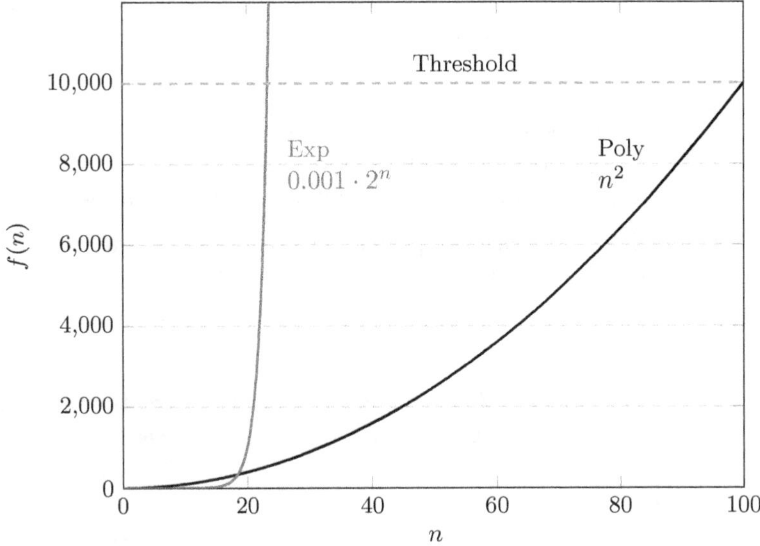

Figure 1.3: Exponential vs. Polynomial Scaling

1   Tractable vs. Intractable Problems

In computational complexity, a problem is called *tractable* if it can be solved in *polynomial time*, that is, in $O(n^k)$ steps for some constant k. Tractable problems are considered efficiently solvable in practice because their runtimes grow moderately with input size. By contrast, *intractable* problems are those for which only exponential-time or worse algorithms are known, such as $O(2^n)$ or $O(n!)$. These algorithms become impractical even at modest input sizes, placing the corresponding problems effectively beyond reach for large-scale instances.

2   Reasons Behind Polynomial vs. Exponential Scaling

Exponential-time algorithms often arise from problems that require *exhaustive search* across 2^n or more possibilities, such as subset enumeration (§ 1.1.1.3) or Boolean satisfiability (§ 1.1.1.4). These problems are treated as having little or no exploitable structure, so the algorithm must consider nearly every possibility.

By contrast, polynomial-time algorithms usually succeed by exploiting *internal structure*—regularities, constraints, or recursive patterns—that enable efficient reduction of the search space. Merge sort uses divide-and-conquer to transform sorting into smaller subproblems; the FFT leverages symmetries in complex roots of unity; Shor's algorithm exploits the periodic structure hidden in modular arithmetic. In each case, recognizing and exploiting structure turns a problem that might appear intractable into one that can be solved efficiently.

The discovery of efficient algorithms is therefore not just a matter of clever programming but of uncovering hidden structures—symmetries, periodicities, or mathematical relationships—that make shortcuts possible. Laws of physics and theorems of mathematics continually reveal such structures in the natural world. Although these do not always translate directly into computational constructs, they often inspire new algorithmic strategies. This process of identifying and exploiting structure lies at the heart of algorithm design, whether in classical computing or in the emerging field of quantum computing.

Chapter Takeaways

1. An algorithm is a well-defined procedure for transforming inputs into outputs, and algorithm design choices determine scalability in practice.

2. Time and space complexity provide machine-independent ways to compare algorithms as the input size n grows.

3. Asymptotic notation ($O(\cdot)$, $\Omega(\cdot)$, $\Theta(\cdot)$) captures dominant growth rates while suppressing constants and lower-order terms.

4. Polynomial scaling is typically considered tractable, while exponential scaling quickly becomes infeasible as n increases.

Problem Set 1

1.1 Complexity analysis depends on a cost model. In this chapter we often treat elementary operations (comparisons, additions, memory accesses) as constant time. In practice, this assumption can fail.

Choose *two* of the following settings:

- arithmetic on very large integers,

- reading data from disk instead of from main memory,

- poor cache locality when scanning a large array.

For each one:

(a) explain in simple terms why the "constant time per step" assumption may fail;

(b) describe which resource becomes the main bottleneck (for example, computation time, memory usage, or data-transfer speed);

(c) say qualitatively how the runtime prediction might change.

Your answer may be informal; no advanced knowledge of computer architecture is required.

1.2 Many exponential-time algorithms arise from exhaustive search over 2^n possibilities. Choose one example from this chapter: subset generation, SAT, or trial-division factoring.

Explain what kind of *structure* would make a polynomial-time algorithm more plausible. For example, structure might mean symmetry, a restricted input form, extra promises about the input, or a hidden pattern that allows many cases to be ruled out at once. Your answer may be informal.

1.3 Worst-case bounds provide guarantees, but typical performance can be very different. Choose one algorithm from this chapter, such as linear search, binary search, bubble sort, merge sort, or brute-force SAT. Then answer the following:

    (a) Describe a plausible type of input on which the algorithm would usually perform better than in the worst case.

    (b) State clearly what assumption you are making about the input.

    (c) Suggest a simple experiment you could run in Python to test whether this assumption is reasonable.

A brief informal discussion is sufficient; no advanced programming or statistics is required.

1.4 $*$ When empirically verifying a scaling law such as $T(n) = O(n \log n)$ or $T(n) = O(n^2)$, the way you plot and fit data matters. Describe a robust procedure to test whether measured runtimes are closer to $n \log n$ or n^2 (for example, using log–log plots, ratio tests such as $T(n)/n$ and $T(n)/(n \log n)$, or regression). Discuss at least two sources of experimental error or bias.

1.5 $*$ A claimed speedup is meaningful only if we compare like with like. Give two examples of how a speedup claim could be misleading if the comparison is not stated carefully (for example, comparing worst case to average case, or comparing time while ignoring memory, preprocessing, or output size). Relate your discussion to one example from this chapter.

1.6 Implement the merge sort and bubble sort algorithms (see § 1.1.1.1) in Python from scratch (i.e., without using built-in sorting libraries). Empirically verify their scaling behavior, corresponding to $O(n \log n)$ for merge sort and $O(n^2)$ for bubble sort.

1.7 The subset generation problem has complexity $O(2^n)$ (see § 1.1.1.3). Estimate the maximum input size n that can be handled within 1 second, 1 hour, and 1 year on a computer performing 10^9 operations per second.

1.8 For the Boolean satisfiability problem (SAT, see § 1.1.1.4), discuss why brute-force search requires exponential time in the worst case. Estimate how many variable assignments must be checked for $n = 20$ and $n = 100$.

1.9 For each of the following functions, determine the dominant term as x grows large and express the growth rate using Big O notation.

    (a) $5 + 0.001x^3 + 0.025x$ (f) $100x + 0.01x^2$

    (b) $1 + 0.1x^{0.3} + 0.05\sqrt{x}$ (g) $2x + x^{0.5}\log_2 x + 0.5x^{1.25}$

    (c) $500x + 100x^{1.5} + 50x\log_{10} x$ (h) $0.01x\log_2 x + x(\log_2 x)^2$

    (d) $x^2\log_2 x + x(\log_2 x)^2$ (i) $100x\log_3 x + x^3 + 100x$

    (e) $3\log_2 x + \log_2\log_2\log_2 x$ (j) $0.003\log_4 x + \log_2\log_2 x$

1.10 Is each of the following statements true or false? Explain. If false, provide a counter example.

(a) $O(f + g) = O(f) + O(g)$

(b) $O(f \cdot g) = O(f)O(g)$

(c) $O(f/g) = O(f)/O(g)$

1.11 Consider the exponential function $E(n) = 2^n$ and the polynomial function $P(n) = 10^6 n^3$. Estimate the largest n for which $E(n) < P(n)$, and explain why $E(n)$ eventually dominates $P(n)$ regardless of constant factors.

1.12 In practice, constant factors matter for small inputs. Consider two algorithms:

$$A(n) = 0.0001 \cdot n^3, \qquad B(n) = 100 \cdot n^2.$$

For what values of n is $A(n) < B(n)$? Which algorithm is asymptotically more efficient?

1.13 $*$ Show that if $f(n) = O(g(n))$ and $g(n) = O(h(n))$, then $f(n) = O(h(n))$. This property is called the *transitivity of Big-O*. Does an analogous property hold for Ω and Θ? Justify.

1.14 $*$ Give an example of a problem that is solvable in polynomial time but still impractical in real-world settings due to a high polynomial degree or large constant factors. Discuss why asymptotic analysis alone is not always sufficient to judge feasibility.

1.15 $*$ For each complexity class below, provide a simple example algorithm (not mentioned in the chapter) that fits the class:

(a) $O(\log n)$ (d) $O(n^2)$

(b) $O(n)$ (e) $O(2^n)$

(c) $O(n \log n)$

1.16 $*$ Prove that $n!$ grows faster than 2^n, i.e., $n! = \Omega(2^n)$.

2. Quantum Algorithms: A New Paradigm

Contents

Classical algorithms largely abstract away the physical details of the machines that run them. Whether implemented on silicon transistors or mechanical devices, the underlying model is the same: information is stored as bits, processed by logic gates, and updated by deterministic or probabilistic rules. Classical algorithms have driven remarkable advances across science and engineering, including recent breakthroughs in artificial intelligence. Yet many important problems remain for which no efficient classical algorithms are known. Fundamental limits, rooted in algorithmic complexity and problem structure, constrain what classical computation can achieve.

Quantum algorithms take a different approach. They process information according to the principles of quantum mechanics—superposition, entanglement, and interference—embedded in the laws of physics themselves. By exploiting structures that classical methods do not naturally access, quantum algorithms can sometimes achieve efficiencies far beyond what is possible classically.

This chapter examines:

- Feynman's insight that simulating physics may require quantum systems,

- the native power of quantum algorithms,

- the operational features that distinguish quantum computation from classical computation, and

- the promises and challenges of realizing this new paradigm.

We begin by reviewing the limits of classical algorithms, which motivate the search for new models of computation and frame the emergence of the Quantum Church–Turing Thesis.

 Algorithms do not exist in isolation: they are always defined relative to a model of computation. In Chapter 1, we worked within the classical model, where information is stored in bits and processed by deterministic or probabilistic logic gates. To understand the potential and limitations of algorithms in the quantum setting, we must broaden our notion of computation. Quantum computation provides the foundation upon which quantum algorithms are built.

2.1 Limits of Classical Algorithms

While classical algorithms have achieved remarkable feats, their power is fundamentally limited by complexity constraints and problem structure. This section highlights representative limits and introduces the Quantum Church–Turing Thesis as a framework for understanding how quantum algorithms extend the classical picture.

2.1.1 Examples of Classically Intractable Problems

Some problems remain infeasible for classical computation not because current programs are poorly implemented, but because no efficient classical algorithms are known for them, and in many cases none are believed to exist. Below is a selection of such problems, with a more comprehensive list provided in Appendix D.

- Integer factorization: Given a large composite number, find its prime factors. No known classical algorithm runs in polynomial time, underpinning the security of RSA encryption. (See § 1.1.1.5.)

- Discrete logarithm problem: Given a prime p, a base b, and an element h, find x such that $b^x \equiv h \pmod{p}$. This problem is classically hard and forms the basis of several cryptographic protocols.

- Protein structure prediction: Predicting protein structure from amino acid sequence has long been a major computational challenge. Although AI-based methods such as AlphaFold have achieved dramatic progress, broader structure-and-dynamics problems in biology remain computationally demanding.

- Boolean satisfiability (SAT): Determine whether a Boolean formula has a satisfying assignment. As the canonical NP-complete problem, it exemplifies the challenges of combinatorial search in logic and verification. (See § 1.1.1.4.)

- Quantum many-body simulation: Classically simulating strongly interacting quantum systems, such as correlated electrons in materials, typically scales exponentially with system size due to the growth of quantum state spaces.

- Vehicle routing problem (VRP): Determine the optimal set of routes for a fleet of delivery vehicles. This generalization of the Traveling Salesman Problem is computationally hard and central to logistics.

- Full quantum state tomography: Reconstructing the full state of a quantum system from measurements becomes infeasible as system size grows, due to exponential resource requirements.

2.1.2 Evolution of the Church–Turing Thesis

The limitations of classical algorithms and the potential of quantum ones are reflected in the evolution of the Church–Turing Thesis, from its original form to later efficiency-based and quantum versions, as illustrated in Fig. 2.1. The key idea is that these statements address not only *what* can be computed, but also *how efficiently* it can be computed.

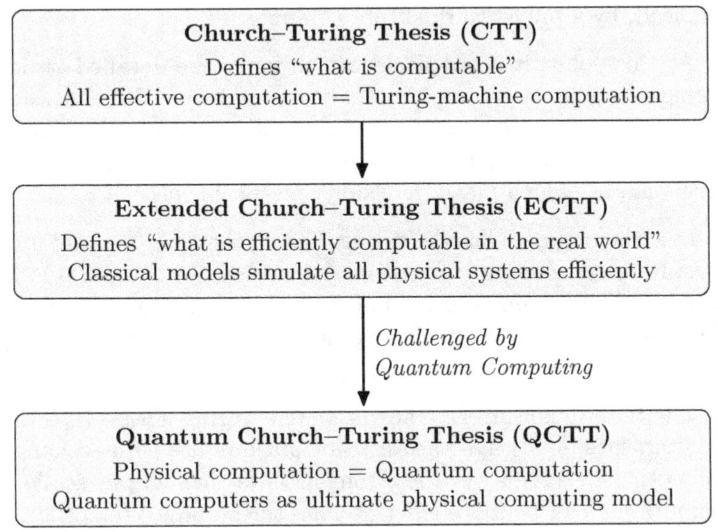

Figure 2.1: Evolution of the Church–Turing Thesis

In the 1930s, several abstract models of computation—including Turing machines, lambda calculus, and recursive functions—were developed and shown to be equivalent in computational power. These models underpin the original Church–Turing Thesis, which states that anything that can be computed by an algorithm can also be computed by a Turing machine. Here, a Turing machine is an idealized mathematical model of computation, not a practical device.

The word "thesis" in Church–Turing Thesis means a general claim or hypothesis, not a proven theorem. The statement cannot be proved mathematically in full generality, because it relies on the informal notion of what counts as an "algorithm."

Quantum algorithms do not invalidate the original Church–Turing Thesis. A quantum computer does not compute functions that are uncomputable for a classical Turing machine. Rather, the point is that some problems appear to be solvable *much more efficiently* on a quantum computer than on any known classical one.

The Extended Church–Turing Thesis (ECTT) strengthens the discussion by adding an efficiency claim. Roughly speaking, it says that any computation that can be carried out by a physically realizable device can also be simulated efficiently, up to polynomial overhead, by a probabilistic classical Turing machine. Thus, while the original Church–Turing Thesis is about computability in principle, the ECTT is about efficient simulation by classical computation.

Quantum algorithms such as Shor's algorithm challenge this stronger thesis. Shor's algorithm factors integers in polynomial time on a quantum computer, while no polynomial-time classical algorithm is known. If such quantum speedups are physically realizable at large scale, then the ECTT cannot hold in full generality.

This motivates a quantum version of the picture. The Quantum Church–Turing Thesis (QCTT) says, informally, that quantum computation is the right general model for efficient computation in a quantum-mechanical world. Equivalently, any physically realizable process should be efficiently simulable, up to polynomial overhead, by a universal quantum computer [4, 5].

At this point, it is not necessary to know the detailed structure of a quantum Turing machine. It is simply a quantum analogue of the classical Turing machine, introduced as an idealized mathematical model of quantum computation. For practical purposes in this book, you may think instead of the universal quantum circuit model, which plays the same conceptual role.

Like the original Church–Turing Thesis, the QCTT is *not* a claim that quantum computers can compute functions beyond those computable by classical Turing machines. The claim is instead about efficiency: quantum computers may provide the most natural and efficient general model for simulating physical processes governed by quantum mechanics.

The term Quantum-Extended Church–Turing Thesis (QECTT) also appears in the literature [6]. Usage varies: some authors use it interchangeably with QCTT, while others reserve it for the stronger claim that no physically realizable model of computation can be more efficient than the standard quantum model.

Limits of the Computational Model

First, physical quantities such as position, density, and velocity are often modeled by real numbers or continuous fields. Standard computational models, however, whether classical or quantum, operate on finite symbolic descriptions and therefore represent such quantities only to finite precision. In this sense, the Church–Turing framework applies most directly to computable approximations of physical systems, rather than to an exact physical continuum.

Second, the standard models of quantum computation used in this book are rooted in nonrelativistic quantum mechanics. They successfully describe many important quantum algorithms and simulation tasks, but they may not capture all aspects of nature equally well. Some physical phenomena are more fundamentally described by relativistic quantum field theory, and the computational implications of such theories are not yet fully understood. Thus, while quantum computers appear to provide a natural model for a wide range of physical processes, it remains possible that the full computational structure of nature is broader than what is captured by current quantum computing models. This does not mean that nature is known to use a "better algorithm," but rather that our present computational models may not yet encompass every physically relevant regime.

Exercise 2.1

(a) Summarize the key differences among the original Church–Turing Thesis, the Extended Church–Turing Thesis, and the Quantum Church–Turing Thesis.

(b) For each version, explain whether Shor's algorithm supports it, challenges it, or leaves it unchanged.

Exercise 2.2

(a) Explain why quantum algorithms are believed to change only what is efficiently computable, rather than what is computable in principle.

(b) State the Quantum Church–Turing Thesis in your own words.

Key Takeaways

1. The original Church–Turing Thesis is about what is computable in principle, not about efficiency.

2. The Extended Church–Turing Thesis is a complexity claim: physical processes can be simulated efficiently on a probabilistic classical Turing machine (up to polynomial overhead).

3. The Quantum Church–Turing Thesis keeps computability unchanged but shifts the model of efficient physical computation from classical to quantum.

2.2 Quantum Algorithms: A New Paradigm

Having framed why classical computation faces intrinsic barriers, we now highlight what changes in the quantum model. Quantum algorithms redefine information processing by leveraging the principles of quantum mechanics—superposition, entanglement, and interference—and by turning these physical principles into algorithmic resources.

2.2.1 Feynman's Vision: Simulating Nature With Quantum Computers

In his influential 1981 lecture, Richard Feynman posed a pivotal question [7]: Why is simulating quantum systems on classical computers so challenging? He noted that the state space of a quantum system grows exponentially with the number of particles, requiring classical resources that scale exponentially with system size, as seen in problems like quantum many-body simulation.

Feynman's insight led to a transformative proposal:

> *Nature isn't classical... and if you want to make a simulation of nature, you'd better make it quantum mechanical.*

Rather than using classical algorithms to simulate quantum systems—a task often intractable—Feynman advocated building quantum computers that evolve according to quantum mechanical principles, naturally emulating the systems they model.

2.2.2 Native Power of Quantum Algorithms

The power of quantum algorithms comes from their alignment with the laws of quantum mechanics. This allows them to exploit mathematical and physical structure in ways that are not naturally available to classical methods. For example:

- Quantum many-body simulation: Predicting the behavior of molecules, materials, or quantum field theories can be exponentially hard for classical algorithms because the quantum state space grows exponentially with system size. Quantum algorithms can represent and evolve such states more directly on quantum hardware. We will explore this topic in Chapter 19.

- Integer factorization: Shor's algorithm factors integers in polynomial time by using the quantum Fourier transform to exploit periodic structure. See Chapter 8.

- Discrete logarithms: Shor's framework also gives an efficient quantum algorithm for the discrete logarithm problem, another classically hard task of major cryptographic importance.

- Other problems: See Appendix D for additional examples.

These examples illustrate a central theme: quantum algorithms gain efficiency by exploiting structure—such as periodicity, linear-algebraic structure, or quantum correlations—rather than by brute-force search over unstructured possibilities. The basic resources behind quantum advantage include superposition, entanglement, and interference.

Key Takeaways

---

1. Quantum advantage is not obtained by "trying all inputs at once," but by using interference to amplify information about a structured global property.

2. Superposition provides coherent access to many basis states, but measurement returns limited classical information per run.

3. Useful speedups arise when the problem has exploitable structure (e.g., periodicity, symmetry, spectral information) that can be converted into a measurable bias.

---

1 Superposition

A quantum bit (qubit) can exist in a linear combination of basis states. A common algorithmic starting point is the uniform superposition

$$|\psi\rangle = \frac{1}{\sqrt{2^n}} \sum_{x \in \{0,1\}^n} |x\rangle. \tag{2.1}$$

This compactly represents 2^n classical inputs at once. However, superposition alone does not yield a speedup: to extract useful information, algorithms must also arrange interference patterns so that measurement outcomes are biased toward the desired answer (see the discussion of interference below).

2 Entanglement

Entanglement enables quantum systems to encode information in a high-dimensional

Hilbert space. A general n-qubit state can be written as

$$|\psi\rangle = \sum_{x \in \{0,1\}^n} c_x \, |x\rangle \, , \tag{2.2}$$

requiring up to 2^n complex amplitudes c_x. In contrast, an unentangled state (a tensor product of single-qubit states) requires only $2n$ complex numbers. This exponential representational capacity is a crucial resource for quantum algorithms, enabling correlations far beyond what classical systems can efficiently capture.

3 Interference

Superposition and entanglement provide access to a vast computational space, but measurement reveals only limited information in any single run. Quantum algorithms employ interference to steer probability amplitudes: constructive interference amplifies correct answers, while destructive interference suppresses incorrect ones. This orchestration of amplitudes is what turns quantum state space into algorithmic power.

■ **Example 2.1 — Deutsch–Jozsa Algorithm.** The Deutsch–Jozsa problem requires determining whether a function $f : \{0,1\}^n \to \{0,1\}$ is constant or balanced. Classically, solving this with certainty may require exponentially many evaluations of f.

In the quantum setting, a uniform superposition is prepared, an oracle applies a phase $(-1)^{f(x)}$, and a Hadamard transform induces interference that concentrates amplitude on basis states encoding the correct global property of f. A single measurement then reveals the answer with certainty, achieving an exponential speedup. (See QCI Book [2], Chapter 11: *Quantum Algorithms: A Sampler* for details.)   ■

Exercise 2.3 Explain why simulating an n-qubit quantum system requires $O(2^n)$ complex numbers in classical memory. Estimate the memory (in gigabytes) required to store the full state of a system with $n = 100$ qubits. Compare this number to the capacity of today's largest supercomputers.

A Practical Interpretation of "Quantum Parallelism"

Preparing a superposition lets a quantum circuit apply an oracle to many basis states coherently in a single unitary evolution. The computational advantage comes only when the algorithm uses interference to convert those coherent phases into a measurable bias toward the correct outcome. Without such amplitude shaping, measurement returns essentially a single sample, and the apparent parallelism does not translate into useful speedup.

2.2.3 Unique Features of Quantum Algorithms

Quantum algorithms exhibit features distinct from classical ones, stemming directly from quantum mechanics and shaping both their strengths and limitations.

1   Reversibility

All operations in a closed quantum system are reversible. Each quantum gate corresponds to a unitary operator U satisfying $U^\dagger U = I$. Unlike classical AND or OR gates, which irreversibly lose information, quantum circuits must be built from reversible gates such as CNOT and Toffoli. Reversibility preserves information until measurement and reflects deeper physical principles such as conservation of information.

2   Measurement

Quantum measurement is non-unitary, projecting a quantum state onto an observable's eigenstate according to the Born rule (see QCI Book [2], Section 6.4: *Measurements of Multi-Qubit Systems*). This collapses superpositions into classical outcomes, introducing intrinsic probabilism.

Moreover, quantum measurements can exhibit contextuality: outcomes may depend on which other compatible measurements are performed. Contextuality is a computational resource in models such as measurement-based quantum computing.

3   No-Cloning Theorem

The no-cloning theorem states that an arbitrary unknown quantum state $|\psi\rangle$ cannot be copied via a unitary operator U such that

$$U(|\psi\rangle \otimes |0\rangle) = |\psi\rangle \otimes |\psi\rangle .$$

This contrasts with classical computing's reliance on copying and imposes constraints on quantum information processing, necessitating alternative techniques such as quantum teleportation.

> **Exercise 2.4** The no-cloning theorem rules out cloning of *arbitrary unknown* states, but it does not forbid copying of classical information encoded in orthonormal states.
>
> (a) Suppose a unitary U satisfies $U(|\psi\rangle |0\rangle) = |\psi\rangle |\psi\rangle$ for two distinct states $|\psi_1\rangle$ and $|\psi_2\rangle$. Show that $\langle\psi_1|\psi_2\rangle \in \{0, 1\}$.
>
> (b) Conclude that any set of perfectly clonable pure states must be mutually orthogonal.
>
> (c) Give an explicit unitary circuit that clones computational basis states $|0\rangle$ and $|1\rangle$ (but does not clone arbitrary superpositions).

4   Fragility and Quantum Error Correction

Quantum states are highly susceptible to decoherence, where environmental interactions disrupt superposition and entanglement. Because the no-cloning theorem forbids simple backups, quantum error correction is essential. Codes such as the Shor code, Steane code, and surface codes use entanglement to detect and correct errors without measuring the state directly. These methods, however, require significant overhead, posing a major challenge for scalable quantum algorithms.

Exercise 2.5

(a) Explain why a quantum system with n qubits can represent 2^n amplitudes.

(b) Compare classical parallelism with quantum superposition, and explain the idea often called "quantum parallelism."

(c) Discuss how the no-cloning theorem impacts the design of quantum algorithms or protocols.

2.2.4 Promises and Challenges

Quantum algorithms offer transformative potential but face significant practical challenges. Below we outline both the promises and the obstacles.

1 Promises of Quantum Algorithms

Quantum algorithms provide new approaches to problems that are inefficient or intractable classically:

- Quantum simulation: Simulating quantum systems (e.g., molecules and materials) is a natural application, as quantum algorithms can represent and evolve quantum states directly.

- Optimization and sampling: Algorithms such as the Quantum Approximate Optimization Algorithm (QAOA) and quantum annealing may improve solution quality for certain combinatorial tasks.

- Machine learning: Quantum linear-algebraic subroutines (e.g., the HHL algorithm) may accelerate tasks involving structured high-dimensional data under appropriate assumptions.

- Cryptanalysis: Shor's algorithm undermines RSA and elliptic curve cryptography, motivating post-quantum cryptography.

Quantum advantage—solving a problem faster than the best known classical algorithm—has been demonstrated in specialized settings (e.g., boson sampling). Extending such advantages to broadly useful practical problems remains an open challenge [8, 9].

2 Challenges in Practice

Despite their promise, quantum algorithms face major hurdles:

- Noise and decoherence: Quantum systems are sensitive to noise, requiring robust error correction. Current fault-tolerant schemes can demand hundreds to thousands of physical qubits per logical qubit.

- Scalability: Modern devices (often called NISQ systems) support only hundreds or, in some platforms, a few thousand physical qubits—far below the scale needed for algorithms such as Shor's on cryptographically relevant inputs.

- Algorithm design: Quantum algorithms must be purpose-built to exploit quantum structure; there is no general recipe to "quantize" a classical algorithm.

- Verification: Verifying quantum computations is challenging when classical simulation is infeasible, motivating new benchmarking and validation methods.

3   Cautious Optimism

Quantum algorithms are still in their early stages, but their theoretical foundations suggest transformative potential in cryptography, materials science, and beyond. Continued advances in hardware scalability, error correction, and algorithmic design will be essential to realizing this promise.

Chapter Takeaways

1. Many important problems appear classically intractable because known algorithms scale exponentially in the worst case.

2. Quantum algorithms change the computational model: superposition, interference, and measurement enable new ways to process information.

3. Quantum computation is not expected to change what is computable, but it can change what is efficiently computable (QCTT).

4. The potential advantage is not "quantum parallelism" alone; it comes from engineered interference that amplifies correct outcomes and suppresses others.

5. Promising speedups are conditional on realistic input access, implementable circuits, and error rates compatible with the algorithm's depth.

Problem Set 2

2.1 Quantifying "State Explosion":

    (a) A general n-qubit pure state requires 2^n complex amplitudes. Assuming 16 bytes per complex number, estimate the memory needed for $n = 50$, $n = 75$, and $n = 100$.

    (b) Repeat the estimate for storing only product states (tensor products of single-qubit states). How does the scaling differ?

    (c) Explain how this motivates Feynman's argument for quantum simulation.

2.2 Quantum Church-Turing Thesis: The Church-Turing Thesis has evolved to account for computational efficiency and the physical realizability of computing devices.

    (a) Briefly explain the difference between the Classical Church-Turing Thesis and the Extended Church-Turing Thesis (ECTT).

    (b) What evidence challenges the ECTT, and how does this motivate the Quantum Church-Turing Thesis (QCTT)?

    (c) Why doesn't quantum computing violate the original Church-Turing Thesis?

2.3 Classical vs. Quantum Computation:

    (a) Give three distinct features that differentiate quantum computation from classical computation.

    (b) Consider a problem that is difficult for classical computers. Explain how a quantum computer might approach this problem differently to achieve a speedup.

2.4 The Promise of Quantum Computing:

    (a) Discuss the "promises" of quantum computation mentioned in § 2.2.4. For each promise, briefly explain its potential impact on a specific field (e.g., medicine, finance, materials science).

    (b) What are some of the "challenges" that need to be overcome to fully realize these promises?

2.5 Classical vs. Quantum Parallelism: In classical parallel computing, evaluating a function on all 2^n possible n-bit inputs requires 2^n processors.

    (a) Explain how an n-qubit quantum state in uniform superposition encodes all 2^n inputs at once.

    (b) Why does this not automatically mean quantum computers can evaluate all 2^n outputs simultaneously?

2.6 ✳ The Limits of Quantum Speedup:

    (a) Are all classically hard problems amenable to quantum speedup? Discuss problems that are believed not to benefit from quantum algorithms and why.

    (b) Consider the "No Free Lunch" theorem in optimization which states that no optimization algorithm is universally better than others when performance is averaged over all possible problems, i.e., any optimization algorithm that performs well on some problems must perform poorly on others. How might this concept apply to the search for quantum algorithms that provide speedups for all types of problems?

2.7 ✳ Choose one application area discussed in Appendix D (e.g., combinatorial optimization, quantum simulation, or AI-augmented modeling). Investigate its current status in quantum computing research. In your presentation, address the following:

- What is the classical complexity of key problems in this domain?

- What quantum algorithms have been proposed or implemented?

- What are the main challenges and recent breakthroughs?

- What impact could quantum speedups have on the industry or science?

2.8 Reversible Computation and Uncomputation:

    (a) Let $f : \{0,1\}^n \to \{0,1\}^m$. Show that $(x, y) \mapsto (x, y \oplus f(x))$ is a permutation on $\{0,1\}^{n+m}$.

    (b) Consider a circuit that computes $f(x)$ into a work register but leaves "garbage" in ancillas. Describe a general uncomputation strategy that resets the garbage to $|0\rangle$.

    (c) Give a concrete example with $f(x_1, x_2) = x_1 \vee x_2$: design a reversible circuit and show how to uncompute any temporary ancillas.

2.9 ✳ Reality Check on Promises and Challenges: Pick one current experimental platform (superconducting qubits, trapped ions, neutral atoms, or photonics).

    (a) Identify one near-term algorithmic target discussed in this chapter (simulation, optimization/sampling, linear-algebraic subroutines, or cryptanalysis).

    (b) Explain which of the practical challenges in § 2.2.4 most directly limit that target (noise, scalability, verification, or algorithm design).

    (c) Propose two concrete metrics you would track to assess progress (e.g., logical error rate, circuit depth at fixed fidelity, problem size, or verification cost).

3. Quantum Circuit Model and Query Model

Contents

The previous chapter introduced quantum algorithms as a new computational paradigm. To move from concepts to concrete implementations, we now formalize quantum computation using the *quantum circuit model*—a framework that specifies how algorithms are built from gates, applied to qubits, and measured to produce classical outcomes.

Just as classical algorithms can be rendered as logic circuits of AND, OR, and NOT gates, quantum algorithms are expressed as circuits of unitary gates and measurements. This chapter also introduces the *query model* and explains how query algorithms can be realized as circuits.

In this chapter, we introduce the structure of quantum circuits and how they represent quantum algorithms, review the basic building blocks—qubits, gates, and measurements—define circuit complexity measures such as size and depth, and present the query model together with the circuit-level realization of query oracles.

3.1 Quantum States, Gates, and Measurements

We begin with a brief review of three essential ingredients of quantum circuits: quantum states, gates, and measurements. These topics are covered in detail in the QCI Book [2]; here we recap the most relevant aspects as preparation for the circuit model.

 This chapter builds on the following chapters from other books in the series:

- QCI Book [2], Chapter 5: *Single-Qubit Quantum Gates*
- QCI Book [2], Chapter 6: *Multi-Qubit Systems*
- QCI Book [2], Chapter 7: *Multi-Qubit Quantum Gates*

3.1.1 Encoding Information in Quantum States

Encoding information into quantum states is the first step of quantum computation. Each encoding method leverages distinct quantum resources—superposition, entanglement, and interference—to enable computational advantages. The choice of encoding depends on the problem structure, the algorithm's requirements, and the physical constraints of the quantum hardware.

Below we summarize the most common approaches.

1 Basis State Encoding

In this encoding, classical bits (0 or 1) are encoded directly into the computational basis states of qubits, $|0\rangle$ and $|1\rangle$. For n qubits, an n-bit string such as "101" is encoded as the tensor product state $|1\rangle\,|0\rangle\,|1\rangle$, also written as $|101\rangle$. This simple and widely used method initializes registers in algorithms like Shor's factoring algorithm, where integers are loaded into quantum states in binary form.

2 Superposition Encoding

Information is spread across a superposition of basis states, so that multiple values are represented simultaneously. A common example is the uniform superposition

$$|\psi\rangle = \frac{1}{\sqrt{2^n}} \sum_{x\in\{0,1\}^n} |x\rangle,$$

which represents all 2^n basis states at once. This encoding underlies Grover's algorithm, where the initial superposition enables simultaneous evaluation of many candidate solutions. (More details in § A.2.)

3 Amplitude Encoding

Classical data, such as a vector $\boldsymbol{v} = (v_0, v_1, \ldots, v_{2^n-1})$, can be stored directly in the amplitudes of a quantum state:

$$|\psi\rangle = \sum_{x=0}^{2^n-1} v_x\,|x\rangle, \quad \sum_x |v_x|^2 = 1.$$

This compactly encodes 2^n complex numbers into n qubits, making it attractive for quantum machine learning and algorithms like the Harrow–Hassidim–Lloyd (HHL) algorithm for solving linear systems. At the same time, preparing an arbitrary amplitude-encoded state can be costly; in many applications, the encoding is only efficient when the data has special structure or is generated coherently as part of the computation.

4 Entanglement-Based Encoding

Entangled states, such as the Bell state

$$\tfrac{1}{\sqrt{2}}(|00\rangle + |11\rangle),$$

are used to represent correlated data or to facilitate communication tasks. This type of encoding is fundamental in quantum teleportation, superdense coding, and distributed quantum information processing, where correlations between qubits carry information more efficiently than classical methods.

5 Phase Encoding

Information may also be stored in the relative phases of quantum states, such as

$$|\psi\rangle = \cos\theta \, |0\rangle + e^{i\phi} \sin\theta \, |1\rangle .$$

Here the parameters θ and ϕ encode data into amplitude ratios and phases. Phase encoding is central in quantum phase estimation, a building block of Shor's algorithm, where hidden periodicities are revealed through interference.

Key Takeaways

1. Encoding is a design choice: basis-state, superposition, amplitude, entanglement-based, and phase encodings emphasize different resources.

2. Amplitude encoding is compact in qubit count, but arbitrary state preparation can be costly; it is efficient mainly when data has exploitable structure.

3. Measurement returns classical samples, not amplitudes; quantum algorithms succeed by using interference to concentrate useful information into observable outcomes.

3.1.2 Quantum Gates

Quantum gates are the building blocks of quantum circuits. Each gate is a unitary operator acting on one or more qubits, applied under external control. Quick reference tables on common gates are available in § A.1.

1 Single-Qubit Gates

Single-qubit gates act on the two-dimensional Hilbert space of a qubit, implementing rotations on the Bloch sphere. Examples include the Pauli gates (X, Y, Z), the Hadamard gate (H), and arbitrary rotation gates $(R_x(\theta), R_y(\theta), R_z(\theta))$.

2 Multi-Qubit Gates

Multi-qubit gates implement interactions between qubits. The most important is the controlled-NOT (CNOT), which can generate entanglement. Other examples include the controlled-Z and the Toffoli (CCNOT) gate.

3 Layers and Circuit Structure

Quantum circuits are composed of layers of gates. Gates acting on different qubits in the same layer can be applied in parallel. This defines the *depth* of the circuit, while the number of qubits defines its *width*. Together, these determine core computational resources of the circuit.

4 Clifford vs. Non-Clifford Gates

A useful distinction is between *Clifford gates* and *non-Clifford gates*. The Clifford group is generated by $\{H, S, \text{CNOT}\}$ and includes operations such as the Pauli gates and controlled-Z. Circuits built entirely from Clifford gates are known as *Clifford circuits*. By the Gottesman–Knill theorem, these can be simulated efficiently on classical computers, meaning they do not provide quantum speedups by themselves.

Non-Clifford gates, such as the T gate (a $\pi/8$ phase rotation), extend computational power beyond the Clifford group. When combined with Clifford gates, they form a universal gate set, capable of approximating any unitary transformation to arbitrary accuracy. Thus, non-Clifford gates are essential for achieving quantum advantage.

This distinction is not only theoretical but also practical. Clifford gates are often relatively easy to implement fault-tolerantly, for instance as transversal gates in many quantum error-correcting codes. Non-Clifford gates, however, are typically expensive to realize fault-tolerantly. In most architectures, they require specialized techniques such as magic state distillation and injection (see § A.5). As a result, measures like T-count and T-depth are critical benchmarks for estimating the cost of fault-tolerant quantum algorithms.

Key Takeaways

---

1. Clifford-only circuits are classically efficiently simulable (Gottesman–Knill), so Clifford gates alone do not yield general quantum advantage.

2. Adding a non-Clifford gate (e.g., T) makes the gate set universal when combined with Clifford gates and entangling operations such as CNOT.

3. In fault-tolerant settings, non-Clifford resources dominate cost, so T-count and T-depth are often the most informative complexity metrics.

3.1.3 Universal Gate Sets

A finite collection of gates is called a *universal gate set* if any unitary operation on n qubits can be approximated to arbitrary accuracy by a sequence of gates from the set. Formally, for any unitary operator U and any $\varepsilon > 0$, there exists a circuit using only gates from the set that approximates U within error ε in operator norm.

Table 3.1 lists several common universal gate sets.

The *Solovay–Kitaev theorem* ensures that approximating arbitrary unitaries with a universal set requires only a polylogarithmic overhead in gate count (as a function of the target precision), making such approximations efficient in practice.

Universal Gate Set	Description
$\{H, T, \text{CNOT}\}$	A minimal Clifford+T set. Universal because H and T generate arbitrary single-qubit rotations, while CNOT provides entanglement.
$\{H, S, T, \text{CNOT}\}$	Extends Clifford+T with the S gate ($\pi/2$ phase). Useful for simplifying synthesis and error correction.
$\{R_x(\theta), R_y(\theta), \text{CNOT}\}$	A parameterized gate set used in many hardware-level platforms. Arbitrary single-qubit gates can be built from R_x and R_y.
$\{X, Y, Z, H, S, T, \text{CNOT}\}$	A pedagogical set including all common Pauli and phase gates. Redundant but convenient for teaching and simulation.
Hardware Basis Gates	Gate sets supported natively by devices. For example: IBM's $\{R_z(\theta), X_{\pi/2}, \text{CX}\}$; Quantinuum's $\{R_x(\theta), R_z(\theta), \text{CX}\}$.

Table 3.1: Common Universal Quantum Gate Sets

3.1.4 Quantum Measurements

Measurement is the interface between quantum and classical information. It is the only process by which outcomes of quantum computation can be read out.

1 Collapse of Superposition and Born's Rule

If a quantum system is in the state

$$|\psi\rangle = \sum_{x \in \{0,1\}^n} c_x |x\rangle ,$$

then measurement in the computational basis yields outcome x with probability $|c_x|^2$. Upon measurement, the system collapses to the corresponding basis state $|x\rangle$, meaning the original superposition—and the information it encodes—is irreversibly lost. This rule, known as Born's rule, governs all quantum measurement outcomes.

The outcome of a measurement in the computational basis is a classical string x, with each outcome occurring with probability $|c_x|^2$. Regardless of how information is encoded in the quantum state, measurement only reveals outcomes sampled from this probability distribution. The complex amplitudes c_x themselves are not directly observable; they can only be inferred statistically through repeated measurements.

2 Partial Measurements

In a multi-qubit system, it is often unnecessary to measure all qubits simultaneously. A partial measurement refers to observing only a subset of qubits, which collapses their states while projecting the remaining qubits into a new state consistent with the measurement outcome. This technique is essential in many quantum algorithms, such as Simon's and Shor's algorithms. A more detailed discussion of this paradigm is provided in § A.3.

3   Holevo's Bound: Limits on Classical Information Extraction

An n-qubit quantum system encodes its state using 2^n complex amplitudes, suggesting an exponential descriptive richness. However, Holevo's theorem places a fundamental limit on what can be extracted as classical information from an ensemble of quantum states: in a single use of an n-qubit system, the accessible classical information is at most n bits. In other words, measurement provides only a bounded classical "readout bandwidth," even though the underlying state vector lives in an exponentially large space.

Quantum algorithms exploit interference to concentrate problem-relevant information into these accessible bits. For example, Shor's and Deutsch–Jozsa algorithms use superposition and interference to ensure that the measurement outcome encodes the desired answer, despite the limited quantum-to-classical readout.

Key Takeaways

1. A measurement in the computational basis samples x with probability $|c_x|^2$ and collapses the state accordingly.

2. Partial measurement collapses only the measured subsystem, projecting the remaining qubits into a conditional post-measurement state.

3. Despite the 2^n-dimensional state space, a single readout of n qubits yields at most n classical bits; algorithms rely on interference to make those bits informative.

3.2   Quantum Circuit Model

The circuit model provides a precise, universal language for quantum computation. In this framework, a quantum algorithm is realized as a finite sequence of gates acting on qubits, followed by measurements that produce classical outcomes. This section introduces the main ingredients of the circuit model and explains how high-level algorithm descriptions are compiled into circuits that can run on hardware.

3.2.1   Foundations of the Circuit Model

1   Components of Quantum Circuits

Quantum circuits serve both as computational blueprints and as a descriptive language for the execution of a quantum algorithm. Typical components include:

- Qubits: Basic units of quantum information, represented by state vectors in a complex Hilbert space.

- Quantum gates: Unitary transformations on one or more qubits (e.g., Hadamard, Pauli, CNOT).

- Mid-circuit measurements and feedforward: Measurements taken during execution whose outcomes may classically control subsequent operations.

- Final measurements: Measurements at the end of the circuit producing classical bits as output.

Figure 3.1 shows an example circuit implementing Shor's three-qubit phase-flip error-correction code (adapted from QCI Book [2], Section 13.3: *Introduction to Error Correction Codes*). This small but rich example already captures the essential ingredients of the circuit model: it involves both qubits and classical bits, uses single- and two-qubit gates, and includes mid-circuit measurements.

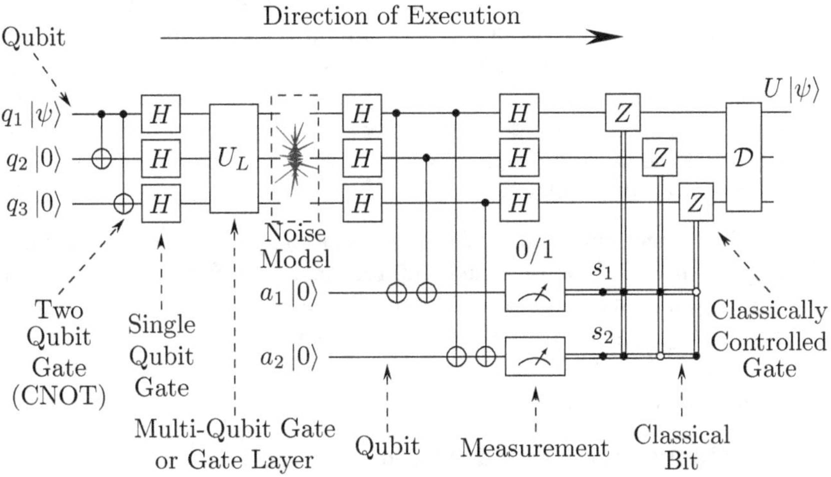

Figure 3.1: Quantum Circuit Example (Phase-Flip Shor Code)

Computation in place: In classical hardware, gates and wires are physical objects. In quantum circuits, the diagram is abstract: the wires represent the time evolution of qubits, and the gates denote externally applied control operations on those qubits. Physically, only the qubits, control signals, and readouts exist. See QCI Book [2], Section 5.3: *From Gate Sequences to Quantum Circuits*.

2 Circuit Model of Quantum Computation

As discussed in § 2.2, a quantum algorithm comprises a finite sequence of unitary transformations and measurements applied to n qubits to solve computational problems. The quantum circuit model formalizes this process by representing algorithms as sequences of quantum gates that implement unitary transformations, followed by measurements. It also provides the foundation for quantum complexity classes such as BQP (see Chapter 4).

This model is universal: any quantum computation can be compiled using a small set of universal gates, analogous to classical AND, OR, and NOT gates (see Table 3.1 for common sets). Quantum circuits therefore make algorithm structure explicit and provide a concrete basis for defining resource costs.

The standard model assumes unitary evolution with measurements at the end. Extensions such as mid-circuit measurements and classically controlled operations (feedforward) can be simulated within the unitary framework using ancilla qubits and controlled gates, preserving the model's universality.

Exercise 3.1 Write the unitary matrix for the two-qubit circuit $\text{CNOT}\,(H \otimes I)$ (Hadamard on the control followed by CNOT).

3 Why Quantum Needs So Many Qubits

A classical CPU (e.g., 64-bit wide) can process large integers by dividing them into manageable chunks and operating on those chunks sequentially or in parallel using modular logic circuits. In contrast, the state of n qubits is a single vector in a 2^n-dimensional space and is often highly entangled, so it generally cannot be decomposed into independent parts that can be processed separately without changing the computation. As a result, many quantum algorithms require coherent control over all n qubits at once.

Real-world quantum hardware further complicates this picture due to noise and decoherence. To sustain long computations, quantum error correction encodes each logical qubit into many physical qubits, often requiring hundreds or thousands of physical qubits per logical qubit. This demand, together with the need for fault-tolerant architectures, explains why practical quantum computation is expected to require very large qubit counts.

4 Quantum Annealing and the Circuit Model

Quantum annealing is based on a different computational viewpoint, commonly framed as *adiabatic quantum computing* (AQC). Instead of applying a sequence of discrete gates, one prepares the ground state of an initial Hamiltonian and then continuously deforms it into a final Hamiltonian whose ground state encodes the solution. If the interpolation is sufficiently slow relative to the minimum spectral gap, the adiabatic theorem implies that the system remains close to the instantaneous ground state, so measuring at the end yields the desired solution with high probability.

Although this looks different from the circuit model, adiabatic quantum computing and the standard circuit model are *polynomially equivalent*: either model can efficiently simulate the other [10, 11]. In this sense, AQC is not merely a heuristic, but a universal model of quantum computation.

Practical *quantum annealers* (such as current D-Wave devices) are best viewed as specialized realizations of this paradigm for optimization problems, typically with restricted Hamiltonian families and without full fault tolerance. Their performance depends strongly on noise, control errors, embedding overhead, and the choice of annealing schedule, but they remain an important experimental platform for studying optimization and continuous-time quantum dynamics.

5 Analog Quantum Simulation and the Circuit Model

A related continuous-time paradigm is *analog quantum simulation*. Here, one engineers a quantum device whose native Hamiltonian closely matches a target model (for example, a many-body lattice Hamiltonian), and then learns about the target by preparing initial states, letting the system evolve, and measuring observables. This approach avoids compiling dynamics into gate sequences, and it can be highly effective for domain-specific tasks.

At first glance, analog simulation seems outside the circuit model. However, the *Quantum Church–Turing Thesis* suggests that any physical quantum process can, in principle, be efficiently simulated by a fault-tolerant gate-based quantum computer [4,

7]. Thus, the circuit model provides a universal digital framework, while analog simulators provide direct physical realizations of particular Hamiltonian dynamics. The trade-off is practical: analog devices can exploit natural interactions with relatively low overhead, but they typically offer weaker and more device-dependent error guarantees than fully error-corrected digital circuits.

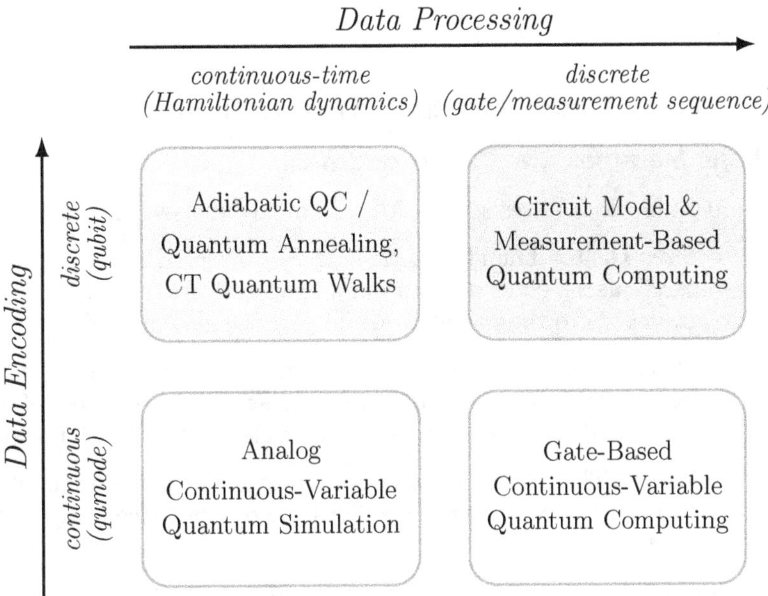

Figure 3.2: Quantum Computing Paradigms by Encoding and Processing

Fig. 3.2 summarizes these paradigms by separating (i) *data encoding* (discrete qubits versus continuous-variable qumodes) and (ii) *data processing* (discrete gate/measurement sequences versus continuous-time Hamiltonian dynamics). It also highlights where adiabatic computation and quantum annealing sit relative to the circuit model, and where analog quantum simulation fits as a domain-specific continuous-time approach.

6 ✳ Uniform Circuit Families

In theoretical computer science, we are often concerned not with isolated circuits, but with *families* of circuits that handle inputs of varying lengths. A circuit family $\{C_n\}$ consists of one circuit for each input size n, where each C_n acts on n input qubits (plus ancillas if needed) and outputs a result consistent with the desired computation.

However, to keep the model computationally grounded, we impose a crucial condition known as *uniformity*. A circuit family is said to be uniform if there exists a classical algorithm that outputs a description of the circuit C_n in time polynomial in n. This ensures that the circuit family does not encode hidden information or uncomputable behavior and can be feasibly generated as part of a well-defined algorithm.

Uniformity plays the same role in quantum complexity theory as it does in classical models: it guarantees that the circuit model faithfully captures what it means to compute something efficiently. Without uniformity, one could artificially embed solutions to hard problems directly into the circuit structure, violating the spirit of algorithmic computation.

Thus, a uniform quantum circuit family represents an algorithm that can scale with input size, making it the formal backbone of complexity classes like BQP, which we explore in Chapter 4. In this way, quantum circuits serve both as executable instructions on a quantum processor and as abstract objects in the landscape of computational complexity.

3.2.2 Quantum Circuit Complexity Measures

1 Basic Measures: Size, Depth, and Width

A quantum circuit's resource cost is commonly analyzed using three basic measures:

- Size: The total number of *elementary gates* used in the circuit, where "elementary" means gates drawn from a fixed gate set (e.g., $\{H, T, \mathrm{CNOT}\}$). This is analogous to the operation count of a classical algorithm.

- Depth: The number of *gate layers*, where each layer consists of gates that can be executed in parallel because they act on disjoint sets of qubits. Depth is often used as a proxy for time under ideal parallelization assumptions.

- Width: The total number of qubits required, including input qubits, ancillas, and workspace qubits. This plays the role of space complexity in the circuit model.

Theoretically, quantum speedups are often stated as reductions in asymptotic computational resources. In many textbook examples, the most salient reduction is in a time-like resource, which can be modeled by circuit depth under parallelization assumptions. For instance, Shor's algorithm factors integers in polynomial time (and hence polynomial depth in standard constructions), whereas all known classical factoring algorithms require superpolynomial time; Grover's algorithm reduces the number of oracle queries for unstructured search from $O(N)$ to $O(\sqrt{N})$.

Practically, both depth and width are constrained by hardware. Width is limited by the number of available qubits, while depth is limited by coherence time and gate fidelity: if a circuit runs too long, accumulated noise overwhelms the computation. Designing circuits that balance these constraints—reducing depth without an unmanageable increase in width—is therefore central in the current Noisy Intermediate-Scale Quantum (NISQ) era.

It is also important to distinguish *abstract* circuit measures from *device-level* execution time. Gate counts and idealized depth are defined for a chosen gate set under perfect parallelism, but real hardware imposes additional constraints: limited connectivity can force SWAP insertion and increase two-qubit depth; native gates differ in duration and error; compilation may rewrite circuits to match calibrated pulse schedules; and measurement, reset, and classical feedforward introduce latencies that do not behave like ordinary unitary layers. As a result, smaller "size" or even smaller "depth" in an abstract cost model may not translate directly to shorter wall-clock runtime on current devices. Thus, implementation costs should state the assumed connectivity and native gate set, and clarify whether depth refers to total depth, two-qubit depth, or a timing-aware schedule.

Exercise 3.2 Consider an n-qubit circuit that consists of

- one layer of n single-qubit Hadamards,

- then n CNOTs arranged as a circular chain $\text{CNOT}_{1,2}$, $\text{CNOT}_{2,3}$, ..., $\text{CNOT}_{n-1,n}$, $\text{CNOT}_{n,1}$,

- then one final layer of n single-qubit Hadamards.

Compute the circuit width, size, and depth for (i) all-to-all connectivity, and (ii) a linear nearest-neighbor (LNN) device.

Exercise 3.3

(a) Explain why the quantum circuit model is a good abstraction for understanding quantum algorithms, even though physical devices are not literally "wires" and "gates" in the classical sense.

(b) Describe the role of initialization, unitary operations, and measurement in a quantum circuit.

2 ✳ Other Measures

Beyond size, width, and depth, several additional measures are important both theoretically and practically. These are summarized in Table 3.2. The relevant choice of measures depends on the setting: in NISQ devices, connectivity and noise often dominate; in fault-tolerant settings, non-Clifford resources and logical-to-physical overheads become central.

Measure	Informal Description	Practical Importance
Ancilla Qubits	Number of extra workspace qubits introduced during computation.	More ancillas can reduce depth but increase width and error-correction costs.
Gate Set / Cost Model	Distinguishes between Clifford gates (often cheaper) and non-Clifford gates (often more expensive, e.g., T).	Fault-tolerant cost is frequently dominated by non-Clifford operations.
T-Count	Total number of T gates.	A central metric in fault-tolerant quantum computing, since high-fidelity non-Clifford resources are expensive.
T-Depth	Number of sequential layers of T gates (after parallelization).	Determines time overhead of non-Clifford resources under error correction.

Connectivity Overhead	Additional gates needed to respect hardware connectivity (e.g., nearest-neighbor constraints).	A major bottleneck on NISQ devices with limited qubit layouts.
Noise Tolerance	Maximum error rates (per gate or per qubit) that the circuit can tolerate.	Determines whether an algorithm can run reliably on a given device.
Logical vs. Physical Depth	Depth before vs. after encoding into error-corrected logical qubits.	Highlights the gap between abstract complexity and fault-tolerant feasibility.
Measurement Complexity	Number and type of measurements (including mid-circuit measurements) and associated classical control.	Affects timing, feedforward latency, and hybrid quantum–classical workflows.
Communication Complexity	Number of entangling operations across physically separated qubits or modules.	Critical for modular and distributed architectures (e.g., photonic interconnects).

Table 3.2: Common Measures of Quantum Circuit Complexity

Key Takeaways

1. Size, depth, and width capture the basic gate, time-like, and space-like resource costs of a circuit.

2. On real devices, two-qubit gates typically dominate error and time budgets, so two-qubit depth and two-qubit gate count are often the practical bottlenecks.

3. Hardware constraints (connectivity, native gate set, noise) can introduce substantial compilation overheads, making "algorithmic" complexity and "executed" complexity differ.

3.2.3 From Applications to Circuits

The quantum circuit model not only underpins much of the theoretical study of quantum computation, but also serves as the practical foundation for programming and executing algorithms on today's quantum devices. In practice, quantum programs pass through multiple layers of representation, from high-level algorithm design to hardware-native gate sequences.

1 From Classical Computer Code to Logic Circuits

When we run a Python program or use an app such as a web browser, we interact with high-level code that is designed for humans to read and write. The computer, however, ultimately executes low-level machine instructions, such as moving data, performing arithmetic, and branching based on conditions.

To bridge this gap, high-level code is *compiled* (or interpreted) into *machine code*, a sequence of binary instructions. These instructions are executed by the CPU (central processing unit), which realizes a small, universal set of operations using logic gates such as AND, OR, and NOT. From this perspective, a program execution can be viewed as the activation of a large, dynamic logic circuit, built from billions of transistors, unfolding over time.

2 From Quantum Algorithms to Hardware

Quantum computers have a similar layered structure: a quantum algorithm is expressed as a circuit at a high level, then compiled and optimized into a hardware-native sequence of operations, and finally executed on a physical device or a simulator. Table 3.3 outlines a typical hierarchy from applications to hardware, indicating how quantum programs are developed, compiled, optimized, and executed.

Layer	Description	Examples
1. Application	Defines the real-world problem or task to be solved using quantum computation.	Factoring, simulating molecules, optimization problems, machine learning
2. Algorithm	A mathematical procedure designed to solve the problem using quantum principles.	Shor's algorithm, Grover's algorithm, QAOA, VQE
3. Programming Platform	Tools used to construct, simulate, and submit quantum circuits.	Qiskit, Cirq, PennyLane, Braket
4. Logical Circuit	Abstract, hardware-agnostic circuit using gates, blocks (e.g., QFT), measurements, and control flow.	Circuit with QFT, H, X, Z, CNOT, and measurement
5. Compilation	Translates high-level quantum programs or logical circuits into a gate-level form.	Qiskit Transpiler, t|ket⟩ compiler, Quilc
6. Optimization	Refines the circuit to reduce gate count and depth and to adapt to hardware constraints such as connectivity and noise.	Gate fusion, qubit routing, pulse-level scheduling
7. Native (ISA) Circuit	Circuit expressed in the hardware's native gate set, defined by its ISA (instruction set architecture).	Circuit based on the $\{R_x(\theta), R_z(\theta), \text{CX}\}$ set
8. Execution Platform	Executes and manages quantum circuits on real hardware or simulators, often through cloud-based services.	AWS Braket, Azure Quantum, IBM Quantum Platform

| 9. Hardware | The physical quantum processor where quantum states and gates are realized. | Superconducting qubits, trapped ions, neutral atom arrays |

Table 3.3: Hierarchy of Quantum Program Execution

Note 1: Some frameworks include basic compilation features, blurring the line between the programming and compilation layers.

Note 2: Optimization can occur at multiple stages, but here it refers specifically to gate-level improvements targeting hardware constraints.

Exercise 3.4 Use a simulator (e.g., Qiskit) to prepare a three-qubit GHZ state. Verify entanglement by measuring in the X basis and comparing correlations to the Z-basis outcomes.

Exercise 3.5 ✳ Using a simulator toolchain (e.g., Qiskit), build a circuit that prepares $|\text{GHZ}_n\rangle$ using $\{H, \text{CNOT}\}$ for $n = 8$.

(a) Report the depth and two-qubit gate count before transpilation.

(b) Choose a specific coupling map corresponding to a linear nearest-neighbor architecture and transpile the circuit. Report the depth and two-qubit gate count after transpilation.

(c) Repeat (b) for a different GHZ construction (e.g., a "tree" construction) and compare the overheads.

State clearly which transpiler options (optimization level, routing strategy, etc.) you used.

Key Takeaways

1. A quantum program passes through layers: application $\rightarrow$ algorithm $\rightarrow$ logical circuit $\rightarrow$ compilation/optimization $\rightarrow$ native circuit $\rightarrow$ execution.

2. Compilation maps a hardware-agnostic circuit to a device's native gate set and connectivity constraints; optimization reduces depth and gate count under those constraints.

3. Reported resource metrics should specify the level (logical vs transpiled) and the metric (total depth vs two-qubit depth, total gates vs two-qubit gates).

3.3 ✳ Query Model

In § 3.2, we established the quantum circuit model as a formal framework for quantum computation. However, several canonical algorithms—including the Deutsch–Jozsa (DJ), Simon, and Grover algorithms—are most naturally analyzed in the query model (or oracle model), where access to a hidden function is idealized as a black box. In this section, we compare the query and circuit models and show that any algorithm defined in the query model can be implemented within the circuit model, reinforcing the circuit model as a universal framework for quantum algorithms.

Deutsch–Jozsa (DJ), Simon, Shor, and Grover Algorithms

Historically, these algorithms were among the first to demonstrate provable speedups over classical approaches for well-defined problem settings. They serve as archetypal examples of quantum speedup. The Deutsch–Jozsa, Simon, and Grover algorithms are commonly presented in the query (oracle) framework, whereas Shor's algorithm is typically analyzed in the circuit model, even though it contains oracle-style subroutines such as order finding.

Algorithm	Year	Key Contribution and Impact
Deutsch / Deutsch–Jozsa	1985 / 1992	Demonstrated quantum parallelism and a separation in the oracle model for a promise problem. While the exponential separation disappears against randomized classical algorithms, it provided an important conceptual breakthrough.
Simon	1994	Gave the first exponential speedup (in the oracle model) over all classical algorithms, including randomized ones. Introduced a hidden-period structure that foreshadowed Shor's period-finding approach.
Shor	1994	Showed exponential speedups for factoring and discrete logarithms, with major cryptographic implications. This result catalyzed widespread interest in quantum computing.
Grover	1996	Achieved a quadratic speedup for unstructured search, applicable to a wide range of problems. Illustrated how quantum mechanics can improve brute-force search.

Table 3.4: Early Quantum Algorithms and Their Contributions

Key Takeaways

1. The query (oracle) model counts the number of black-box calls to f and treats other computation as secondary.

2. Query complexity isolates the informational advantage of quantum superposition and interference.

3. Any query-model algorithm can be implemented in the circuit model, provided the oracle U_f itself can be realized efficiently.

3.3.1 ✳ Understanding the Query Model

Let us begin with a high-level comparison of the query and circuit models of quantum computation:

- Circuit model: A universal model based on sequences of unitary gates, typically followed by measurements at the end. It aligns closely with physical implementations and is the standard framework for defining complexity classes.

- Query model (or oracle model): An abstract model that provides black-box access to a hidden function f. It separates the cost of querying f from other computational operations, enabling a clean analysis of information access and query complexity.

The query model is especially useful for:

- Analyzing resource complexity in terms of the number of oracle calls, independent of gate-level details.

- Highlighting the distinction between classical and quantum access to information, thereby clarifying the source of quantum advantage.

- Providing a structured setting for proving lower bounds, so that speedups can be attributed to quantum information processing rather than to model-dependent artifacts.

We now illustrate the utility of the query model using the Deutsch–Jozsa and Simon algorithms.

Deutsch–Jozsa (DJ) Algorithm

1. A phase oracle U_f encodes $f(x)$ into the phase: $U_f \left| x \right\rangle = (-1)^{f(x)} \left| x \right\rangle$.

2. Hadamards create and recombine superpositions so that global phase patterns become observable interference.

3. Deutsch–Jozsa solves the promised constant-vs-balanced problem with 1 quantum query, while deterministic classical algorithms may require up to $2^{n-1} + 1$ queries.

■ **Example 3.1 — Deutsch–Jozsa (DJ) Algorithm in the Query Model.** The DJ algorithm is illustrated in Fig. 3.3. While we do not repeat its full analysis here, we focus on the role of the oracle. Readers interested in additional details are referred to QCI Book [2], Section 11.1: *The Deutsch-Jozsa Algorithm.*

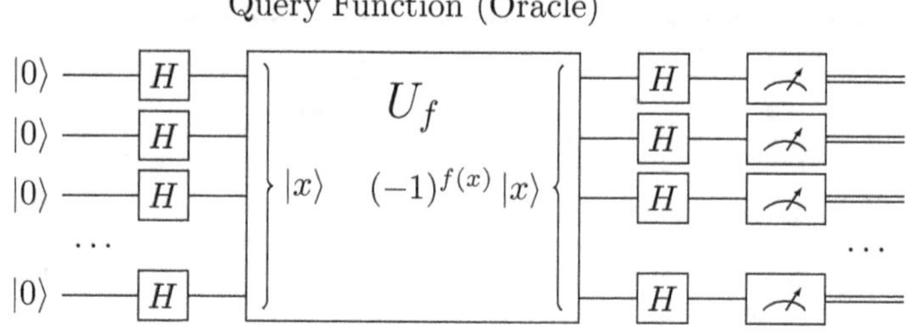

Figure 3.3: Query Oracle in the Deutsch–Jozsa Algorithm

In the query model, the function $f : \{0,1\}^n \to \{0,1\}$ is accessed via a phase oracle U_f that acts as

$$U_f \left| x \right\rangle = (-1)^{f(x)} \left| x \right\rangle . \tag{3.1}$$

This oracle encodes the function value into the phase of the computational basis state. The algorithm places this oracle between two layers of Hadamard gates: the first Hadamard layer prepares a uniform superposition over all x, the oracle imprints $f(x)$ into the phase, and the second Hadamard layer enables interference that reveals whether f is constant or balanced. The promise problem is solved with a single oracle query, whereas classical deterministic algorithms may require up to $2^{n-1}+1$ queries in the worst case.   ∎

■ **Example 3.2 — Simon's Algorithm in the Query Model.** Whereas the DJ algorithm demonstrates a separation in a promise setting, its advantage disappears when randomized classical algorithms are permitted. In contrast, Simon's algorithm achieves an exponential speedup over all classical (including randomized) algorithms in the oracle model and introduced a hidden-period structure that is foundational to Shor's period-finding approach.

Simon's problem is defined as follows. We are given a function

$$f_s : \{0,1\}^n \to \{0,1\}^n,$$

where $s \in \{0,1\}^n$ is an unknown nonzero bit string. The key promise is that there is a hidden XOR shift s such that

$$f_s(x) = f_s(x \oplus s)$$

for every input x. Here, $\oplus$ denotes bitwise XOR.

In fact, each output comes from exactly one such pair. That is, two inputs give the same output only when they differ by s. So the inputs are partitioned into pairs of the form

$$(x, \; x \oplus s),$$

and each pair maps to a single output value. Since there are 2^n input strings, there are 2^{n-1} such pairs.

The task is to recover the hidden string s using as few queries to the function as possible.

As shown in Fig. 3.4, Simon's algorithm uses two n-qubit registers, both initialized to $|0\rangle^{\otimes n}$. Applying Hadamard gates to the first register creates the uniform superposition

$$\frac{1}{\sqrt{2^n}} \sum_{x \in \{0,1\}^n} |x\rangle .$$

The oracle U_{f_s} is then applied as a black-box transformation:

$$U_{f_s} |x\rangle |0\rangle = |x\rangle |f_s(x)\rangle . \tag{3.2}$$

By linearity, its action on the superposition is

$$\frac{1}{\sqrt{2^n}} \sum_{x \in \{0,1\}^n} |x\rangle |0\rangle \longmapsto \frac{1}{\sqrt{2^n}} \sum_{x \in \{0,1\}^n} |x\rangle |f_s(x)\rangle .$$

This correlates the two registers. Because Simon's promise says that $f_s(x) = f_s(x \oplus s)$ and that these are the only collisions, measuring the second register and obtaining some value $f_s(\tilde{x})$ forces the first register into the equal superposition

$$\frac{1}{\sqrt{2}} \big(|\tilde{x}\rangle + |\tilde{x} \oplus s\rangle \big).$$

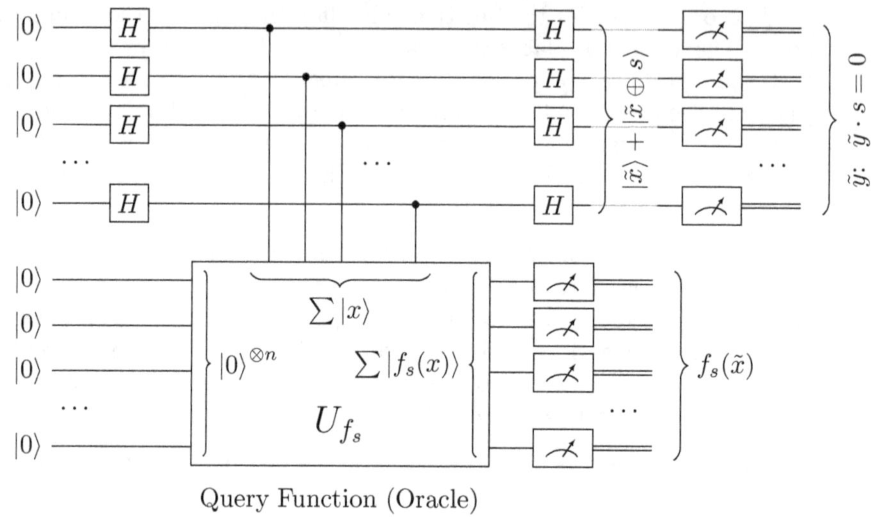

Query Function (Oracle)

Figure 3.4: Query Oracle in Simon's Algorithm

Next, Hadamard gates are applied to the first register. The measurement outcome is then a bit string $\tilde{y}$ satisfying

$$\tilde{y} \cdot s = 0 \bmod 2,$$

where $\tilde{y} \cdot s$ denotes the bitwise inner product modulo 2, equivalently the standard inner product over $\mathbb{F}_2$.

Repeating the circuit $O(n)$ times produces enough independent linear equations of the form $\tilde{y} \cdot s = 0 \bmod 2$ to recover s with high probability by solving a linear system over $\mathbb{F}_2$. In the oracle model, any classical randomized algorithm requires exponentially many queries to determine s with non-negligible success probability. Simon's algorithm was therefore the first example of an exponential quantum advantage in a black-box setting. ∎

Simon's Algorithm

1. Each run of Simon's algorithm produces a bit string $\tilde{y}$ satisfying $\tilde{y} \cdot s = 0 \bmod 2$, where s is the secret string.

2. Repeating the circuit yields enough independent linear constraints to recover s via linear algebra over $\mathbb{F}_2$.

3. The quantum speedup comes from coherent oracle access plus interference; the final step is classical post-processing.

3.3.2 ✳Realizing Query Oracles in the Circuit Model

In the query model, an oracle represents restricted access to information: the algorithm may query a function, dataset, or physical process without directly inspecting its internal structure. This viewpoint is useful both theoretically, for comparing classical and quantum query complexity, and practically, in settings where computation proceeds through subroutine calls, database access, simulation routines, or experimental devices.

To use such an oracle in a quantum algorithm, however, it must be realized in the circuit model as a concrete unitary transformation that acts coherently on superposition states. Thus, while the query model isolates the cost of information access, an actual circuit implementation must also account for the gate cost of building the oracle.

For a quantum algorithm to deliver an efficient speedup in the usual sense, the oracle circuit U_f must itself be implementable with polynomially many elementary gates. If realizing U_f requires exponentially many gates, then the overall algorithm is no longer efficient, even if its query complexity is small.

We next use Simon's algorithm to illustrate how an abstract oracle can be realized as a reversible quantum circuit with coherent access. For details on implementing the oracle in the Deutsch–Jozsa algorithm, refer to QCI Book [2], Section 11.1: *The Deutsch-Jozsa Algorithm.*

Key Takeaways

1. A standard reversible oracle implements $U_f |x\rangle |y\rangle = |x\rangle |y \oplus f(x)\rangle$, enabling coherent evaluation on superpositions.

2. Phase oracles and standard oracles are interconvertible with constant overhead using ancillas and simple gates (phase kickback).

3. Low query complexity alone does not guarantee an efficient algorithm; the circuit cost of implementing U_f must also be polynomial in n.

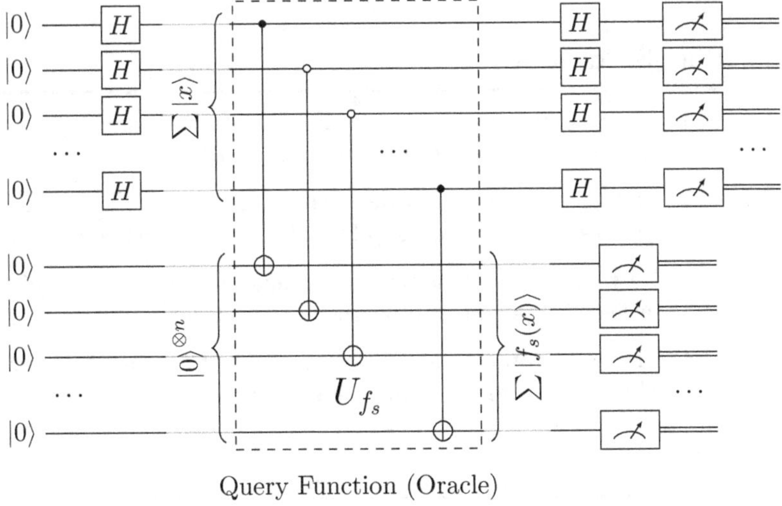

Query Function (Oracle)

Figure 3.5: Oracle Circuit in Simon's Algorithm

As discussed in Example 3.2, Simon's function is of the form $f_s : \{0,1\}^n \to \{0,1\}^n$ parameterized by a secret string s. The circuit in Fig. 3.5 implements $f_s(x)$ for a specific choice $s = 011\ldots 0$ using CNOT and $\overline{\text{CNOT}}$ gates. This construction realizes a reversible XOR of the function value into the second register, demonstrating how an abstract oracle can be expressed as a concrete unitary circuit.

More generally, for a function $f : \{0,1\}^n \to \{0,1\}^m$, a standard reversible oracle is implemented as a unitary U_f acting as

$$U_f \left|x\right\rangle \left|y\right\rangle = \left|x\right\rangle \left|y \oplus f(x)\right\rangle. \tag{3.3}$$

Here, $\left|x\right\rangle$ is the n-qubit input register and $\left|y\right\rangle$ is an m-qubit auxiliary register. The function value $f(x)$ is XORed into $\left|y\right\rangle$, while $\left|x\right\rangle$ remains unchanged, ensuring reversibility and coherent access in superposition.

In Simon's algorithm, we have $m = n$, $\left|y\right\rangle = \left|0\right\rangle^{\otimes n}$, and Eq. 3.3 reduces to Eq. 3.2.

Exercise 3.6 This exercise guides you through implementing and analyzing the Deutsch–Jozsa algorithm.

(a) Construct specific Deutsch–Jozsa oracle functions for input sizes of 2, 4, 8, 12, and 16 bits.

(b) Implement a classical deterministic algorithm to solve the problem in Python or another programming language of your choice.

(c) Implement a classical randomized algorithm to solve the problem in Python or another programming language.

(d) Implement the Deutsch–Jozsa quantum algorithm using Qiskit or another quantum programming platform.

(e) For each approach, analyze and compare the number of oracle queries as a function of input size.

Chapter Takeaways

1. Quantum information is encoded in states, transformed by unitary gates, and accessed through measurement statistics.

2. The circuit model provides a concrete language for algorithms: a finite gate sequence acting on qubits plus measurements and classical control.

3. Universal gate sets allow approximate implementation of arbitrary unitaries with polynomial overhead in accuracy.

4. The query (oracle) model isolates the cost of black-box access, clarifying where quantum speedups come from and how they scale.

Problem Set 3

3.1 The table below lists eigenstates and preparation circuits for the Pauli operators. For instance, the $+1$ eigenstate of the Pauli X operator is $\left|+\right\rangle = \frac{1}{\sqrt{2}}(\left|0\right\rangle + \left|1\right\rangle)$, which can be prepared by applying a Hadamard gate to a qubit initialized in the state $|0\rangle$. Your task is to complete the missing entries (marked with "?") in the table.

Pauli	+1 Eigenstate	Prep Circuit	−1 Eigenstate	Prep Circuit
Z	$|0\rangle$	None	$|1\rangle$	?
X	$\frac{1}{\sqrt{2}}(|0\rangle + |1\rangle)$	H	$\frac{1}{\sqrt{2}}(|0\rangle - |1\rangle)$	?
Y	$\frac{1}{\sqrt{2}}(|0\rangle + i|1\rangle)$	?	$\frac{1}{\sqrt{2}}(|0\rangle - i|1\rangle)$	?

3.2 Encoding practice: Consider the vector $\boldsymbol{v} = (1, 1, 0, 0)/\sqrt{2}$.

    (a) Write the amplitude-encoded quantum state $|\psi\rangle$ for $\boldsymbol{v}$.

    (b) Give a simple circuit that prepares this state from $|00\rangle$.

3.3 This exercise helps you refresh your understanding of quantum gate and circuit analysis. If you are unfamiliar with this topic, refer to QCI Book [2], Chapter 7: *Multi-Qubit Quantum Gates*.

Demonstrate that the magic T-state injection circuit shown in Fig. A.5 of § A produces $T|\psi\rangle$ (up to a global phase), regardless of the measurement outcome.

3.4 This exercise is designed to reinforce your understanding of the basics of quantum measurement. If you are unfamiliar with this topic, refer to QCI Book [2], Section 6.4: *Measurements of Multi-Qubit Systems*.

Consider a collection of qubits initialized in the multi-qubit state $|\psi\rangle$. As shown below, the qubits are first transformed by a unitary U, then measured in the computational basis, and finally transformed by $U^\dagger$:

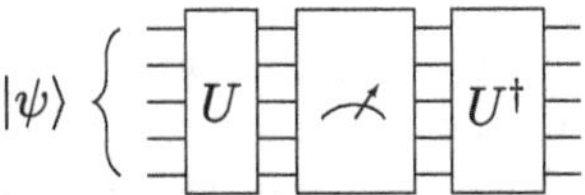

    (a) What are the possible measurement outcomes?

    (b) What is the final state of the system for a given measurement outcome?

3.5 We want to apply a CNOT gate between q_1 and q_n on a linear nearest-neighbor architecture (example shown for $n = 6$ in the figure). Since only adjacent qubits can interact, a sequence of SWAP gates must be used to bring the qubits together, as illustrated.

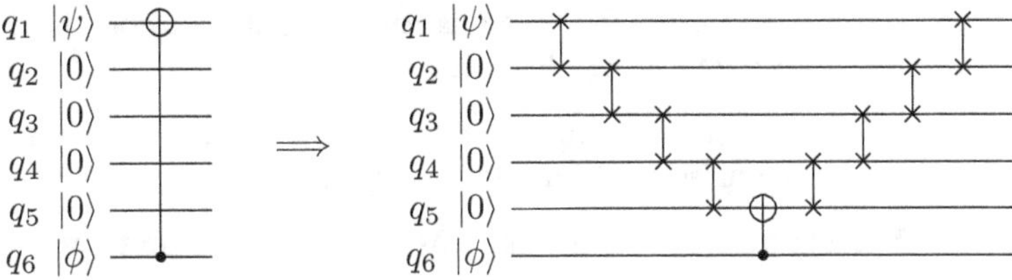

Assume each SWAP uses 3 elementary two-qubit gates and the original qubit order must be restored. For a system of n qubits, what are the circuit width, size, and depth under this assumption?

3.6 Let $f : \{0,1\}^n \to \{0,1\}$ and let U_f be the standard reversible oracle

$$U_f \,|x\rangle\,|y\rangle = |x\rangle\,|y \oplus f(x)\rangle\,.$$

(a) Show that if the second register is initialized in $|-\rangle = \frac{1}{\sqrt{2}}(|0\rangle - |1\rangle)$, then

$$U_f \,|x\rangle\,|-\rangle = (-1)^{f(x)}\,|x\rangle\,|-\rangle\,.$$

(b) Conclude that a phase oracle $|x\rangle \mapsto (-1)^{f(x)}\,|x\rangle$ can be implemented using one call to the standard oracle U_f plus single-qubit gates.

3.7 Query model warm-up: Let $f : \{0,1\}^n \to \{0,1\}$ be a Boolean function.

(a) Write down the action of the quantum oracle U_f on basis states.

(b) Show how U_f acts on a uniform superposition of inputs.

3.8 Bernstein–Vazirani as a query problem: Let $s \in \{0,1\}^n$ and define $f_s(x) = s \cdot x \bmod 2$.

(a) Write the standard oracle U_{f_s} and the phase oracle form for this function.

(b) Show that one quantum query suffices to recover s with certainty.

(c) Prove that any classical deterministic algorithm requires n queries in the worst case.

3.9 Oracle cost vs. query cost: In the query model, an algorithm may use Q oracle calls and "free" additional computation.

(a) Explain, with a concrete example, how an algorithm can have small query complexity but still be inefficient in the circuit model if U_f is expensive to implement.

(b) For a family of functions f_n whose classical circuit size is $\Theta(n^2)$, give a reasonable upper bound (order of growth) on the size of a reversible circuit for U_{f_n}, allowing ancillas.

3.10 $*$ This exercise guides you through implementing and analyzing Simon's algorithm.

(a) Construct specific Simon oracle functions for input sizes of 2, 4, 6, 8, and 10 bits, each with a nontrivial hidden period s.

(b) Implement a classical brute-force algorithm to determine the hidden string s in Python or another programming language of your choice.

(c) Implement Simon's quantum algorithm using Qiskit or another quantum programming platform.

(d) For each approach, analyze and compare the number of oracle queries required as a function of input size.

3.11 $*$ Figure 3.6 shows several circuit families for preparing the n-qubit GHZ state.

Conventions: Use gates $\{H, \text{CNOT}\}$. Depth counts parallel layers of two-qubit gates; a layer of single-qubit gates counts as depth 1 if used. When asked for "size," count two-qubit gates; state clearly if you include single-qubit gates. "Nearest-neighbor" means a linear array (LNN). "All-to-all" means any pair can interact. "Mid-circuit" measurement and "reset" each contribute one depth layer.

(a) For each implementation in Fig. 3.6, show that the output state is

$$|\mathrm{GHZ}\rangle = \frac{1}{\sqrt{2}}\big(|0\,0\cdots 0\rangle + |1\,1\cdots 1\rangle\big).$$

(b) Part (i) of Fig. 3.6 uses only LNN connectivity. Part (ii) reduces the depth by parallelizing CNOTs. For part (ii), give the circuit width, size, and depth as functions of n.

(c) Part (iii) arranges the CNOTs so they can be morphed into the tree in part (iv) (assume all-to-all connectivity). Prove that the depth of part (iv) grows logarithmically in n.

(d) Part (v) achieves constant depth on an LNN device using mid-circuit measurements, feedforward (classically controlled X), and qubit reset/reuse. For an n-qubit GHZ, report width, size, and depth under the conventions above, counting one layer for the measurements and one for resets. (The figure shows $n = 7$ as an example.) In the classical side-processing, the updates on the classical bits are XORs, that is, additions modulo 2, so the stored bits track parities of measurement outcomes rather than ordinary logical ORs.

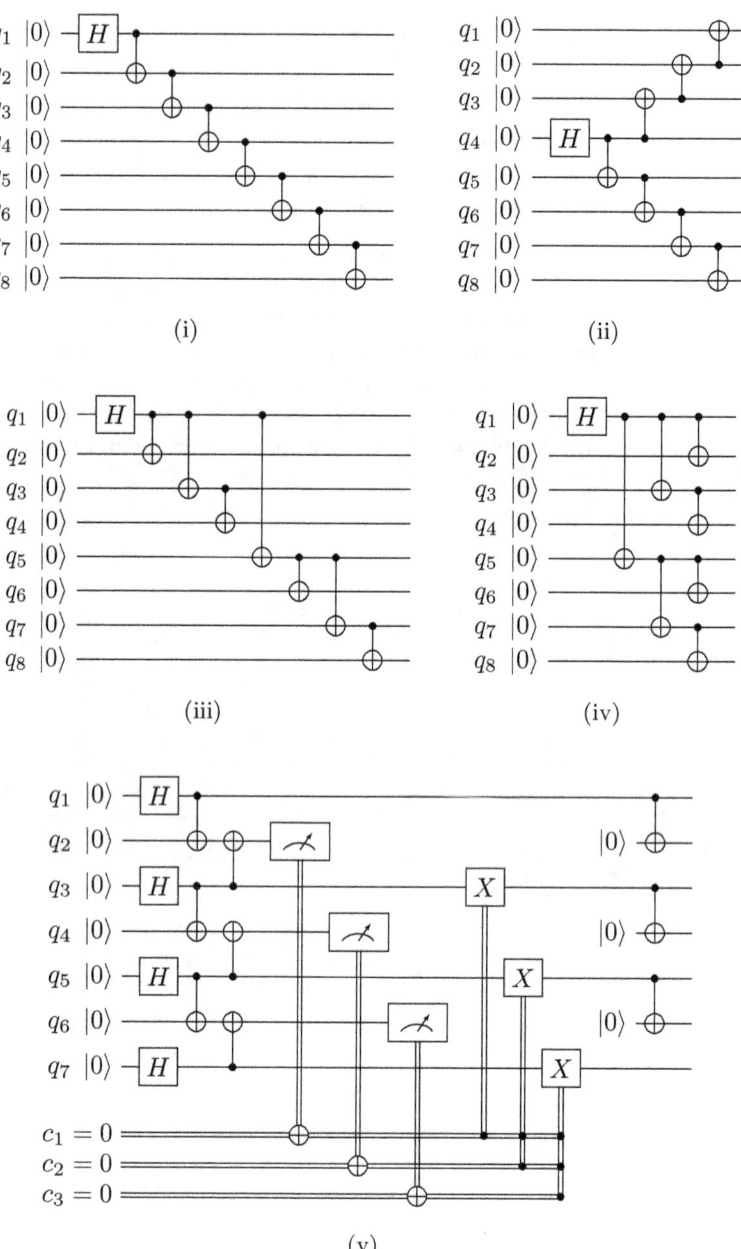

Figure 3.6: GHZ State Preparation Circuits

4. Classification of Computational Complexity

Contents

In the previous chapter we saw that the efficiency of an algorithm can vary dramatically depending on the problem structure. We now turn to a broader question: which problems can be solved efficiently at all? This leads us to the study of *computational complexity classes*, which organize problems by the resources required to solve them. Of particular interest will be the class BQP, which captures the problems solvable in polynomial time using quantum algorithms.

4.1 Introducing Complexity Classes

The theory of computational complexity groups problems into classes according to the resources needed to solve them—most commonly time (the number of steps) and space (the amount of memory). Complexity classes help us reason about which problems are tractable, which appear intractable, and how different problems relate to one another.

4.1.1 Common Complexity Classes

Some of the most important complexity classes are summarized in Table 4.1, arranged roughly from the most easily solvable to the most resource-intensive.

Class	Informal Description
P	Problems that a deterministic algorithm can solve in a number of steps that grows at most polynomially with the input size.
NP	Problems where, if the correct answer is "yes," a candidate solution (a certificate) can be verified in polynomial time, even though finding that solution may require much longer.
NP-Complete	The hardest problems in NP. Every problem in NP can be reduced in polynomial time to one of these. Efficiently solving any NP-complete problem would imply $P = NP$.
BPP	Problems that a probabilistic algorithm can solve in polynomial time using randomness, with error probability at most $1/3$ (the error can be reduced arbitrarily by repetition).
BQP	Problems that a quantum algorithm can solve in polynomial time, also with error probability at most $1/3$ (again reducible by repetition).
QSAMPLING	Problems where the goal is to sample from a probability distribution that a quantum process can generate efficiently but classical algorithms are believed to struggle with (e.g., boson sampling, random quantum circuits).
QMA	The quantum analogue of NP. If the answer is "yes," a prover can send a quantum state as a proof, which a quantum verifier can check in polynomial time with bounded error (e.g., $\leq 1/3$). If the answer is "no," every proof is rejected with high probability.
PSPACE	Problems solvable with only a polynomial amount of memory, even if the runtime may be exponential.
EXP	Problems solvable in exponential time. Typically far beyond practical reach for large inputs.
Computable	All decision problems for which some algorithm exists that always halts with the correct "yes" or "no" answer (Turing-decidable problems).

Table 4.1: Informal Descriptions of Common Complexity Classes

Related Concepts

A few concepts used in Table 4.1 are worth clarifying:

- Decision problem: A problem with a yes/no answer (e.g., "Does this number have a factor smaller than 1000?").

- Deterministic algorithm: An algorithm that always follows the same steps and produces the same output on the same input.

- Probabilistic algorithm: An algorithm that makes random choices (coin flips), so different runs may produce different results.

- Polynomial time: An algorithm runs in polynomial time if its number of steps is bounded by n^k for some constant k, where n is the input size. This is contrasted with exponential time, where runtime grows like 2^n or worse (see § 1.2.4).

Exercise 4.1 For each of the following problems, write an explicit decision-problem formulation (yes/no) that captures its essence:

(a) finding a shortest path between two cities in a road network,

(b) computing the greatest common divisor of two integers,

(c) sampling from the output distribution of a quantum circuit.

Key Takeaways

1. P models efficient deterministic computation; NP models problems whose proposed solutions can be verified efficiently.

2. NP-complete problems are the hardest problems in NP under polynomial-time reductions; a fast algorithm for one would imply $P = NP$.

3. Bounded-error classes such as BPP and BQP allow constant error that can be reduced by repetition (probability amplification).

4.1.2 P and NP Problems

P (Polynomial time) is the class of decision problems solvable by deterministic classical algorithms in polynomial time. These are considered tractable, under the working assumption that polynomial-time algorithms are practical for real-world input sizes.

NP (Nondeterministic Polynomial time) contains problems where a "yes" answer has a certificate that can be verified in polynomial time by a verifier. The subtlety is that finding such a certificate may require exponential search.

Classic NP problems include SAT, subset sum, and graph coloring. Integer factorization also lies in NP, though it is not known to be either in P or NP-complete. See Appendix D for more examples.

■ **Example 4.1 — Student Housing: A Real-World NP Problem.** Imagine assigning dormitory space to 100 students out of 400 applicants, subject to a list of incompatible pairs who should not be housed together.

Checking a proposed group of 100 students is straightforward: one simply verifies that no incompatible pairs are present. However, finding such a group is far more challenging. The number of possible ways to choose 100 students from 400 is $\binom{400}{100}$, an astronomically large number. A brute-force search through all possibilities is infeasible.

This captures the essence of an NP problem: verifying a candidate solution is efficient, but finding one may require intractable computation. Whether every problem in NP is also in P—that is, whether P $=$ NP—remains one of the most important open questions in computer science.   ∎

■ **Example 4.2 — Integer Factorization is in** NP. Consider the decision version of the problem: given integers n and m, does n have a nontrivial factor less than m?

This is indeed a decision problem, since the answer is either "yes" or "no." If a factor f is provided, checking that $1 < f < m$ and $n \bmod f = 0$ can be done efficiently. Thus, integer factorization is in NP.   ∎

4.1.3   Limits Beyond Time: Space and Randomness

Time is the most common resource in complexity theory, but it is not the only one.

Space plays a distinct role. The class PSPACE consists of problems solvable using only polynomial memory, even if they may require exponential time. This shows that time and space can trade off as computational resources.

When randomness is allowed in computation, one obtains additional complexity classes. For instance, BPP (Bounded-Error Probabilistic Polynomial Time) is the class of decision problems solvable by a randomized polynomial-time algorithm whose error probability is bounded away from 1/2. A standard convention is to require that the algorithm output the correct answer with probability at least 2/3 for every input.

The constant 2/3 is not special. Any fixed success probability greater than 1/2 would lead to the same complexity class, because repeated independent runs and majority vote can reduce the error exponentially while increasing the runtime by only a polynomial factor.

Error Reduction by Repetition (Probability Amplification)

The constant error bound in BPP may seem restrictive, but it can be improved by repetition and majority vote. Suppose a randomized algorithm is correct on each run with probability $p > \frac{1}{2}$, and suppose we run it k times independently and return the majority outcome. Then the probability that the majority vote is wrong decreases exponentially in k (a standard Chernoff/Hoeffding-type phenomenon).

For example, if each run is correct with probability at least 2/3, then for $k = 3$ the probability that at least 2 runs are wrong is at most

$$\binom{3}{2}\left(\frac{1}{3}\right)^2\left(\frac{2}{3}\right) + \left(\frac{1}{3}\right)^3 = \frac{7}{27} < 0.26.$$

For larger k, the tail probability decreases rapidly; in particular, there exists a constant $c > 0$ such that the error probability is bounded by e^{-ck}.

Thus, as long as each run succeeds with probability strictly greater than 1/2, repetition and majority vote can reduce the error arbitrarily close to zero. To make

> the error less than ϵ, it suffices to repeat the algorithm $O(\log(1/\epsilon))$ times, which is still only a polynomial overhead.

These refinements—space-bounded and randomized computation—prepare us to understand the role of quantum mechanics, where superposition and interference introduce still more powerful models of efficient computation.

4.1.4 NP-Complete Problems

NP-complete problems are the hardest problems in NP: a problem is NP-complete if it lies in NP and every problem in NP can be reduced to it by a polynomial-time many-one reduction. If any one NP-complete problem were solvable in polynomial time, then *every* problem in NP would be solvable in polynomial time (i.e., P = NP). Prominent NP-complete problems include:

- Boolean satisfiability (SAT): Given a Boolean formula, determine whether there exists a variable assignment that makes the formula true (see § 1.1.1.4). This was the first problem proven NP-complete (Cook–Levin theorem).

- Subset sum: Given a set of integers and a target T, determine whether some subset sums exactly to T.

- Hamiltonian cycle: Given a graph, determine whether it contains a cycle that visits every vertex exactly once.

- Graph coloring: Given a graph and an integer k, determine whether the vertices can be colored with at most k colors so that adjacent vertices receive different colors.

Reduction is the formal notion behind "at least as hard as." A language A reduces to a language B (written $A \leq_p B$) if there is a polynomial-time computable mapping that transforms any instance of A into an instance of B with the same yes/no answer. If $A \leq_p B$ and B is easy, then A is easy as well.

> **Exercise 4.2** Give a concrete certificate (witness) for each of the following "yes" instances, and explain how it can be verified in polynomial time:
>
> (a) SAT,
>
> (b) Hamiltonian cycle,
>
> (c) graph 3-coloring.

4.1.5 NP-Hard Problems

NP-hard problems are at least as hard as the NP-complete problems but need not themselves be in NP (they may not even be decision problems). Formally, a problem X is NP-hard if there exists an NP-complete problem Y such that $Y \leq_p X$. In particular, every NP-complete problem is also NP-hard, but not every NP-hard problem is in NP.

Classical examples include optimization versions of NP-complete problems (e.g., Traveling Salesman Problem by cost minimization) and certain undecidable problems that are NP-hard under appropriate reductions. For instance, the Halting Problem asks: given a program P and input I, does P halt on I? It is undecidable and (under standard polynomial-time many-one reductions) is at least as hard as every problem in NP; in this sense it is NP-hard.

Exercise 4.3 Classify each item below as belonging to P, NP, NP-hard/NP-complete, or Undecidable. Briefly justify your answer and state any decision formulation you use.

(a) Primality testing (Given an integer n, determine whether n is prime.)

(b) Sudoku validity (Given a completed grid, determine whether it satisfies the Sudoku rules.)

(c) Integer factorization (Decision Version: "Does n have a nontrivial factor $\leq m?$")

Key Takeaways

1. NP-complete means "in NP and as hard as all of NP" under polynomial-time reductions.

2. NP-hard means "at least as hard as NP" but the problem need not be in NP (it may be an optimization problem or even undecidable).

3. NP-hardness is a reduction-based relation, not a single region in a containment diagram.

4.2 Relationships Among Major Complexity Classes

Complexity theory is full of open questions about how major classes relate to one another. Some containments are known, but many separations remain conjectural. To visualize the current beliefs and conjectures, it is helpful to use a conceptual diagram.

Figure 4.1 provides a Venn-style illustration, showing how classes such as P, NP, BPP, and BQP are often thought to fit together relative to broader classes such as PSPACE and EXP. We emphasize that the diagram reflects current intuition rather than proven separations.

4.2.1 Standard Inclusions

The following containments are known; whether they are strict (for example, whether $P \neq NP$) is open:

$$P \subseteq NP \subseteq PSPACE \subseteq EXP.$$

In addition,

$$P \subseteq BPP \subseteq PSPACE.$$

We will relate BQP to these classes in § 4.3.

Exercise 4.4 Explain why $NP \subseteq PSPACE$. Your explanation should make clear what is being searched over and why the search can be done using only polynomial

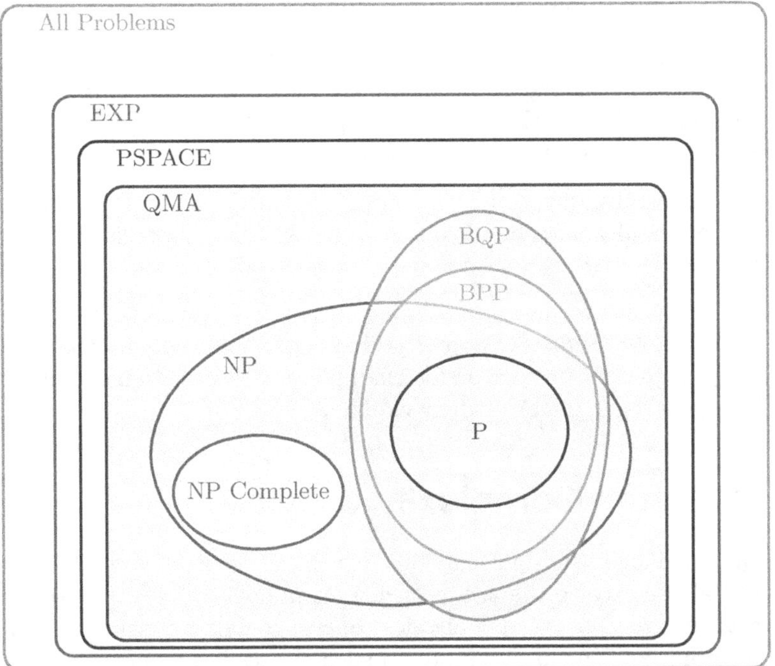

Figure 4.1: Conjectured Relationships Among Complexity Classes

Note: There is no single "NP-hard" region in this diagram. NP-hardness is a *relation*: a problem is NP-hard if every problem in NP reduces to it in polynomial time, regardless of whether it lies in NP, is decidable, or is even a decision problem.

space (even if it takes exponential time).

4.2.2 The P vs. NP Question

At the heart of complexity theory lies one of the most important open problems:

Is every problem whose solution can be verified in polynomial time also solvable in polynomial time?

This is the famed $P \stackrel{?}{=} NP$ question. If $P \neq NP$, then some problems are inherently intractable for classical algorithms, even though proposed solutions can be checked efficiently.

4.2.3 The P vs. BPP Question

Whether $BPP = P$ is also unknown. If equality holds, every efficiently solvable randomized algorithm would have a deterministic counterpart with similar efficiency. Many researchers conjecture $BPP = P$, motivated by derandomization techniques and pseudorandom generators suggesting that randomness can often be replaced by deterministic processes without losing efficiency.

■ **Example 4.3 — P vs. NP: Word Problems vs. Multiple Choice Problems.** A useful analogy for the P vs. NP question compares word problems with multiple choice problems. A word problem asks you to produce the answer from scratch, while a multiple choice problem provides candidate answers that you can test.

If "easy" means solvable in polynomial time, then P consists of problems whose answers can be found from scratch efficiently—easy word problems. In contrast, NP contains problems where checking a proposed answer is easy—easy multiple choice problems—even if finding such an answer appears difficult.

The central question is whether there exist problems that are easy to check but fundamentally hard to solve from scratch. This is precisely the open question of whether $P = NP$.   ■

> **Exercise 4.5** Suppose $P = NP$. What would this imply for the security of common cryptographic protocols (for example, those based on integer factorization or discrete logarithms)? Briefly explain.

4.3 Quantum Complexity Classes

Quantum algorithms not only broaden the scope of what can be computed efficiently but also motivate the study of new complexity classes. These quantum-specific classes extend or parallel classical ones, providing a richer map of computational possibilities. The formal study of such classes is grounded in the quantum circuit model, which offers a uniform framework for defining and analyzing them.

4.3.1 The Class BQP

The central complexity class of quantum algorithms is BQP (Bounded-Error Quantum Polynomial Time). This class contains decision problems that a quantum computer can solve in polynomial time, with error probability at most $1/3$ for all inputs. As with the classical class BPP, this error can be reduced arbitrarily by repetition and majority vote (a standard form of probability amplification).

Problems in BQP are considered efficiently solvable on a quantum computer, assuming access to a fault-tolerant device. This class captures the formal power of quantum algorithms such as Shor's factoring algorithm and Grover's unstructured search algorithm. These illustrate different kinds of quantum advantage: Shor's algorithm achieves a superpolynomial speedup over the best known classical methods for factoring, whereas Grover's algorithm achieves a quadratic speedup for unstructured search.

> **Exercise 4.6** In the definition of BQP, the constant error bound $1/3$ is arbitrary.
>
> (a) State a definition of BQP using completeness (probability on yes-instances) $\geq 1 - \delta$ and soundness $\leq \delta$ for a fixed constant $\delta < 1/2$.
>
> (b) Explain why this definition yields the same class as the $1/3$ definition (you may use repetition and majority vote).

4.3.2 Classical–Quantum Containment Hierarchy

Quantum complexity classes are related to classical ones through the following standard inclusions:

$$P \subseteq BPP \subseteq BQP \subseteq PSPACE.$$

All containments shown above are known to hold; whether any of them are strict is open.

Here, BQP is known to contain all of P, and it is widely believed to be strictly larger. Its relationship to NP remains one of the major open questions in quantum complexity theory. In particular, it is unknown whether NP $\subseteq$ BQP. Most evidence suggests that NP-complete problems cannot, in general, be solved efficiently by quantum computers, although this remains unproven.

Here, BQP is known to contain all of P, and it is widely believed to be strictly larger. Its relationship to NP remains one of the major open questions in quantum complexity theory. In particular, it is unknown whether NP $\subseteq$ BQP. Most evidence suggests that NP-*complete* problems cannot, in general, be solved efficiently by quantum computers, although this remains unproven.

The inclusion P $\subseteq$ BQP means that quantum computers are at least as powerful as classical ones. This inclusion rests on the ability of quantum circuits to simulate classical computation using reversible gates such as the Toffoli gate. A full proof of this fact is given in § 4.3.5.

Exercise 4.7 Define the quantum complexity class BQP. How does it relate intuitively to P and NP?

Exercise 4.8 Why is BQP generally believed to be distinct from both P and NP? Summarize the current understanding of its relationship to NP.

Key Takeaways

1. The containments P $\subseteq$ BPP $\subseteq$ BQP $\subseteq$ PSPACE are known to hold.

2. Whether BQP is strictly larger than P or whether NP $\subseteq$ BQP are major open questions.

3. When interpreting such statements, always separate what is proven (containment) from what is conjectured (strictness/separation).

4.3.3 QMA: Quantum Merlin–Arthur

A common misconception is that BQP is the quantum analogue of NP. In fact, they represent different ideas:

- NP describes verification: if a candidate solution is given, a deterministic algorithm can check it in polynomial time.

- BQP describes computation: a quantum algorithm solves the problem directly in polynomial time with bounded error.

The true quantum analogue of NP is QMA. A decision problem belongs to QMA if, whenever the correct answer is "yes," there exists a quantum state that serves as a proof (a quantum witness) and convinces a polynomial-time quantum verifier with high probability. If the answer is "no," then no possible proof—quantum or otherwise—can convince the verifier except with small probability (typically at most 1/3).

The names "Merlin" and "Arthur" reflect this interactive proof model: Merlin is an all-powerful but untrusted prover, and Arthur is a polynomial-time quantum verifier. QMA captures problems where quantum information may be essential to verification, including quantum versions of constraint satisfaction and problems arising in the simulation of physical systems.

1　Quantum Money as a QMA Problem

A striking illustration of QMA comes from the idea of *quantum money*, first proposed by Stephen Wiesner in the 1970s (see Section 11.3: *Quantum Bomb and Quantum Money*).

Suppose a bank issues banknotes, each consisting of a unique serial number (classical data) and a corresponding quantum state $|\psi\rangle$ known only to the bank. To verify a note, the bank runs a quantum verification algorithm that checks whether the attached state matches the secret $|\psi\rangle$ for that serial number.

- If the banknote is genuine, the verifier accepts with high probability.

- If it is counterfeit, then no quantum state supplied by a counterfeiter will pass verification except with small probability.

This fits the QMA framework: the prover (Merlin) provides a quantum witness (the banknote's quantum state), and the verifier (Arthur) uses a polynomial-time quantum algorithm to check its validity.

The power of quantum mechanics ensures that genuine notes cannot be copied due to the no-cloning theorem, and counterfeit states cannot consistently fool the verifier. Thus, quantum money highlights how QMA captures problems where verification requires intrinsically quantum information.

2　Physical Simulation as a QMA Problem

Many problems in quantum physics can be framed as decision problems suitable for the class QMA. A central example is the *local Hamiltonian problem*, often regarded as the quantum analogue of SAT.

Consider a quantum system described by a Hamiltonian $H = \sum_i H_i$, where each H_i acts on only a few qubits ("local terms"). The decision problem is:

> *Given thresholds E_0 and E_1 with $E_1 > E_0$, determine whether the ground state energy of H is $\leq E_0$ (a "yes" instance) or $\geq E_1$ (a "no" instance), promised that one of these is the case.*

This fits the QMA framework:

- If the answer is "yes," a prover (Merlin) can send a candidate ground state $|\psi\rangle$ as a quantum witness.

- A verifier (Arthur), running a polynomial-time quantum algorithm, can efficiently estimate $\langle\psi| H |\psi\rangle$ by measuring the local terms H_i. If the energy is $\leq E_0$, Arthur accepts with high probability.

- If the answer is "no," then every state has energy $\geq E_1$, so no possible witness can fool the verifier into acceptance except with small probability.

Because verifying such properties may require quantum states as proofs, not just classical strings, the local Hamiltonian problem is not in NP (as far as we know) but belongs to QMA. In fact, it is QMA-complete: just as SAT is the canonical NP-complete problem, the local Hamiltonian problem serves as the canonical hard problem for QMA. This establishes a deep connection between quantum complexity theory and the physics of many-body quantum systems.

Exercise 4.9 Promise problems arise naturally in quantum complexity.

(a) Explain why the local Hamiltonian problem is stated with a promise gap $E_1 - E_0 > 0$.

(b) What could go wrong (conceptually) if we asked the algorithm to decide the case where the ground energy might lie between E_0 and E_1?

(c) Give an everyday (non-physics) example of a decision task that becomes well-posed only after adding a promise (a guaranteed gap or margin).

4.3.4 Relationship Between BQP and QMA

1 Containment

Every BQP problem is also in QMA:

$$BQP \subseteq QMA.$$

This follows because QMA is a verification model. If a quantum computer can decide the problem directly, then Arthur can ignore the witness (or accept a trivial one) and simply run the BQP algorithm.

2 Strictness?

It is not known whether the inclusion is strict:

$$BQP \overset{?}{=} QMA.$$

Most researchers believe that QMA is strictly larger, since QMA contains natural problems that are not believed to be in BQP, such as the local Hamiltonian problem. The local Hamiltonian problem is QMA-complete, meaning it is as hard as any problem in QMA.

3 Analogy with Classical Classes

The relationship between BQP and QMA mirrors the classical relationship between P and NP:

$$P \subseteq NP, \quad BQP \subseteq QMA.$$

Just as NP is believed to be larger than P, QMA is believed to be larger than BQP.

4.3.5 ✳ Proof That BQP Contains All of P

A fundamental property of quantum computers is that they are at least as powerful as classical computers, formally expressed by the inclusion $P \subseteq BQP$. This inclusion follows from the fact that classical computation can be simulated within the quantum model using reversible gates.

Key Takeaways

1. Quantum circuits can simulate classical computation because classical logic can be made reversible.

2. The Toffoli gate is universal for reversible classical computation, so any classical circuit can be embedded into a quantum circuit with polynomial overhead.

3. Uncomputation is used to erase garbage (ancilla) information so the output is clean and not entangled with intermediate workspace.

1 Expressing Classical Gates Using the Toffoli Gate

Classical logic gates such as AND, OR, and NOT are typically irreversible: they discard information and cannot be inverted. Quantum gates, by contrast, must be reversible, since they represent unitary operations. To reconcile this difference, classical computation can be reformulated using reversible gates. A key example is the *Toffoli gate* (also called the CCNOT gate), which is universal for reversible classical computation.

Gate	Symbol	Boolean Representation
NOT (X)	$\begin{aligned}&\lvert 1\rangle \longrightarrow \lvert 1\rangle \\ &\lvert 1\rangle \longrightarrow \lvert 1\rangle \\ &\lvert z\rangle \longrightarrow \lvert \bar{z}\rangle\end{aligned}$	$\lvert z\rangle \mapsto \lvert \bar{z}\rangle$
AND	$\begin{aligned}&\lvert x\rangle \longrightarrow \lvert x\rangle \\ &\lvert y\rangle \longrightarrow \lvert y\rangle \\ &\lvert 0\rangle \longrightarrow \lvert (xy)\rangle\end{aligned}$	$\lvert xy\rangle \mapsto z = \lvert (xy)\rangle$
FANOUT	$\begin{aligned}&\lvert 1\rangle \longrightarrow \lvert 1\rangle \\ &\lvert y\rangle \longrightarrow \lvert y\rangle \\ &\lvert 0\rangle \longrightarrow \lvert y\rangle\end{aligned}$	Only for classical bits represented by basis states; not for superposition states.
CNOT	$\begin{aligned}&\lvert 1\rangle \longrightarrow \lvert 1\rangle \\ &\lvert y\rangle \longrightarrow \lvert y\rangle \\ &\lvert z\rangle \longrightarrow \lvert (y \oplus z)\rangle\end{aligned}$	$\lvert yz\rangle \mapsto \lvert y(y \oplus z)\rangle$

Table 4.2: Expressing Common Gates Using Toffoli Gate

Basic classical operations such as NOT, AND, and FANOUT can all be implemented using Toffoli gates, as summarized in Table 4.2 (adapted from QCI Book [2], Chapter 7: *Multi-Qubit Quantum Gates*). An OR gate can also be built indirectly via De Morgan's law:

$$OR(x, y) = NOT(AND(NOT(x), NOT(y))).$$

These constructions often require ancillary qubits to maintain reversibility, which is essential in quantum circuits due to unitarity.

Since any classical circuit can be expressed using NOT, AND, and FANOUT, and these can all be built from Toffoli gates, it follows that any classical logic circuit can be simulated by a quantum circuit. This proves that quantum computers can efficiently emulate classical ones.

2 Uncomputation

A subtle issue arises when simulating classical computation within a quantum circuit: intermediate values stored in ancillary qubits, often called *garbage bits*, are not part of the intended output. While harmless in a classical setting, in a quantum circuit they remain entangled with the output and can corrupt later operations or measurements. If not removed, this residual entanglement undermines the correctness of the simulation.

The solution is a technique called *uncomputation*. The procedure is to compute the output, copy it into a clean register, and then reverse the computation to reset all ancilla qubits to their initial $|0\rangle$ state. This disentangles the output from the workspace and restores a clean quantum state.

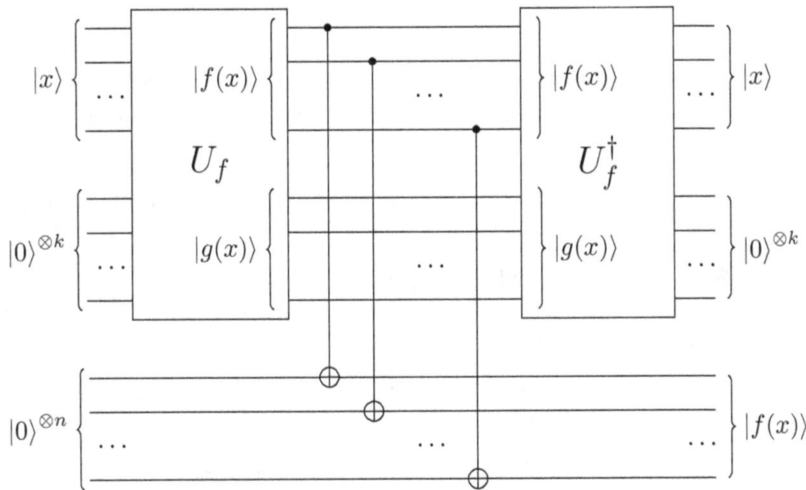

Figure 4.2: Uncomputation Circuit

In Fig. 4.2, the unitary U_f computes $f(x)$ from input $|x\rangle$, yielding $|f(x)\rangle$ along with garbage $|g(x)\rangle$ in ancilla qubits. The result $|f(x)\rangle$ is copied into a clean register using CNOT gates. Then the inverse operation $U_f^\dagger$ erases the garbage, returning ancilla qubits to $|0\rangle^{\otimes k}$. The final state is

$$|x\rangle\, |0\rangle^{\otimes k}\, |f(x)\rangle\,,$$

in which all ancilla qubits have been reset and are disentangled from the result. This cleanup step is crucial for completing the formal argument that $P \subseteq BQP$.

Chapter Takeaways

1. Complexity classes organize problems by the resources needed to solve or verify them (time, space, randomness, and interaction).

2. The P vs. NP question highlights the difference between efficiently *solving* and efficiently *verifying*.

3. Quantum computation introduces classes such as BQP and QMA, enabling a refined comparison between classical and quantum computa-

tional power.

4. Containment relationships (known and conjectured) guide where quantum advantage is plausible and where it is unlikely.

5. $P \subseteq BQP$: a quantum computer can efficiently simulate any classical polynomial-time computation.

Problem Set 4

4.1 Assume $P \neq NP$, $BQP \supset NP$, and that no classical polynomial-time algorithm exists for integer factoring. For each of the following complexity classes, determine whether integer factoring is known to lie in the class, and briefly justify your answer:

$$P,\ NP,\ \text{NP-Complete},\ \text{NP-Hard},\ BQP,\ QMA,\ PSPACE,\ EXP.$$

4.2 Show that $P \subseteq NP$ by explicitly explaining how a deterministic polynomial-time algorithm can also serve as a polynomial-time verifier. Why does this inclusion not imply $P = NP$?

4.3 $*$ The Local Hamiltonian problem is QMA-complete. Compare it to Boolean SAT (which is NP-complete). In each case:

  (a) State the decision version of the problem.

  (b) Identify what the certificate or witness looks like.

  (c) Describe how the verifier checks it in polynomial time.

4.4 Suppose you have a randomized polynomial-time algorithm that solves a decision problem with success probability 0.6.

  (a) If you repeat the algorithm 5 times and take the majority outcome, what is the probability that the overall result is wrong?

  (b) Generalize your answer to k repetitions.

  (c) Explain how this supports the claim that BPP algorithms can achieve arbitrarily small error probability.

4.5 Consider the following problems. For each, classify it (as best as currently known) into one or more complexity classes:

  (a) Deciding whether a given graph has a Hamiltonian cycle.

  (b) Determining whether a number n is prime.

  (c) Simulating the ground state energy of a quantum many-body system.

  (d) Computing the permanent of a matrix with non-negative entries.

4.6 The P vs. NP question asks whether every problem with efficiently verifiable solutions also has efficiently computable solutions.

   (a) Explain why an efficient algorithm for any NP-complete problem would imply P = NP.

   (b) Give a real-world example of a task (outside mathematics or computer science) that naturally fits the NP framework.

4.7 Let a randomized algorithm output the correct yes/no answer with probability at least $p = 1/2 + \gamma$ for every input, where $\gamma > 0$.

   (a) Show that repeating the algorithm k times and taking the majority vote reduces the error probability to at most $e^{-2k\gamma^2}$ (you may cite Hoeffding's inequality).

   (b) How large must k be to make the error probability at most 10^{-6} when $\gamma = 0.1$?

4.8 Give a short, self-contained explanation of why P $\subseteq$ BQP is not merely "obvious." Your explanation should mention:

   (a) reversibility and why quantum gates must be reversible,

   (b) how Toffoli gates allow classical computation to be embedded into quantum circuits,

   (c) why uncomputation is needed to remove garbage.

4.9 One natural QMA example beyond local Hamiltonians is *group non-membership*, where Merlin provides a quantum witness that helps Arthur verify that an element is not in a subgroup.

   (a) State the group non-membership decision problem.

   (b) Explain, at a high level, why a *quantum* witness can help even when no efficient classical certificate is known.

   (c) Compare the "witness" in this problem to the witness in local Hamiltonians.

4.10 $*$ Implement in Qiskit (or an alternative platform) a reversible circuit that computes the Boolean function

$$f(x_1, x_2, x_3) = (x_1 \wedge x_2) \oplus x_3$$

into a target qubit, using only CNOT and Toffoli gates and any necessary ancillas.

   (a) Verify correctness on all 2^3 inputs.

   (b) Add an uncomputation step so that all ancillas return to $|0\rangle$, and verify this property by simulation.

4.11 $*$ The class QSAMPLING is informally described as sampling tasks that quantum devices can generate efficiently.

   (a) Give an example of a sampling task associated with a quantum circuit family (for instance, output sampling from a random quantum circuit).

   (b) Propose a related *decision* problem whose input is a circuit description and whose answer is yes/no, intended to capture some aspect of the sampling hardness.

(c) Discuss (informally) what evidence one would want in order to claim a separation between classical and quantum algorithms for your decision formulation.

5. FTQC vs. NISQ Algorithms

Contents

The landscape of quantum algorithms is shaped by the capabilities and limitations of the underlying hardware. Most present-day devices belong to the *Noisy Intermediate-Scale Quantum (NISQ)* era, where computations are constrained by noise, limited coherence, and modest qubit counts. In contrast, the long-term goal is *Fault-Tolerant*

Quantum Computing (FTQC), in which quantum error correction enables reliable, large-scale computations on logical qubits.

This chapter explores the distinction between these two eras and their algorithmic consequences. We begin by examining the defining features of NISQ and FTQC systems, together with the transition milestones that bridge them. We then review the fundamental concepts of decoherence, noise, and errors, and compare the roles of quantum error correction and quantum error mitigation. The discussion proceeds to highlight how algorithm design differs across eras, tracing the evolution from near-term prototypes to scalable, error-resilient algorithms. Finally, we consider the broader quest for quantum utility and advantage, including current challenges, benchmarks, and possible directions for future research.

By contrasting NISQ and FTQC algorithms, this chapter provides a framework for understanding both the limitations of today's hardware and the opportunities that lie ahead as quantum systems move toward fault tolerance.

5.1   Introduction to Quantum Computing Eras

Quantum computing has entered a transitional period marked by the coexistence of two paradigms: the *Noisy Intermediate-Scale Quantum (NISQ)* era and the aspirational stage of *Fault-Tolerant Quantum Computing (FTQC)*. Each represents a milestone on the path toward practical, scalable quantum computation, with distinct opportunities, challenges, and implications for algorithm design.

5.1.1   NISQ (Noisy Intermediate-Scale Quantum)

The NISQ era refers to the current generation of quantum devices (roughly the mid-2020s), typically featuring between 50 and 1,000 physical qubits. Examples include IBM's 156-qubit Heron processor, Google's 105-qubit Willow processor, and QuEra's 256-qubit Aquila system. While remarkable in scale, such devices remain limited by relatively high error rates, often on the order of 0.01–0.1% per two-qubit gate, and coherence times that extend only from microseconds to milliseconds.

NISQ machines cannot yet support full-fledged, large-scale quantum error correction. As a result, algorithms must be designed to run on shallow circuits that minimize error accumulation. Despite these limitations, NISQ devices play a crucial role in demonstrating early signs of quantum utility—proof-of-concept benefits in areas such as optimization, chemistry, and quantum-enhanced machine learning. They also expose the central challenges of decoherence, gate imperfections, and readout errors, which define much of the present-day research frontier.

5.1.2   FTQC (Fault-Tolerant Quantum Computing)

Looking further ahead, the goal of FTQC is to build scalable quantum computers with very large numbers of physical qubits, capable of sustaining logical qubits whose error rates are low enough to support deep computations reliably (often quoted targets are $\lesssim 10^{-10}$ per logical operation for certain applications). This vision is grounded in the quantum threshold theorem, which asserts that if physical error rates can be pushed below a code-dependent threshold, then arbitrarily long computations become feasible with only polynomial resource overhead.

FTQC systems will rely heavily on quantum error correction (QEC) codes such as the surface code, which typically encodes a single logical qubit into hundreds to thousands of physical qubits, depending on the target logical error rate and the noise model. The payoff is transformative: once realized, fault-tolerant architectures will enable universal quantum computation on problems far beyond the reach of classical hardware, from cryptographic breakthroughs to large-scale simulations in chemistry and physics. Achieving this milestone will require significant advances in qubit quality, control engineering, and algorithmic resource optimization, and is widely estimated to be at least 5–10 years away (as of early 2026).

5.1.3 Transition Phases and Their Implications

Between NISQ devices and fully fault-tolerant systems lies a transition phase characterized by partial error correction and gradually increasing algorithmic depth. This intermediate regime, sometimes referred to as early error correction (EEC), becomes accessible once hardware quality and error-correction cycles are good enough to sustain small numbers of logical qubits with useful logical error rates (for example, aiming at $\sim 10^{-6}$–10^{-8} logical error per operation for early demonstrations). At this stage, modest numbers of logical qubits can be realized, enabling early demonstrations of practical quantum utility beyond NISQ yet short of universal computation [12].

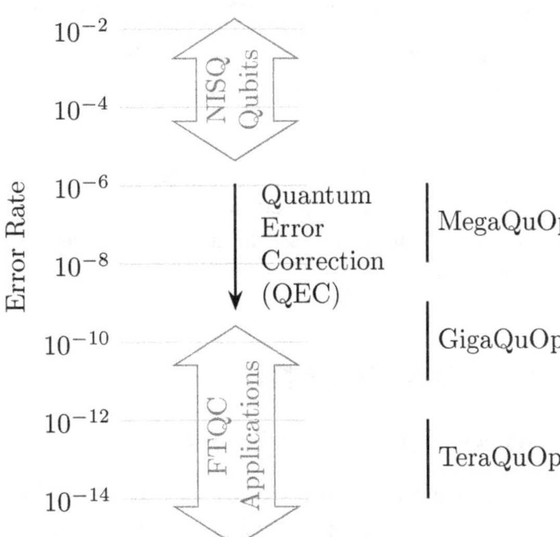

Figure 5.1: Transition from NISQ to FTQC

A critical milestone in this progression is the *MegaQuOp* regime, in which one million error-corrected quantum operations can be performed reliably. As illustrated in Fig. 5.1, reaching this level can allow demonstrations involving on the order of 10^2 logical qubits and circuit depths around 10^4, provided logical error rates are suppressed to roughly 10^{-7} (or closer to 10^{-6} with the assistance of error mitigation). These advances pave the way for subsequent thresholds such as GigaQuOp (10^9 operations) and TeraQuOp (10^{12} operations), where the scope of feasible applications expands dramatically.

The implications of this staged evolution extend beyond physics and engineering.

In the NISQ era, industry and academia focus on developing noise-resilient, low-depth algorithms to extract value from imperfect devices. As systems transition toward FTQC, algorithmic strategies will shift toward deeper, more resource-intensive methods, demanding careful accounting of qubit overhead, logical error budgets, and fault-tolerant depth. On a broader scale, this transition will influence policy, investment, and education, driving near-term commercial pilots while laying the groundwork for breakthroughs in areas such as drug discovery, secure communications, and materials science.

> **Exercise 5.1** A resource estimate for a fault-tolerant algorithm must track more than the number of logical qubits. Other practically important components include the logical circuit depth, the non-Clifford gate count, the throughput of magic-state factories, and connectivity- or routing-induced overhead. Investigate and explain how any one of these could become a bottleneck even when the number of logical qubits is modest.

Key Takeaways

1. NISQ devices are limited by noise and coherence, so algorithms must be shallow and often hybrid quantum–classical.

2. FTQC relies on quantum error correction to create logical qubits that support deep circuits with controllable failure probability.

3. Between NISQ and FTQC lies an intermediate regime where small numbers of logical qubits enable early, error-corrected demonstrations of utility.

4. Transition milestones (e.g., MegaQuOp) summarize the ability to run long sequences of error-corrected logical operations reliably. Reaching higher milestones translates hardware progress into algorithmic capability: deeper circuits and larger problem instances become feasible.

5.2 Fundamental Concepts

5.2.1 Decoherence, Noise, and Errors

In the study of quantum computation, the terms *decoherence*, *noise*, and *error* are sometimes used interchangeably, yet they carry distinct meanings that are important to separate.

Decoherence refers to the gradual loss of quantum coherence when a quantum system interacts with its surrounding environment. In such cases, a pure state—capable of exhibiting superposition and entanglement—evolves into a mixed state, in which these distinctly quantum features are degraded or lost. Decoherence is most naturally described using the density matrix formalism, and it represents one of the primary obstacles to storing and manipulating quantum information over meaningful time scales.

Noise is a broader term that encompasses unwanted disturbances acting on a quantum system. Sources of noise include decoherence, imperfect gate control, fluctuations in the external environment, and measurement inaccuracies. Mathematically, noise is often modeled by quantum channels, represented by sets of Kraus operators, which capture the probabilistic evolution of states under such disturbances.

Finally, an error is the deviation between the actual outcome of a quantum operation and the intended ideal result. Errors can arise as a consequence of noise but are classified differently depending on their characteristics. Coherent errors result from systematic miscalibrations, such as over-rotation in a quantum gate, and can often be reduced with improved control and calibration. By contrast, incoherent errors are random and less predictable, making them harder to suppress directly. Together, decoherence, noise, and errors form the central challenge in building reliable quantum computers.

For more details on this topic, refer to QCI Book [2], Chapter 12: *Density Operators and Quantum Channels* and Chapter 13: *Quantum Error Correction: A Primer.*

5.2.2 Quantum Error Correction (QEC): Principles and Necessity

The fragility of quantum states makes quantum error correction (QEC) indispensable for scalable quantum computing. The basic principle of QEC is to encode a single logical qubit into many physical qubits so that errors can be detected and corrected without collapsing the encoded quantum information. Early examples include the 9-qubit Shor code, while more practical modern schemes rely on topological codes such as the surface code, which typically require on the order of 10^2–10^3 physical qubits per logical qubit for application-relevant logical error rates.

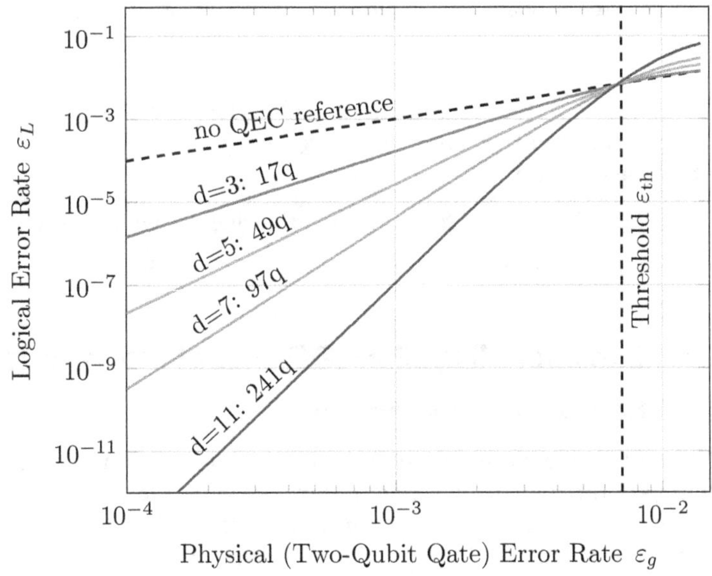

Figure 5.2: Quantum Error Correction and Threshold

Note: The threshold ε_{th} and the illustrative logical-error scaling shown here are adapted from a specific demonstration in [13]. These values are not universal constants; they depend on the code family, noise model, decoding method, and hardware assumptions, and may change as the literature and technology evolve.

The mathematical foundation of QEC is built on the stabilizer formalism and the threshold theorem. (See Appendix B for a more formal discussion of quantum error correction and fault-tolerance.) In broad terms, if the physical error rate p lies below a threshold p_{th} (which depends on the code and noise model), then increasing the *code distance* d suppresses the logical error rate rapidly. Roughly speaking, the code distance measures the minimum number of physical errors needed to cause an undetectable logical error; larger d therefore means stronger protection.

As illustrated in Fig. 5.2, a common heuristic scaling is that the logical error rate behaves roughly like $(p/p_{\text{th}})^{(d+1)/2}$ up to prefactors. The main message is that once the physical error rate is below threshold, larger code distance can strongly suppress logical errors, though this comes at the cost of greater physical-qubit overhead. In the figure, labels such as "$d = 3$: 17q" and "$d = 7$: 97q" indicate illustrative examples in which a distance-3 code uses 17 physical qubits per logical qubit, whereas a distance-7 code uses 97.

The necessity of QEC becomes clear when considering error accumulation. Without QEC, the depth of quantum circuits is fundamentally limited, as errors compound too quickly. With QEC, however, the effective error per logical operation can be reduced to extremely low levels, enabling deep algorithms such as factoring large RSA keys or simulating strongly correlated quantum materials.

 In the context of QEC, an "error rate" typically refers to a probability of failure per operation (or per cycle), not a rate per unit time. For example, it may describe the likelihood that a single-qubit gate, a two-qubit gate, a measurement, or a full error-correction cycle fails. In Fig. 5.1, the "error rate" refers to an effective logical error probability per operation after error correction has been applied.

Exercise 5.2 A common heuristic scaling for logical error is $\varepsilon_L \sim (p/p_{\text{th}})^{(d+1)/2}$ (up to prefactors), where p is the physical error rate, p_{th} is a threshold, and d is the code distance.

(a) Explain why this form suggests that increasing d is only useful when $p < p_{\text{th}}$.

(b) For fixed $p < p_{\text{th}}$, how does $\log \varepsilon_L$ scale with d?

5.2.3 Quantum Error Mitigation (QEM): Techniques and Limitations

1 Why Mitigation Matters in Practice

While QEC offers strong suppression of errors, it comes with high overheads in qubits, operations, and architectural complexity. For this reason, present-day NISQ devices often rely on *quantum error mitigation (QEM)*, a collection of techniques that attempt to reduce the impact of noise without implementing full error correction.

QEM is also expected to remain valuable in the *early FTQC* regime, where logical qubits will be scarce and costly, and logical operations may retain residual noise. Practical workflows may therefore combine partial error correction with mitigation and inference, such as post-selection (or "abort") rules in decoding, mitigation applied at the logical level, and statistical estimators that extract more accurate outputs from imperfect data [14, 15].

2 Inference Overhead and Hardware-Aware Costs

More broadly, on near-term and early fault-tolerant devices, the decisive costs are often not logical gate counts but the *inference overhead* surrounding the circuit: the number of circuit repetitions required for a target confidence level, the need to recalibrate or re-characterize noise as devices drift, and the impact of non-local noise correlations that violate simple independent-error models. Making these costs explicit is essential for realistic resource estimates and for interpreting benchmark claims.

3 Representative Mitigation Strategies

QEM techniques typically operate through additional circuit executions and classical post-processing. One common method is *zero-noise extrapolation (ZNE)*, in which circuits are executed at artificially increased noise levels and extrapolated back toward a zero-noise estimate. Another approach, *probabilistic error cancellation (PEC)*, involves characterizing noise channels and probabilistically inverting them during classical post-processing. Additional strategies include readout error mitigation, which corrects measurement biases (often via calibration and matrix inversion), and dynamical decoupling, which applies carefully designed pulse sequences to suppress decoherence during idle times. Increasingly, machine-learning-based approaches are also explored as a way to learn noise-aware correction maps or to adapt mitigation parameters from calibration data.

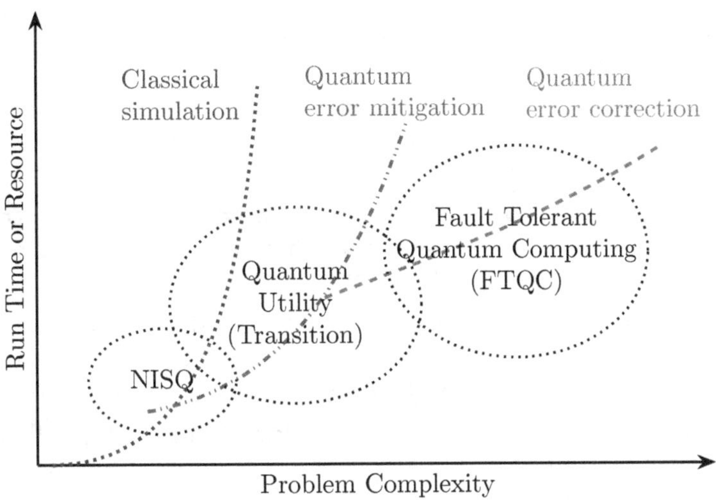

Figure 5.3: NISQ to Fault Tolerance Progression

4 Scalability Limits and When Mitigation Helps

The central limitation of QEM is scalability. The overhead required for accurate mitigation often grows rapidly with the number of qubits and the circuit depth, and in some regimes it can become effectively exponential. Moreover, not all error types are equally well handled: time-dependent drift and incoherent stochastic errors can be particularly problematic. Despite these limitations, QEM plays a central role in extracting useful signals from NISQ devices, especially in variational algorithms such as the Variational Quantum Eigensolver (VQE). As shown in Fig. 5.3, QEM provides a practical stepping stone in the progression from noisy devices to fully fault-tolerant quantum computers.

> **Exercise 5.3** In Qiskit-style workflows, it is common to combine multiple mitigation and suppression options (for example, measurement error mitigation, gate twirling, dynamical decoupling, and zero-noise extrapolation (ZNE)). Investigate and explain why "turning on more options" might fail to improve results for a given circuit family.

5.2.4   Comparison of QEC and QEM in Practical Applications

A comparison between QEC and QEM highlights their complementary roles in the development of quantum computing. QEM is indispensable in the near term, where shallow circuits can still yield insights into optimization, chemistry, and machine learning. QEC, on the other hand, is the enabling technology for large-scale, deep, universal computation. In practice, however, the boundary is not strict: in early FTQC, mitigation can remain important even after logical qubits exist, because logical resources are limited and residual logical errors and bias can still be reduced through post-processing and statistical inference. The trade-offs are summarized in Table 5.1.

Aspect	Quantum Error Correction (QEC)	Quantum Error Mitigation (QEM)
Era Suitability	FTQC (long-term)	NISQ (near-term)
Mechanism	Active encoding, syndrome extraction, and correction	Post-processing and noise-aware inference
Error Suppression	Strong (below threshold; improves with code distance)	Limited (problem- and method-dependent)
Resource Overhead	High ($\sim 10^2$–10^3 physical qubits per logical qubit, plus cycles)	Low-to-medium (extra circuit runs and classical computation)
Scalability	Designed for deep circuits	Typically limited to shallow circuits (often ~ 10–10^2 layers)
Examples in Use	Surface code for fault-tolerant gates	ZNE and readout mitigation in quantum chemistry
Limitations	Hardware- and architecture-intensive	Does not by itself enable universal, deep computation

Table 5.1: Comparison of QEC and QEM in Practical Applications

> **Key Takeaways**
>
> 1. QEC protects quantum information by encoding logical qubits into many physical qubits and repeatedly extracting syndromes.
>
> 2. Below a code-dependent threshold, increasing the code distance can strongly suppress logical error, enabling deep computations.
>
> 3. QEM reduces the impact of noise using extra circuit runs and classical post-processing, without encoding into logical qubits. Unlike QEC, it typically does not scale to large circuits: the sampling and calibration overhead can grow rapidly with circuit size and desired accuracy.

5.3 Characteristics of Algorithms by Era

5.3.1 NISQ-Era Algorithms: Design Constraints and Examples

Algorithms designed for the NISQ era must contend with severe hardware limitations. Chief among these are constraints on circuit depth, since noise accumulates rapidly as gates are applied. As a result, many NISQ algorithms are intentionally shallow and adopt hybrid quantum–classical workflows, where parameterized quantum circuits are optimized by classical routines. These approaches emphasize robustness and empirical performance on imperfect devices, rather than fully rigorous guarantees of correctness.

A hallmark of NISQ-era algorithms is the use of parameterized ansatz circuits, often tailored to a specific problem family rather than universal applicability. They are also designed to be measurement-efficient, seeking to extract useful information with a feasible number of circuit repetitions (shots). The emphasis is often less on provable asymptotic speedups and more on demonstrating practical *quantum utility*—evidence that noisy quantum hardware can provide competitive or otherwise scientifically useful performance under realistic resource constraints.

High-level examples include optimization and scheduling, where the Quantum Approximate Optimization Algorithm (QAOA) has been explored, and quantum simulation tasks such as estimating molecular ground-state energies via variational methods. Sampling-oriented workloads, including random circuit sampling and related models, provide another important class of near-term demonstrations. Across these domains, the design principles are similar: hybrid workflows that partition tasks between quantum and classical resources, careful error budgeting to reduce the impact of noise, and benchmarking against state-of-the-art classical baselines.

5.3.2 FTQC-Era Algorithms: Scalability and Error Resilience

In contrast, algorithms designed for the FTQC era assume access to error-corrected logical qubits and the ability to execute deep circuits reliably. Such algorithms fully exploit superposition, entanglement, and universality of the quantum gate set, yielding quantum speedups that are unattainable for known classical algorithms under standard complexity assumptions.

A canonical example is unstructured search: Grover's algorithm finds a marked item using $O(\sqrt{N})$ oracle queries, compared with $\Theta(N)$ queries classically. Other hallmark FTQC applications include factoring large integers via Shor's algorithm and accurate, large-scale simulations in quantum chemistry and materials science, such as computing molecular energies to chemical accuracy or predicting properties of strongly correlated materials beyond the reach of classical methods. In these settings, algorithm design is expressed in terms of logical resources—logical qubits, logical gate counts, and fault-tolerant circuit depth—rather than in terms of the raw physical device alone.

A defining feature of FTQC algorithms is that error resilience is incorporated into the computation model. Logical operations are realized through fault-tolerant primitives (syndrome extraction cycles, encoded operations, and carefully compiled non-Clifford resources), and the cost of maintaining logical fidelity must be accounted for explicitly in resource estimates. Transpilation from high-level algorithm descriptions to fault-tolerant instruction sets is therefore a central requirement, ensuring that algorithms can run efficiently on architectures protected by codes such as the surface code.

5.3.3 Current Status and Trends in Algorithm Development

The relationship between NISQ-era and FTQC-era algorithms is not one of strict separation, but rather of staged evolution. Many NISQ methods can be viewed as prototypes, heuristics, or reduced-depth surrogates for workflows that may later be realized more cleanly in the fault-tolerant setting. For example, variational and sampling-based methods, although currently limited by noise and optimization instability, may still inform the development of more exact, error-corrected solvers and resource-efficient subroutines.

As of early 2026, the field presents a mixed picture. NISQ algorithm families such as QAOA, VQE, and related hybrid methods remain scientifically important and widely studied, but they have not yet established a broad, decisive practical advantage over the best classical methods across major problem classes. At the same time, the most visible near-term successes have come in more specialized forms: sampling-style demonstrations designed to stress classical simulation, and a growing number of utility-oriented or verifiable experiments tied more closely to concrete scientific tasks. These should be understood as milestones in a developing landscape rather than as final evidence that the algorithmic questions of the NISQ era have been settled.

Several broad trends are now visible. First, modularity is increasingly emphasized, with reusable primitives forming the building blocks for larger workflows. Second, algorithm–hardware co-design has become central, since connectivity, native gate sets, calibration stability, and dominant noise channels strongly affect practical performance. Third, greater attention is being paid to end-to-end workflow costs, including measurement overhead, calibration drift, classical outer-loop optimization, and verification against strong classical baselines. In this sense, the practical performance of a quantum algorithm is often determined as much by the surrounding workflow as by the nominal circuit itself.

A further trend is continuity across eras. As hardware improves and limited logical qubits become available, some ideas developed in the NISQ setting may persist in modified form, while others will be replaced by more analytic, resource-accounted FTQC methods. Retrofitting a NISQ success into an FTQC workflow, however, is rarely straightforward: higher precision, deeper circuits, and explicit management of logical resources are typically required. The broad trend is therefore one of partial inheritance rather than simple carryover.

Trend	NISQ Focus	FTQC Evolution
Circuit Depth	Shallow (noise-limited)	Deep (error-corrected)
Hybrid Nature	Strong classical involvement	Optional (quantum-dominant in many tasks)
Performance Metric	Practical utility on hardware	Complexity-theoretic speedups and logical-resource scaling
Optimization	Noise-aware heuristics	Resource-efficient fault-tolerant design
Validation	Empirical benchmarking and mitigation-aware inference	Verified logical performance and formal resource estimates

Table 5.2: Current Status and Evolutionary Trends in Algorithm Development

5.3.4 Analytic vs. Hybrid Algorithmic Paradigms

A useful way to contrast NISQ-era and FTQC-era algorithms is through the methodological paradigms they emphasize. In the fault-tolerant setting, many cornerstone algorithms are *analytic* in the sense that they are built from explicit constructions with stated approximation guarantees. Typical examples include product-formula simulation, the linear combination of unitaries (LCU) method, and qubitization with quantum signal processing (QSP) or quantum singular value transformation (QSVT). These frameworks provide controllable error bounds and resource estimates, and they are naturally expressed in terms of logical qubits and fault-tolerant gate complexity.

Near-term algorithms, by contrast, are often *hybrid and heuristic.* They are designed around short, hardware-feasible circuits combined with classical post-processing. A prominent subclass is *variational* algorithms, which use parameterized ansatz circuits optimized by classical routines. More broadly, hybrid methods may also include non-variational workflows, such as sampling-based protocols or error-mitigated estimation pipelines, where the emphasis is on empirical performance under realistic noise rather than worst-case asymptotic guarantees.

As summarized in Table 5.3, analytic/provable methods capture much of the long-term FTQC algorithmic landscape, while hybrid/heuristic approaches dominate the NISQ setting. They represent two ends of a spectrum: explicit constructions with tunable accuracy for the long term, and adaptive, hardware-facing strategies for extracting utility in the near term.

Feature	Analytic / Provable Methods (FTQC)	Hybrid / Heuristic Methods (NISQ)
Circuit Depth	Deep, scalable under error correction	Shallow, noise-limited
Guarantees	Explicit error bounds and asymptotic scaling	Problem- and method-dependent; limited worst-case guarantees
Resource Model	Logical qubits and fault-tolerant non-Clifford resources	Physical qubits with hardware-adapted circuits and post-processing
Typical Examples	QPE, Shor, Grover, HHL, product formulas, LCU, QSP/QSVT	VQE, QAOA, variational quantum simulation, sampling-based protocols

Table 5.3: Methodological Contrast Between FTQC and NISQ Algorithms

Key Takeaways

1. NISQ algorithms are typically hybrid and hardware-facing: shallow circuits, many measurements, and classical optimization or inference.

2. FTQC algorithms are typically analytic and resource-accounted: deep fault-tolerant circuits with stated scaling in logical resources.

3. The present landscape is transitional: many NISQ algorithm families remain research directions, while utility-style demonstrations are becoming more credible and application-linked.

4. The central shift is from *device-limited* design (NISQ) to *error-budgeted* design (FTQC), where logical error and depth are explicit parameters.

5.4 Quest for Quantum Advantage in the NISQ Era

This section introduces the language of quantum utility and quantum advantage only insofar as it is needed for the NISQ–FTQC comparison in the present chapter. It is not intended as a full treatment. We return to these questions in much greater depth in Chapter 23, where advantage claims are analyzed more systematically in terms of definitions, classical baselines, verification, and resource accounting.

5.4.1 Defining Quantum Advantage and Its Measurement

The pursuit of quantum advantage is central to the development of quantum algorithms. In the NISQ setting, the term *quantum utility* is often used to emphasize practical performance on real hardware: a quantum device carries out a task of genuine scientific or computational interest under realistic resource constraints, even if no provable asymptotic speedup is known.

By contrast, *quantum advantage* (historically called "quantum supremacy") usually refers to a clearer and more decisive separation between quantum and classical performance for a specific, well-defined task under an explicitly stated comparison model. Early demonstrations focused on sampling tasks, such as random circuit sampling, designed to be hard for classical simulation. More broadly, quantum advantage is now understood as a multidimensional notion that depends on the task, the classical baseline, the verification method, and the performance metric being used [8, 9, 16].

Measurement of such milestones requires rigorous benchmarking. Comparisons may be drawn in terms of runtime (quantum execution vs. classical simulation), resource usage (qubits and gate counts vs. classical FLOPs), or task-specific benchmarks (for example, families of chemistry or optimization instances). Variants of advantage are often distinguished as weak versus strong (task-specific vs. broader algorithmic regimes), or heuristic versus certified (empirical demonstrations vs. provable guarantees).

Exercise 5.4 Discuss why an experiment showing "quantum supremacy" (e.g., random circuit sampling) might not necessarily demonstrate quantum utility.

5.4.2 Current Challenges in Achieving Quantum Utility

Despite progress, several challenges hinder the achievement of meaningful quantum utility. On the hardware side, noise continues to obscure quantum signals, making it difficult to separate genuine improvement from statistical fluctuations. The scalability of FTQC remains a major engineering challenge, with large numbers of physical qubits required for many applications.

Algorithmically, the field still lacks universally agreed "killer apps" that clearly require quantum resources under realistic cost models. Many candidate problems, such as optimization, still admit highly effective classical approximations, raising questions about when robust utility will emerge. Verification poses another hurdle, since distinguishing quantum outputs from sophisticated classical spoofing or from hardware noise remains nontrivial, leading to critiques of "supremacy without utility." Broader concerns include energy costs, accessibility, and the ethical implications of large-scale quantum computing.

These concerns have motivated systematic efforts to formalize notions of advantage, including explicit criteria for verification, robustness to noise, and fair classical comparisons [8, 9, 16].

5.4.3 Potential Benchmarks for Success

Defining benchmarks for progress provides tangible goals for both NISQ and FTQC research. In the near term, success is best measured by clear, reproducible improvements over strong classical baselines on meaningful instance families under transparent resource accounting. As of spring 2026, the most credible milestones include verifiable or carefully benchmarked utility-style demonstrations on real hardware, increasingly strong logical-qubit and error-correction results, and task-specific experiments that connect more directly to scientific applications rather than to contrived sampling tasks alone.

In the longer term, hallmark demonstrations would include cryptographically relevant factoring instances via Shor's algorithm, chemically accurate quantum simulations for challenging molecules, and fault-tolerant computations sustained over algorithmically significant depths. These are still future benchmarks rather than present-day achievements. Current roadmaps from major platforms instead point first toward intermediate goals such as larger and more reliable logical-qubit arrays, deeper error-corrected circuits, and application-linked demonstrations of quantum utility before truly large-scale FTQC becomes routine.

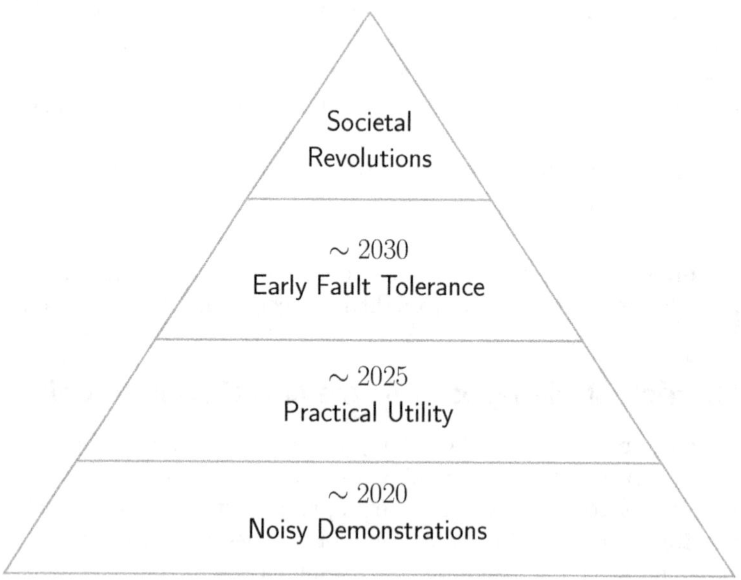

Figure 5.4: Pyramid of Quantum Utility Levels

A useful visualization is the pyramid of utility levels in Fig. 5.4. Its date labels should be read as approximate historical markers rather than fixed deadlines. By 2026, the lower two layers are no longer merely aspirational: noisy demonstrations are well established, and the field has begun to report stronger forms of utility and verifiable advantage. At the same time, the upper layers—early fault tolerance at broad algorithmic scale and wider societal transformation—remain goals for the future rather than settled realities.

Other widely cited benchmarks include increasing quantum volume or related hardware-scale metrics, realizing logical qubits that outperform physical ones under repeated error correction, and demonstrating error-corrected operations over longer algorithmic depths. Such benchmarks reflect a growing consensus that progress toward quantum advantage should be evaluated across a spectrum of tasks and regimes rather than through isolated demonstrations. This broader perspective is one of the main themes of Chapter 23.

Exercise 5.5 A benchmark claim of quantum advantage requires a careful statement of the comparison model. To be scientifically meaningful, such a claim should specify at least the task definition, the verification method, and the classical baseline. Investigate and provide (from the literature or a plausible toy scenario) a concrete example showing how changing the classical baseline could change the conclusion of the benchmark.

> **Key Takeaways**
>
> 1. In the NISQ setting, *quantum utility* usually emphasizes practical perfor-
>   mance under realistic hardware constraints, whereas *quantum advantage*
>   refers to a stronger and more explicit separation from classical methods.
>
> 2. Advantage claims are comparative: they depend on the task definition,
>   the classical baseline, the verification method, and the performance
>   metric.
>
> 3. The field has made progress toward stronger utility-style and verifi-
>   able demonstrations, but broadly accepted, application-wide near-term
>   advantages remain limited.

5.5 Possible Directions and Future Outlook

5.5.1 Research Priorities for NISQ and FTQC Convergence

Bridging the gap between NISQ and FTQC requires research efforts on multiple fronts.
One promising avenue is the integration of hybrid error management, combining QEM
with early forms of QEC to extend circuit depth modestly without incurring full-scale
overhead. Another priority is algorithm co-design, where software frameworks are
developed to support programming models that remain meaningful across eras.
Equally important are benchmarking frameworks that provide standardized tests of
utility and quantify the impact of quantum processors on real-world tasks.

5.5.2 Emerging Technologies and Methodologies

On the hardware side, modular quantum architectures—including distributed quantum networks and heterogeneous platforms—promise improved scalability. Topological qubits offer another path, with built-in protection against certain types of noise.
Methodologically, machine learning is increasingly being applied to optimize error
mitigation protocols and automate compiler-level optimizations. The software ecosys-
tem is also evolving, with open-source tools such as PennyLane providing bridges
between NISQ experimentation and fault-tolerant simulation environments, and
with industrial platforms such as Qiskit and Cirq supporting increasingly detailed
resource analysis.

5.5.3 Long-Term Vision for Quantum Algorithm Development

The long-term vision for quantum algorithms emphasizes universal design frameworks
that adapt across eras while remaining practical on near-term hardware. A central
ambition is to identify tasks that are representative of the power of BQP and to
realize them with transparent assumptions and credible benchmarks. Societal impact
may be profound, with potential applications in climate modeling, drug discovery, AI
acceleration, and secure communications. Yet open questions remain: Will NISQ-era
algorithms produce unexpected advantages before full FTQC is realized? How should
"quantum worth" be quantified in practice?

Ultimately, advancing quantum algorithms requires cross-era thinking. By designing modular workflows with explicit error-management layers, researchers can build methods that remain robust across a spectrum of hardware capabilities, ensuring continuity in progress from NISQ to FTQC.

Direction	Short-Term (NISQ)	Long-Term (FTQC)
Algorithm Focus	Hybrid / heuristic, noise-aware	Analytic / provable, fault-tolerant
Tech Enablers	QEM tools, improving qubits	Full QEC, large-scale logical qubits
Key Milestones	Reproducible utility vs. strong baselines	Broad quantum advantage on practical tasks

Table 5.4: Future Directions in Quantum Algorithm Development

Chapter Takeaways

1. Quantum algorithms are shaped by hardware eras: fault-tolerant (FTQC) goals differ fundamentally from near-term (NISQ) constraints.

2. Noise and limited depth motivate hybrid and variational methods, while FTQC enables deeper algorithms with rigorous guarantees.

3. "Quantum advantage" depends on end-to-end workflow costs, including state preparation, error management, and readout.

4. A realistic outlook balances theoretical speedups with practical resource requirements and application-specific structure.

Problem Set 5

5.1 Choose a concrete hardware limitation (for example, limited coherence time, two-qubit gate error, or measurement error) and explain how it constrains *each* of the following:

(a) circuit depth,

(b) number of shots required to estimate an expectation value,

(c) the choice of ansatz structure in a variational algorithm.

Your answer should explicitly connect the limitation to a design decision.

5.2 The NISQ era is often defined by its hardware limitations.

(a) List three key constraints of NISQ devices that directly shape algorithm design.

(b) Provide an example of a real-world problem where a variational quantum algorithm might offer utility despite these constraints.

5.3 Quantum error correction (QEC) provides exponential error suppression below a threshold.

(a) Explain the meaning of the error threshold theorem in your own words.

(b) Suppose the physical error rate per gate is $p = 10^{-3}$ and the surface code has distance $d = 7$. Estimate the scaling of the logical error rate using the rule $\varepsilon_L \sim p^{d/2}$.

5.4 Quantum error mitigation (QEM) plays a different role from QEC.

(a) Describe one QEM technique (e.g., zero-noise extrapolation) and explain how it reduces error without full error correction.

(b) Why is QEM unsuitable as a long-term solution for universal quantum computing?

5.5 Consider the pyramid of quantum utility levels (see Fig. 5.4).

(a) Identify the main milestones separating the four stages: noisy demonstrations, practical utility, early fault tolerance, and societal revolutions.

(b) Investigate and suggest one possible benchmark experiment that could mark the transition from practical utility to early fault tolerance.

5.6 Algorithmic eras can be contrasted by their structure and scalability.

(a) Explain the main difference between hybrid/heuristic algorithms in the NISQ era and analytic, deep-circuit algorithms in the FTQC era.

(b) Give two reasons why classical–quantum hybrid workflows are essential in NISQ but optional in FTQC.

5.7 The transition from NISQ to FTQC involves significant shifts in algorithm characteristics, including circuit depth and error resilience.

(a) Identify two key design constraints for NISQ-era algorithms and explain how these differ from the requirements for FTQC-era algorithms.

(b) Suggest an evolutionary trend in algorithm development that could facilitate the transition from NISQ to FTQC, providing a rationale based on hardware advancements.

5.8 Assume a target experiment requires n_L logical qubits and depth D_L in logical operations. Suppose the acceptable total failure probability is at most 0.1.

(a) Give an order-of-magnitude estimate for the logical error per operation ε_L required to keep the total failure probability below 0.1, expressed in terms of D_L.

(b) Interpret your result: how does increasing depth tighten the logical error budget?

Part II develops the core quantum algorithmic primitives that reappear across many application domains. We start with Fourier-based methods—the Quantum Fourier Transform (QFT) and Quantum Phase Estimation (QPE)—and then study period finding and Shor's algorithm as a landmark example where these primitives combine into a complete, high-impact procedure.

We then build the modern "polynomial approximation" toolkit: Hamiltonian simulation viewpoints, block-encoding, Linear Combination of Unitaries (LCU), Quantum Signal Processing (QSP), and Quantum Singular Value Transformation (QSVT). These tools support quantum linear system algorithms and also clarify how many seemingly different algorithms share a common backbone. The part concludes by treating adiabatic and variational families and by collecting reusable measurement primitives, since readout is often the dominant cost in practice. By the end of Part II, readers should be able to recognize recurring patterns and assemble primitives into new workflows.

6. Quantum Fourier Transform (QFT)

Contents

Fourier transforms provide a way to decompose a function into its frequency components, converting from the time (or spatial) domain to the frequency domain. The Discrete Fourier Transform (DFT, introduced in § 1.1.1.2) is the discrete counterpart of the continuous Fourier transform. It operates on a finite set of equally spaced samples of a function, producing a finite sequence that represents the frequency spectrum. This process can reveal periodic behaviors and simplify certain computational tasks, making the DFT a cornerstone of modern signal processing, communications, and related fields.

The Quantum Fourier Transform (QFT) is essentially the "quantum version" of the DFT. It is a central primitive in several quantum algorithms, and it is one of the key ingredients behind exponential quantum speedups for problems such as factoring and discrete logarithms.

In essence, the QFT is the DFT defined over a computational basis of size $N = 2^n$, implemented via unitary operations on n qubits. It leverages quantum gates—particularly controlled-phase gates—along with Hadamard gates and a final bit-reversal (swap) pattern to realize the same mathematical transform. This chapter provides a rigorous treatment of the QFT and highlights how it reveals periodic structure in quantum states.

 This chapter builds on the following chapters from other books in the series:

- QC Math [1], Chapter 11: *Functions of Vectors and Matrices*

- QCI Book [2], Chapter 7: *Multi-Qubit Quantum Gates*

6.1　Definition and Properties

6.1.1　Definition

Let n be the number of qubits, so that our Hilbert space has dimension $N = 2^n$. We label the computational basis states by integers $x \in \{0, 1, \ldots, N-1\}$, representing each basis vector as $|x\rangle$. The QFT acts on these basis states as follows:

$$\mathrm{QFT}_N : |x\rangle \mapsto |\widetilde{x}\rangle = \frac{1}{\sqrt{N}} \sum_{k=0}^{N-1} e^{\frac{2\pi i k x}{N}} |k\rangle. \tag{6.1}$$

It is also often written as:

$$\mathrm{QFT}_N : |x\rangle \mapsto |\widetilde{x}\rangle = \frac{1}{\sqrt{N}} \sum_{k=0}^{N-1} \omega_N^{kx} |k\rangle, \tag{6.2}$$

where $\omega_N = e^{\frac{2\pi i}{N}}$ is the primitive N-th root of unity.

In operator (or matrix) form, Eq. 6.1 is equivalent to:

$$\mathrm{QFT}_N = \frac{1}{\sqrt{N}} \sum_{x=0}^{N-1} \sum_{k=0}^{N-1} e^{\frac{2\pi i k x}{N}} |k\rangle\langle x|. \tag{6.3}$$

We often drop the subscript N when the context is clear and simply write QFT or F.

 **DFT vs. QFT.** While QFT can be considered the "quantum version" of the DFT, the two have subtle differences.

1. Classical DFT definitions often use a factor of $\frac{1}{N}$ in one direction and 1 in the other. The QFT on an N-dimensional Hilbert space uses $\frac{1}{\sqrt{N}}$ in both directions so that it is a unitary operator.

2. By convention, the QFT adopts a positive sign in the phase factor $e^{\frac{2\pi i k x}{N}}$. This is opposite to a common DFT convention that uses a negative sign, i.e., $e^{-\frac{2\pi i k x}{N}}$.

Exercise 6.1 Verify directly from Eq. 6.3 that QFT_N is unitary by showing $\text{QFT}_N^\dagger \, \text{QFT}_N = I$. (Hint: use the finite geometric-series identity $\sum_{k=0}^{N-1} \omega_N^{k(x-y)} = N\,\delta_{x,y}$.)

Exercise 6.2 Compute $\text{QFT}_N\,|0\rangle$ and $\text{QFT}_N\,|N-1\rangle$. Interpret your results as "flat" superpositions with different phase patterns.

6.1.2 Basic Properties

The following properties are inherited from the DFT:

- Linearity: Like any quantum operation, QFT is defined linearly on the basis states.

- Symmetry: The QFT matrix is symmetric (though not necessarily Hermitian).

- Unitarity: The QFT is unitary, meaning its inverse is given by its conjugate transpose, $\text{QFT}^{-1} = \text{QFT}^\dagger$.

- Inverse transform: The inverse QFT^{-1} is obtained by replacing $e^{\frac{2\pi i k x}{N}}$ with $e^{-\frac{2\pi i k x}{N}}$. In particular, for computational basis states,

$$\text{QFT}_N^{-1}: \ |k\rangle \ \mapsto \ \frac{1}{\sqrt{N}} \sum_{x=0}^{N-1} e^{-\frac{2\pi i k x}{N}} \, |x\rangle . \tag{6.4}$$

Exercise 6.3 Show that $\text{QFT}_N^4 = I$ for $N = 2$. Does the identity $\text{QFT}_N^4 = I$ hold for general N? Justify your answer.

Exercise 6.4 Show that QFT_N^{-1} is obtained from QFT_N by complex conjugation of all matrix entries. Explain why this is consistent with $\text{QFT}_N^{-1} = \text{QFT}_N^\dagger$.

Exercise 6.5 Let $|\widetilde{k}\rangle := \text{QFT}_N\,|k\rangle$. Show that $\{|\widetilde{k}\rangle\}_{k=0}^{N-1}$ is an orthonormal basis, and compute $\langle \widetilde{k} | x \rangle$ explicitly.

Key Takeaways

1. The QFT is the DFT written as a unitary basis change on an $N = 2^n$ dimensional Hilbert space.

2. The QFT uses the symmetric normalization factor $\frac{1}{\sqrt{N}}$ so that $\text{QFT}^{-1} = \text{QFT}^\dagger$.

3. Different sign conventions ($e^{\pm 2\pi i k x / N}$) correspond to QFT vs. inverse QFT; you must track which convention your circuit implements.

6.1.3   Extracting Periodicity via QFT

Just as the classical Fourier transform uncovers discrete frequency components from periodic signals in the time domain, the QFT can reveal periodic structure embedded in a quantum superposition. The manifestation of periodicity is different in appearance: instead of smooth spectral peaks, the QFT produces sharp amplitude concentrations at specific basis states that encode the underlying periodicities.

1   Transforming Superposition States

The QFT (Eqs. 6.1 and 6.2) specifies its action on computational basis states. By linearity, it extends to any superposition state

$$|\psi\rangle = \sum_{x=0}^{N-1} c_x \,|x\rangle , \tag{6.5}$$

which encodes the complex vector $\boldsymbol{c} = (c_0, \dots, c_{N-1})$ via amplitude encoding (see § 3.1.1).

Under the QFT, such a state transforms into

$$|\widetilde{\psi}\rangle \equiv \mathrm{QFT}\,|\psi\rangle = \sum_{k=0}^{N-1} \widetilde{c}_k \,|k\rangle , \tag{6.6}$$

with

$$\widetilde{c}_k = \frac{1}{\sqrt{N}} \sum_{x=0}^{N-1} c_x \, e^{\frac{2\pi i k x}{N}} , \tag{6.7}$$

which matches the classical DFT up to the normalization convention.

Conversely, if $|\widetilde{\psi}\rangle = \sum_{k=0}^{N-1} \widetilde{c}_k \,|k\rangle$, then

$$\mathrm{QFT}^{-1}\,|\widetilde{\psi}\rangle = \sum_{x=0}^{N-1} c_x \,|x\rangle , \tag{6.8}$$

where

$$c_x = \frac{1}{\sqrt{N}} \sum_{k=0}^{N-1} \widetilde{c}_k \, e^{-\frac{2\pi i k x}{N}} . \tag{6.9}$$

2   Example 1: A Pure Complex Exponential

Consider a genuine discrete complex exponential:

$$|\psi\rangle = \frac{1}{\sqrt{N}} \sum_{x=0}^{N-1} e^{-\frac{2\pi i \omega x}{N}} \,|x\rangle ,$$

where the integer frequency index satisfies $\omega \in \{0, \dots, N-1\}$.

This state is already a Fourier basis vector. Applying the QFT gives

$$|\widetilde{\psi}\rangle = |\omega\rangle .$$

Hence the QFT maps a discrete sinusoid of frequency ω to the corresponding basis state $|\omega\rangle$ with certainty.

3 Example 2: A Three-Harmonic Superposition

Consider the counterpart of the FFT example in QC Math [1], Chapter 11: *Functions of Vectors and Matrices*:

$$c_x = \frac{1}{\sqrt{N}}\frac{15}{\sqrt{259}}\left(e^{-\frac{2\pi i}{N}5x} - \tfrac{1}{3}e^{-\frac{2\pi i}{N}15x} + \tfrac{1}{5}e^{-\frac{2\pi i}{N}25x}\right).$$

This represents a signal with three harmonic components when $|x\rangle$ and $|k\rangle$ are interpreted as time and frequency basis states.

Applying the QFT,

$$|\widetilde{\psi}\rangle = \frac{1}{N}\frac{15}{\sqrt{259}}\sum_{x=0}^{N-1}\sum_{k=0}^{N-1}\left(e^{-\frac{2\pi i}{N}5x} - \tfrac{1}{3}e^{-\frac{2\pi i}{N}15x} + \tfrac{1}{5}e^{-\frac{2\pi i}{N}25x}\right)e^{\frac{2\pi i}{N}kx}|k\rangle$$

$$= \frac{15}{\sqrt{259}}\left(|5\rangle - \tfrac{1}{3}|15\rangle + \tfrac{1}{5}|25\rangle\right).$$

Thus, only the frequencies $k = 5, 15, 25$ have nonzero amplitude.

4 Example 3: A Discrete Periodic Superposition

Consider the periodic superposition

$$|\psi\rangle = \frac{1}{\sqrt{M}}\left(|0\rangle + |r\rangle + |2r\rangle + \cdots + |(M-1)r\rangle\right),$$

where $1 < r < N$ and $M = \lfloor \frac{N}{r} \rfloor$. This emulates a periodic signal of period r sampled on the computational basis.

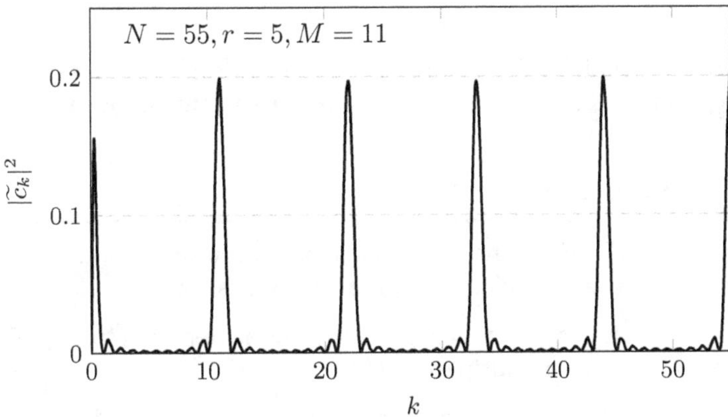

Figure 6.1: Periodicity from Quantum Fourier Transform

Applying the QFT gives

$$|\widetilde{\psi}\rangle = \sum_{k=0}^{N-1}\widetilde{c}_k\,|k\rangle,$$

where

$$\widetilde{c}_k = \frac{1}{\sqrt{MN}}\sum_{x=0}^{M-1}e^{\frac{2\pi ikrx}{N}} = \frac{1}{\sqrt{MN}}\frac{1-e^{\frac{2\pi ikrM}{N}}}{1-e^{\frac{2\pi ikr}{N}}}.$$

Thus,

$$|\widetilde{c}_k|^2 = \frac{1}{MN}\left|\frac{\sin\left(\frac{M\pi kr}{N}\right)}{\sin\left(\frac{\pi kr}{N}\right)}\right|^2.$$

The quantity on the right is a discrete Dirichlet kernel in k. As shown in Fig. 6.1, it has dominant peaks when $\frac{kr}{N}$ is near an integer. If r divides N, then the strongest peaks occur exactly at

$$k = 0,\ \frac{N}{r},\ 2\frac{N}{r},\ \ldots,$$

revealing the underlying period r from the spacing $\frac{N}{r}$. If r does not divide N, the peaks concentrate near these values, and one typically recovers r by a rational-approximation step (as in period finding).

Exercise 6.6 Assume r divides N. In Example 3, show that $\widetilde{c}_k = 0$ whenever k is not a multiple of N/r, and that all nonzero amplitudes have the same magnitude. Conclude that measuring $|\widetilde{\psi}\rangle$ yields a uniformly random multiple of N/r.

Exercise 6.7 Assume r does not divide N. Use the expression for $|\widetilde{c}_k|^2$ in Example 3 to explain why the distribution concentrates near multiples of N/r. Give a qualitative explanation of how this enables period recovery by rational approximation.

Exercise 6.8 Let $s \in \{0, \ldots, N-1\}$. Consider the shifted periodic superposition

$$|\psi_s\rangle = \frac{1}{\sqrt{M}}\big(|s\rangle + |s+r\rangle + \cdots + |s+(M-1)r\rangle\big).$$

Show that $\mathrm{QFT}\,|\psi_s\rangle$ has the same measurement probabilities as $\mathrm{QFT}\,|\psi\rangle$ in Example 3, and identify how s appears in the output phases.

Key Takeaways

1. The QFT maps a periodic pattern in the computational basis to a sparse (peaked) distribution in the Fourier basis.

2. If a state has period r, then measurement after QFT concentrates near integer multiples of N/r.

3. QFT-based algorithms exploit this concentration to recover global structure (such as a period or eigenphase) from a small number of measurements.

6.1.4 Tensor Product Representation

The QFT has a tensor product representation that underpins its circuit implementation. This representation encodes integers and fractions in binary form and expresses the output as a product of single-qubit states with phases determined by progressively finer binary fractions. A key point is that these tensor factors naturally appear in reversed qubit order, which explains the final bit-reversal stage in the standard circuit.

1   Binary Notation of Integers and Fractions

The n-qubit state vector

$$|x_1 x_2 \ldots x_n\rangle \equiv |x_1\rangle \otimes |x_2\rangle \otimes \cdots \otimes |x_n\rangle,$$

can encode an integer x by interpreting $x_1 x_2 \ldots x_n$ as a binary number:

$$x = [x_1 x_2 \ldots x_n] = \sum_{k=1}^{n} x_k \, 2^{n-k}, \quad \text{where } x_k \in \{0,1\}.$$

As a result, the state vector $|x_1 x_2 \ldots x_n\rangle$ is often written as $|x\rangle$. For instance, $|5\rangle = |00\ldots 101\rangle$.

We also define the binary fraction:

$$[x_1 x_2 \ldots x_m \cdot x_{m+1} x_{m+2} \ldots x_n] = \sum_{k=1}^{n} x_k \, 2^{m-k}.$$

For example, $[10.1] = 2 + \frac{1}{2}$, while $[0.0101] = \frac{1}{4} + \frac{1}{16} = \frac{5}{16}$.

> Throughout this text we use the *big-endian* (i.e., big end first) convention: in a register $|x_1 x_2 \cdots x_n\rangle$, the leftmost bit x_1 is the most significant bit (MSB) and the rightmost bit x_n is the least significant bit (LSB).
>
> It is important to note that some software frameworks—most notably Qiskit—use a little-endian convention in which the lowest-index qubit corresponds to the least significant bit. When comparing circuits or measurement outcomes across conventions, care must be taken to translate between qubit indices, bitstring order, and the corresponding integer values.

2   Two-Qubit Example

Consider a two-qubit system ($n = 2$, $N = 4$) as a simple illustration. In this case, Eq. 6.1 reduces to:

$$
\begin{aligned}
2\,\mathrm{QFT}(|x_1 x_2\rangle) &= |00\rangle + e^{2\pi i \frac{2x_1 + x_2}{4}} |01\rangle + e^{2\pi i \frac{(2x_1 + x_2)\,2}{4}} |10\rangle + e^{2\pi i \frac{(2x_1 + x_2)\,3}{4}} |11\rangle \\
&= |00\rangle + e^{2\pi i \frac{2x_1 + x_2}{4}} |01\rangle + e^{2\pi i \frac{4x_1 + 2x_2}{4}} |10\rangle + e^{2\pi i \frac{6x_1 + 3x_2}{4}} |11\rangle \\
&= |00\rangle + e^{2\pi i \frac{2x_1 + x_2}{4}} |01\rangle + e^{2\pi i \frac{2x_2}{4}} |10\rangle + e^{2\pi i \frac{2x_1 + 3x_2}{4}} |11\rangle \\
&= \left(|0\rangle + e^{2\pi i \frac{x_2}{2}} |1\rangle\right) \otimes \left(|0\rangle + e^{2\pi i (\frac{x_1}{2} + \frac{x_2}{4})} |1\rangle\right) \\
&= \left(|0\rangle + e^{2\pi i [0.x_2]} |1\rangle\right) \otimes \left(|0\rangle + e^{2\pi i [0.x_1 x_2]} |1\rangle\right).
\end{aligned}
$$

In the third step, we have used the fact that $e^{2\pi i j} = 1$ for any integer j. Hence, any integer part of $\frac{kx}{N}$ does not affect the phase $e^{2\pi i \frac{kx}{N}}$.

Exercise 6.9 Work out Eq. 6.1 for a three-qubit system, and show that $|\widetilde{x}\rangle$ can be written as a tensor product of three factors, analogous to the two-qubit example above.

3 n-Qubit System

We can generalize the two-qubit result to an n-qubit system:

$$
\begin{aligned}
\mathrm{QFT}(|x_1 x_2 \ldots x_n\rangle) &= \frac{1}{\sqrt{N}} \bigotimes_{k=1}^{n} \left(|0\rangle + e^{2\pi i [0.x_k x_{k+1} \ldots x_n]} |1\rangle \right) \\
&= \frac{1}{\sqrt{N}} \left(|0\rangle + e^{2\pi i [0.x_n]} |1\rangle \right) \otimes \left(|0\rangle + e^{2\pi i [0.x_{n-1} x_n]} |1\rangle \right) \quad (6.10) \\
&\quad \otimes \cdots \otimes \left(|0\rangle + e^{2\pi i [0.x_1 x_2 \ldots x_n]} |1\rangle \right).
\end{aligned}
$$

Key Takeaways

1. The QFT of $|x_1 x_2 \ldots x_n\rangle$ factorizes into n single-qubit states whose phases encode binary fractions reveals from the bits of x.

2. The k-th output qubit carries a phase determined by $[0.x_k x_{k+1} \ldots x_n]$, so finer bits contribute smaller phase rotations.

3. The natural factor order is reversed, which is why the standard QFT circuit ends with a bit-reversal (swap) stage.

Below is an outline of the proof of Eq. 6.10. (Readers lacking sufficient background may skip it.)

✴ *Proof.*

$$
\begin{aligned}
\mathrm{QFT}(|x\rangle) &= \frac{1}{\sqrt{N}} \sum_{y=0}^{N-1} e^{\frac{2\pi i\, x\, y}{N}} |y\rangle \\
&= \frac{1}{\sqrt{N}} \sum_{y_1=0}^{1} \cdots \sum_{y_n=0}^{1} e^{2\pi i\, x \left(\sum_{k=1}^{n} y_k\, 2^{-k} \right)} |y_1 y_2 \ldots y_n\rangle \\
&= \frac{1}{\sqrt{N}} \sum_{y_1=0}^{1} \cdots \sum_{y_n=0}^{1} \bigotimes_{k=1}^{n} \left(e^{2\pi i\, x\, y_k\, 2^{-k}} |y_k\rangle \right) \\
&= \frac{1}{\sqrt{N}} \bigotimes_{k=1}^{n} \left[\sum_{y_k=0}^{1} e^{2\pi i\, x\, y_k\, 2^{-k}} |y_k\rangle \right] \\
&= \frac{1}{\sqrt{N}} \bigotimes_{k=1}^{n} \left(|0\rangle + e^{2\pi i\, x\, 2^{-k}} |1\rangle \right) \\
&= \frac{1}{\sqrt{N}} \bigotimes_{k=1}^{n} \left(|0\rangle + e^{2\pi i\, [x_1 x_2 \ldots x_n]\, 2^{-k}} |1\rangle \right) \\
&= \frac{1}{\sqrt{N}} \bigotimes_{k=1}^{n} \left(|0\rangle + e^{2\pi i [x_1 x_2 \ldots x_{k-1}.x_k x_{k+1} \ldots x_n]} |1\rangle \right)
\end{aligned}
$$

$$= \frac{1}{\sqrt{N}} \bigotimes_{k=1}^{n} \left(|0\rangle + e^{2\pi i[0.x_k x_{k+1} \cdots x_n]} |1\rangle \right).$$

$\square$

Exercise 6.10 In Eq. 6.10, explain carefully why the tensor factors appear in reversed order. If one omits the final bit-reversal swap network, what is the resulting map on basis states?

Exercise 6.11 Prove the identity

$$e^{2\pi i[x_1 x_2 \cdots x_n] \, 2^{-k}} = e^{2\pi i[x_1 x_2 \cdots x_{k-1} . x_k x_{k+1} \cdots x_n]} = e^{2\pi i[0.x_k x_{k+1} \cdots x_n]},$$

and state precisely where the equality uses the fact that $e^{2\pi i j} = 1$ for integers j.

6.2 Circuit Implementation

Overview

1. The standard QFT circuit is built from Hadamard gates plus a cascade of controlled-phase gates whose angles decrease as powers of two.

2. Each output qubit receives one Hadamard and then phases controlled by less significant qubits.

3. A final swap network performs bit reversal; it can often be omitted if classical post-processing relabels the measured bits.

6.2.1 Circuit for an Individual QFT Tensor Factor

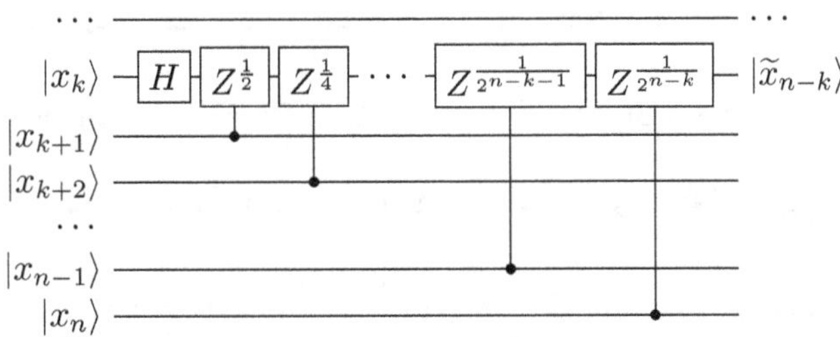

Figure 6.2: Quantum Fourier Transform: A Single Output Factor

Each tensor factor in Eq. 6.10 has the form

$$\frac{1}{\sqrt{2}} \left(|0\rangle + e^{2\pi i[0.x_k x_{k+1} \cdots x_n]} |1\rangle \right),$$

and can be realized by the quantum circuit shown in Fig. 6.2. A common source of confusion is qubit order: the factors appear from $[0.x_n]$ up to $[0.x_1 \ldots x_n]$, so the natural tensor-product order is reversed relative to the input register. The final bit-reversal stage in the full circuit restores the conventional ordering.

A Hadamard gate (H) transforms $|x_k\rangle$ into $\frac{1}{\sqrt{2}}(|0\rangle+|1\rangle)$ if $x_k = 0$, or $\frac{1}{\sqrt{2}}(|0\rangle-|1\rangle)$ if $x_k = 1$. Both cases can be written as

$$\frac{1}{\sqrt{2}}\left(|0\rangle + e^{2\pi i \frac{x_k}{2}}|1\rangle\right) = \frac{1}{\sqrt{2}}\left(|0\rangle + e^{2\pi i[0.x_k]}|1\rangle\right).$$

To progress from $|0\rangle + e^{2\pi i[0.x_k]}|1\rangle$ to $|0\rangle + e^{2\pi i[0.x_k x_{k+1}]}|1\rangle$, we must incorporate an additional phase $e^{2\pi i \frac{x_{k+1}}{4}} = e^{\frac{i\pi}{2}x_{k+1}}$ into the $|1\rangle$ component whenever $x_{k+1} = 1$. This is achieved using a controlled $Z^{\frac{1}{2}}$ gate conditioned on the control qubit $|x_{k+1}\rangle$. Generally,

$$Z^{\frac{1}{m}} = |0\rangle\langle 0| + e^{\frac{\pi i}{m}}|1\rangle\langle 1| = \begin{bmatrix} 1 & 0 \\ 0 & e^{\frac{\pi i}{m}} \end{bmatrix} \tag{6.11}$$

is the m-th root of the Z gate, identical to the phase gate $P(\frac{\pi}{m})$.

Continuing this pattern, the full phase factor $e^{2\pi i[0.x_k x_{k+1} \cdots x_n]}$ is implemented by controlled gates $Z^{\frac{1}{2}}, Z^{\frac{1}{4}}, Z^{\frac{1}{8}}, \ldots$, each controlled by the appropriate more-significant input qubit, as illustrated in Fig. 6.2.

6.2.2 Full QFT Circuit Construction

To construct the entire QFT circuit, we realize each factor of Eq. 6.10 on the respective output qubit, with controlled-phase gates conditioned on the other qubits. Repeating the circuit shown in Fig. 6.2 for every qubit leads to the structure of Fig. 6.3.

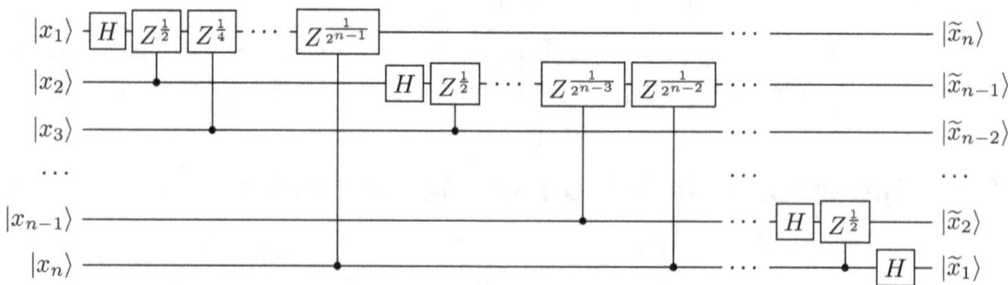

Figure 6.3: Standard Quantum Fourier Transform (QFT) Circuit

Hence, a standard QFT circuit is generally implemented through the following steps:

1. Apply a Hadamard gate (H) to the first (most significant) qubit.

2. Apply controlled-phase gates to the first qubit, with phases $e^{i\frac{\pi}{2}}, e^{i\frac{\pi}{4}}, \ldots, e^{i\frac{\pi}{2^{n-1}}}$ conditioned on the remaining qubits.

3. Repeat the procedure on the second qubit (now using phases down to $e^{i\frac{\pi}{2^{n-2}}}$), then the third, and so on, until all qubits have been processed.

4. Conclude with a bit-reversal operation (a sequence of swap gates) that reverses the order of the qubits. This aligns the output ordering with the standard QFT convention.

In practice, the bit-reversal stage can often be postponed or handled by classical post-processing, depending on the application.

Exercise 6.12 Sketch the QFT circuit for three qubits from memory.

Exercise 6.13 Show explicitly that your three-qubit QFT circuit implements the basis transform in Eq. 6.1 by tracing the states before and after each Hadamard gate.

Exercise 6.14 For $n = 4$, write the list of controlled-phase gates acting on each target qubit in the standard QFT circuit (including their angles). Then repeat the list for the inverse QFT.

Exercise 6.15 Show that the inverse QFT circuit is obtained from the QFT circuit by reversing the gate order and replacing each controlled phase by its inverse. Draw the three-qubit inverse QFT circuit explicitly.

6.2.3 Exponential Speedup of QFT

1 Gate Counts

The n-qubit QFT circuit in Fig. 6.3 employs:

1. One Hadamard gate on each qubit, giving a total of n Hadamards.

2. Controlled-phase gates in the amount $\displaystyle\sum_{k=1}^{n-1} k = \frac{n(n-1)}{2}$.

3. $\lfloor n/2 \rfloor$ swap gates, if the final bit-reversal is implemented.

Hence, the total number of gates is:

$$\text{Total gates} = n + \frac{n(n-1)}{2} + \lfloor n/2 \rfloor. \tag{6.12}$$

The dominant term in $n + \frac{n(n-1)}{2} + \lfloor n/2 \rfloor$ is on the order of n^2, set by the $\frac{n(n-1)}{2}$ controlled-phase gates. Consequently, the overall gate complexity of the QFT is $O(n^2)$.

2 What "Exponential Speedup" Means Here

A classical DFT on an $N = 2^n$ dimensional input requires $O(N^2)$ basic operations if implemented naively. The well-known Fast Fourier Transform (FFT) leverages symmetries in the DFT to reduce this to $O(N \log N)$, which is $O(2^n n)$.

The QFT is mathematically almost identical to the DFT, but its circuit implementation uses only $O(n^2)$ gates. This is an exponential improvement with respect to the number of qubits n when one compares the circuit depth/size needed to implement the linear transform.

At the same time, it is important to interpret this comparison correctly: the QFT does not, by itself, provide all N Fourier coefficients as classical output in one run. Measuring QFT $|\psi\rangle$ yields a single sample from the frequency distribution $|\tilde{c}_k|^2$. The main algorithmic power of the QFT comes from using it as a coherent subroutine inside larger quantum procedures (such as phase estimation and period finding), where one only needs to extract structured information (e.g., a period or an eigenphase) rather than the entire spectrum.

Key Takeaways

1. The QFT implements the Fourier basis change with $O(n^2)$ gates on n qubits, vs. $O(N \log N)$ arithmetic steps for FFT on $N = 2^n$ data.

2. This does not mean one can read out all N Fourier coefficients in one run; measurement yields a sample from the frequency distribution.

3. The QFT's power comes from being used coherently inside larger algorithms (period finding, phase estimation), where only structured information must be extracted.

6.3 ✳ Variations of the QFT Circuit

Using gate-sequence equivalences, the standard QFT circuit in Fig. 6.3 can be reshaped into several alternative forms. These equivalent circuits are useful in practice: they can reduce circuit depth, adapt the QFT to hardware connectivity constraints, and reveal structural symmetries that are later reused in more advanced algorithmic primitives. In what follows, we use a four-qubit QFT to illustrate these transformations. The standard four-qubit QFT circuit, including the final bit-reversal, appears in Fig. 6.4.

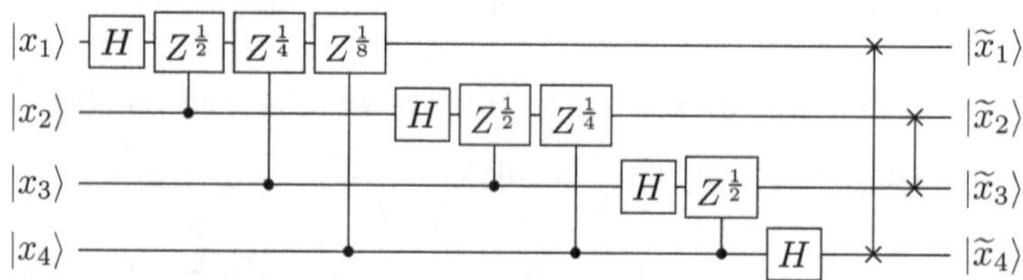

Figure 6.4: Four-Qubit QFT Reference Circuit

R | Refer to QCI Book [2], Section 7.4: *Equivalent Gate Sequences.*

6.3.1   QFT with Swapped Controlled-Phase Gates

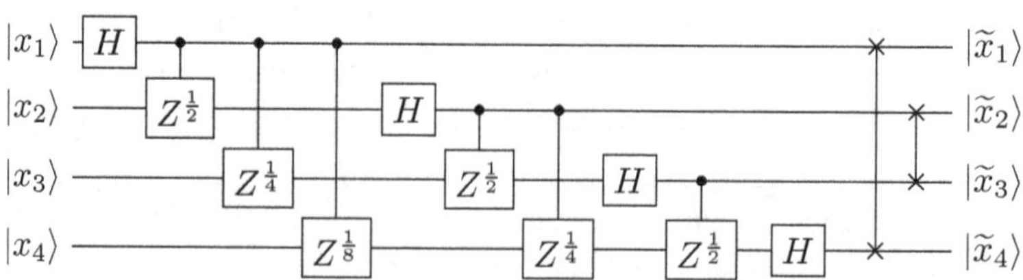

Figure 6.5: QFT with Swapped Controlled-Phase Gates

First, note that controlled-phase gates are symmetric in the roles of the control and target qubits: swapping "control" and "target" does not change the unitary. For

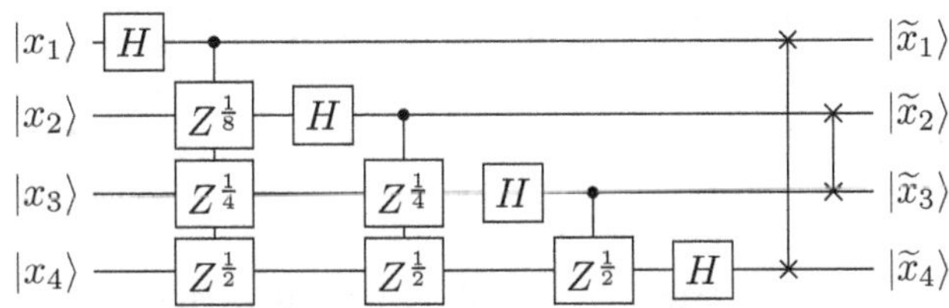

Figure 6.6: QFT with Stacked Controlled-Phase Gates

example, a phase gate on qubit 1 controlled by qubit 2, denoted $\text{CZ}^{\frac{1}{2^l}}{}_{12}$, is equal to the gate on qubit 2 controlled by qubit 1, $\text{CZ}^{\frac{1}{2^l}}{}_{21}$. Concretely,

$$\text{CZ}^{\frac{1}{2^l}}{}_{12} = |00\rangle\langle 00| + |01\rangle\langle 01| + |10\rangle\langle 10| + e^{i\frac{\pi}{2^l}} |11\rangle\langle 11|.$$

Swapping qubits 1 and 2 via $|x_1 x_2\rangle \to |x_2 x_1\rangle$ yields

$$\text{CZ}^{\frac{1}{2^l}}{}_{21} = |00\rangle\langle 00| + |10\rangle\langle 10| + |01\rangle\langle 01| + e^{i\frac{\pi}{2^l}} |11\rangle\langle 11|,$$

which is the same operator.

A QFT circuit that uses swapped $\text{CZ}^{\frac{1}{2^l}}$ gates is shown in Fig. 6.5. In many circuit diagrams, phase gates controlled by the same qubit are grouped together, producing a circuit like Fig. 6.6. This grouping is functionally equivalent to Fig. 6.4, but it often has fewer gate layers and can reduce circuit depth on hardware.

6.3.2 Consolidating Phase Gates with a Common Control Qubit

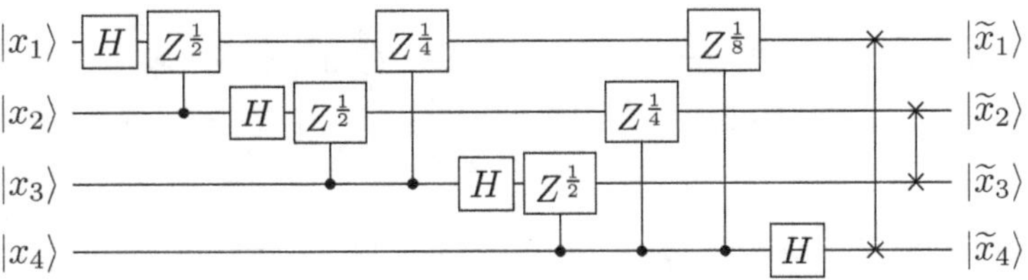

Figure 6.7: QFT with Reordered Controlled-Phase Gates

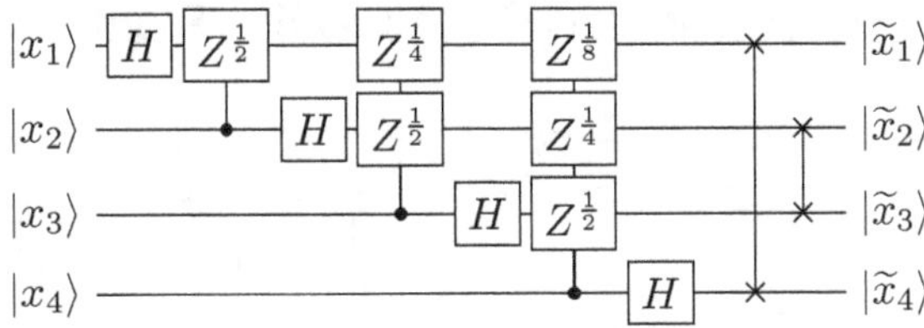

Figure 6.8: QFT with Reordered and Stacked Controlled-Phase Gates

Many controlled-phase gates in the QFT commute because they act on disjoint pairs of qubits. For instance, the H gate and $\text{CZ}^{\frac{1}{2}}{}_{23}$ do not act on either the control or the target qubit of $\text{CZ}^{\frac{1}{4}}{}_{13}$. Hence, we can postpone $\text{CZ}^{\frac{1}{4}}{}_{13}$ until after

H and $\mathrm{CZ}^{\frac{1}{2}}{}_{23}$ have been applied. Similarly, $\mathrm{CZ}^{\frac{1}{8}}{}_{14}$ and $\mathrm{CZ}^{\frac{1}{4}}{}_{24}$ can be postponed. Collectively, these rearrangements yield the circuit shown in Fig. 6.7. By further stacking all phase gates controlled by a single qubit, we arrive at the circuit in Fig. 6.8.

6.3.3 Gathering Hadamard Gates to the End

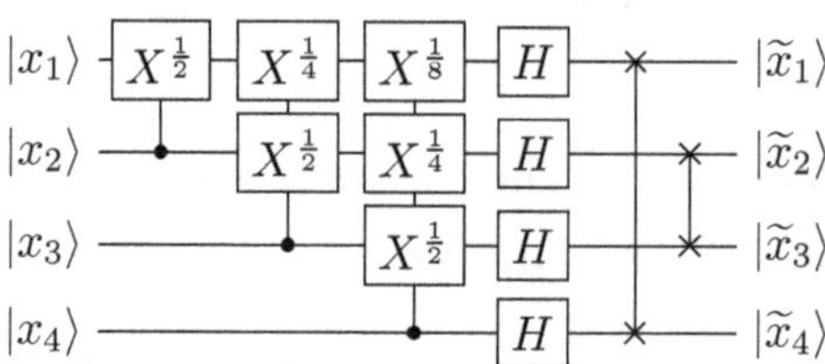

Figure 6.9: QFT with All Hadamard Gates at the End

A useful "gate-pushing" rule is that moving a Hadamard gate past a fractional Z rotation converts it into the corresponding fractional X rotation. In circuit form:

$$-\boxed{H}\,\boxed{Z^{\frac{1}{m}}}- \;\; = \;\; -\boxed{X^{\frac{1}{m}}}\,\boxed{H}-$$

R | See QC Math [1], Section 11.3: *Matrix-Valued Functions*, on analytic functions.

To justify this identity, recall that X is diagonalized by H: indeed, $X = HZH$, where $H^\dagger H = H^2 = I$ and Z is diagonal. Hence, for any analytic function f,

$$f(X) \;=\; H\,f(Z)\,H.$$

Taking $f(t) = t^{1/m}$ gives $X^{\frac{1}{m}} = H\,Z^{\frac{1}{m}}\,H$, and therefore

$$H\,Z^{\frac{1}{m}} \;=\; X^{\frac{1}{m}}\,H,$$

which is the relationship used in the circuit equivalence above.

This equivalence allows the circuit in Fig. 6.8 to be rewritten as Fig. 6.9, where all Hadamards appear at the end.

6.3.4 Gathering Hadamard Gates to the Front

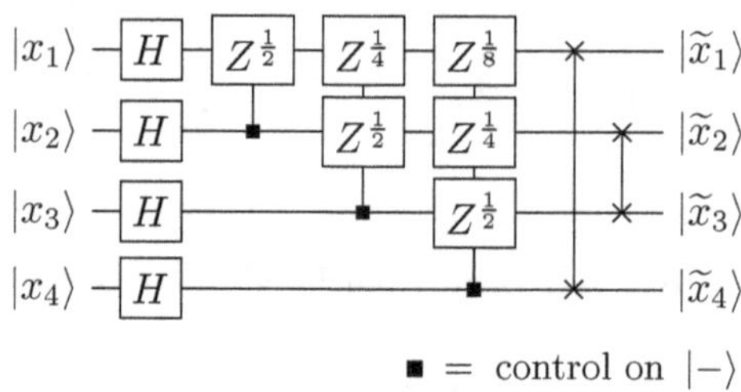

Figure 6.10: QFT with All Hadamard Gates Positioned at the Front

One may also ask whether all Hadamard gates can be pushed to the front. Indeed, using a related equivalence, the circuit in Fig. 6.8 can be rearranged into Fig. 6.10, where every Hadamard gate is applied at the outset:

$$\text{control on } |1\rangle \qquad = \qquad \text{control on } |-\rangle$$

In this front-loaded design, each controlled $Z^{\frac{1}{m}}$ gate is triggered when the control qubit is in the state $|-\rangle \equiv \frac{1}{\sqrt{2}}(|0\rangle - |1\rangle)$ (indicated by a black square), rather than the usual $|1\rangle$ (often indicated by a black circle). This follows from $H|1\rangle = |-\rangle$ and $H|-\rangle = |1\rangle$, which changes which control-state activates the phase.

Exercise 6.16 Choose one of the QFT circuit variations (Fig. 6.7, Fig. 6.8, Fig. 6.10, or Fig. 6.9). Starting from the standard circuit in Fig. 6.4, justify the transformation step by step using commutation relations or gate-pushing identities.

Exercise 6.17 In the "front-loaded" design, explain in one or two sentences why "control on $|1\rangle$" becomes "control on $|-\rangle$".

Key Takeaways

1. Many QFT circuits are equivalent up to commuting gates and simple conjugation identities such as $HZ^\alpha = X^\alpha H$.

2. Reordering and stacking controlled-phase gates can reduce circuit depth without changing the implemented unitary.

6.4 ✳ Extensions to the QFT Algorithms

The QFT is a fundamental operation in quantum computing, and several extensions adapt it to different hardware and algorithmic constraints. Variations such as the Approximate QFT, the Iterative QFT, and Qubit Recycling trade off accuracy, circuit depth, and qubit count in ways that can be advantageous on near-term or connectivity-limited devices.

6.4.1 Approximate Quantum Fourier Transform (AQFT)

A notable strength of the QFT is its $O(n^2)$ gate complexity, which enables its use as a scalable subroutine inside larger algorithms. However, some applications do not require an exact QFT. An approximate QFT (AQFT) truncates or coarsens small-angle phase gates, since these often contribute only marginal precision gains.

Concretely, if the QFT is implemented via controlled-phase gates $Z^{\frac{1}{2^k}}$, then gates with $k > m$ may be discarded or replaced by a coarse approximation. Figure 6.11 illustrates this approach for $n = 6$ and $m = 3$.

Discarding these small-angle phase gates reduces the overall gate count from $O(n^2)$ to $O(nm)$. A thorough analysis (see, e.g., Nielsen & Chuang [3]) shows that

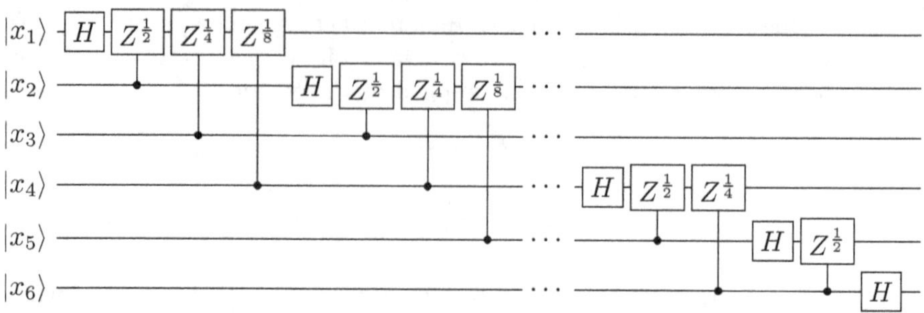

Figure 6.11: Six-Qubit AQFT Circuit

in many QFT-based algorithms, the final success probability remains high when the largest omitted phase is sufficiently small. As a rough guideline, if the smallest phase we keep is comparable to $\frac{2\pi}{n}$, then we can set $\frac{\pi}{2^m} \sim \frac{2\pi}{n}$, giving

$$\frac{\pi}{2^m} \sim \frac{2\pi}{n} \implies m = O(\log n),$$

and thus an AQFT complexity closer to $O(n \log n)$ rather than $O(n^2)$.

> **Exercise 6.18** Pick $n = 8$ and compare the standard QFT to the AQFT with $m = 3$ by listing which controlled-phase gates are removed. Which output qubits lose the finest phase information?

> **Exercise 6.19** Consider the AQFT truncation rule "discard all controlled-phase gates $Z^{1/2^k}$ with $k > m$." For a given n, compute the exact number of remaining controlled-phase gates as a function of m. Verify that it scales as $O(nm)$.

Key Takeaways

1. AQFT discards small-angle controlled-phase gates that contribute only fine-grained phase information.

2. Keeping angles down to about $\frac{\pi}{2^m}$ reduces gate count from $O(n^2)$ to $O(nm)$.

3. Choosing $m = O(\log n)$ often preserves high success probability in QFT-based algorithms while reducing depth substantially.

6.4.2   Iterative and Recycling Implementations of the QFT

A direct, fully parallelized QFT circuit can be challenging when hardware connectivity, coherence time, or circuit depth is limited. Several closely related variants reorganize the same Fourier basis change to better match such constraints. The key idea is to execute the QFT in stages, so that the gates associated with one output qubit are completed before moving on to the next. This staged execution is often called the Iterative Quantum Fourier Transform (IQFT).

In some literature and quantum software libraries, "IQFT" denotes the *inverse* QFT. Here, it refers *iterative* implementation of the QFT.

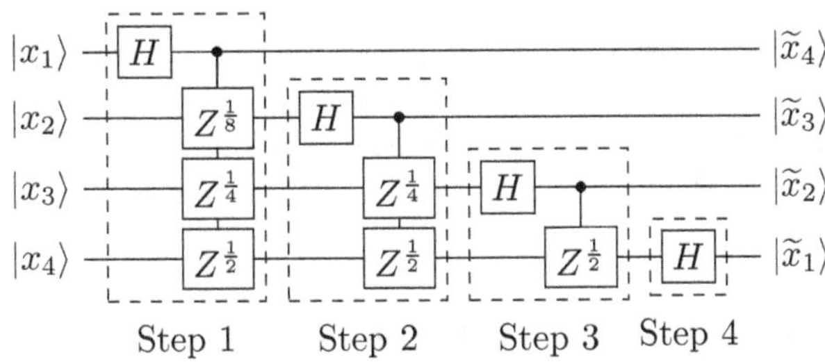

Step 1   Step 2   Step 3   Step 4

Figure 6.12: An Iterative (Staged) Implementation of the QFT

1   Staged IQFT

Figure 6.12 shows a four-qubit example based on a QFT circuit with swapped controlled-phase gates (Fig. 6.6). The circuit is organized into four stages:

1. Apply a Hadamard (H) to qubit 1, then apply controlled-phase gates from qubit 1 to qubits 2, 3, and 4. After this stage, qubit 1 will not be acted on again.

2. Apply H to qubit 2, then the remaining controlled-phase gates from qubit 2 to qubits 3 and 4.

3. Continue similarly for qubit 3.

4. Apply the final H on qubit 4.

No swap network is shown: the bit-reversal is handled by the output wire labeling (equivalently, by classical relabeling of the output bits).

Whereas a standard QFT interleaves Hadamards and controlled-phase gates across all n qubits, the staged IQFT simply reorganizes the same unitary into a schedule that can be more convenient for depth optimization and hardware mapping. Other IQFT variants exist, for example versions combined with truncation (as in the AQFT) to further reduce depth.

2   From Staging to Qubit Recycling

Once a qubit has completed its stage and will not be acted on again, an implementation may choose to measure it, reset it, and reuse it for later stages. This is the idea of qubit recycling: using fewer than n physical qubits to emulate an n-qubit QFT (or inverse QFT), while storing intermediate outcomes classically for later interpretation. Recycling typically entails mid-circuit measurement and reset, and therefore relies on hardware support for such operations.

3   Semiclassical Inverse QFT

A particularly important special case is the semiclassical inverse QFT due to Griffiths and Niu [17]. In this construction, one implements QFT^{-1} by measuring qubits one at a time and using the measurement outcomes to classically control phase corrections on the remaining qubits. This "measure-and-feedforward" structure removes the need for many coherent controlled-phase gates across distant qubits and naturally enables qubit recycling, since a measured qubit can be reset and reused immediately. This variant is widely used in settings such as quantum phase estimation, where only classical bits are needed as output.

Exercise 6.20 Explain how staging the QFT into an IQFT changes the requirements on qubit connectivity compared with the standard QFT. Give one example of a hardware constraint (e.g., linear nearest-neighbor coupling) for which a staged schedule is advantageous.

Exercise 6.21 Propose a strategy to implement an effective n-qubit QFT (or inverse QFT) using only $O(1)$ or $O(\log n)$ reusable qubits plus mid-circuit measurement and reset. Describe what classical information must be stored and how it is used to interpret the final output.

Chapter Takeaways

1. The QFT changes from the computational basis to the Fourier basis, turning *periodic structure* into measurable interference.

2. The standard QFT circuit factors into Hadamards plus controlled-phase rotations; many circuit layouts are equivalent up to gate-commutation and simple conjugation identities.

3. Approximate QFT (AQFT) drops very small-angle rotations, giving a tunable depth–accuracy tradeoff that is often sufficient for algorithmic postprocessing.

4. Iterative QFT / qubit recycling reduces qubit count by measuring and reusing ancillas with classical feedforward, trading parallelism for repeated short circuits.

Problem Set 6

6.1 Verify that the $N \times N$ QFT matrix has entries $F_{k,x} = \frac{1}{\sqrt{N}} \omega_N^{kx}$, and prove that the columns are orthonormal. Conclude that $F^{-1} = F^\dagger$.

6.2 Show that the QFT diagonalizes the cyclic shift operator $S\,|x\rangle = |x + 1 \bmod N\rangle$. Specifically, prove that $|\widetilde{k}\rangle := \mathrm{QFT}\,|k\rangle$ is an eigenvector of S and determine the eigenvalue.

6.3 Define the "phase gradient" state

$$|\phi_\omega\rangle = \frac{1}{\sqrt{N}} \sum_{x=0}^{N-1} e^{\frac{2\pi i \omega x}{N}} |x\rangle.$$

Show that $\mathrm{QFT}\,|\phi_\omega\rangle = |\omega\rangle$ (up to a global phase). Explain how this relates to Example 2 (§ 6.1.3.3).

6.4 Let $|\psi\rangle = \frac{1}{\sqrt{2}}(|0\rangle + |N/2\rangle)$ for even N. Compute $\text{QFT}\,|\psi\rangle$ explicitly and describe the measurement distribution over $k \in \{0, \ldots, N-1\}$.

6.5 In Example 3 (§ 6.1.3.4), assume r divides N and $M = N/r$. Prove that the measurement outcome k after the QFT is always a multiple of N/r, and that each such multiple occurs with equal probability.

6.6 In Example 3 (§ 6.1.3.4), drop the assumption that r divides N. Let k^* be the integer closest to jN/r for some integer j. Use the Dirichlet-kernel expression to argue that $|\widetilde{c}_{k^*}|^2$ is large. Provide a clear qualitative explanation of why the peaks concentrate near multiples of N/r.

6.7 Prove the tensor-product identity Eq. 6.10 by expanding $|x\rangle$ and $|y\rangle$ in binary, as outlined in the text. Your proof should explicitly justify the step where a global integer phase is discarded.

6.8 For $n = 5$, write the standard QFT circuit as a list of gates in time order (Hadamards, controlled-phase gates, swaps). Then write the inverse QFT circuit as a corresponding list.

6.9 Derive an exact expression for the total number of two-qubit controlled-phase gates in the standard n-qubit QFT. Then include the number of swaps needed for bit reversal. State the total gate count in closed form.

6.10 Hadamard gate pushing. Prove the identity $HZ^\alpha = X^\alpha H$ for real α, assuming $Z^\alpha = \text{diag}(1, e^{i\pi\alpha})$. Then derive the corresponding identity $HX^\alpha = Z^\alpha H$.

6.11 AQFT gate complexity. In the AQFT truncation rule that discards all $Z^{1/2^k}$ with $k > m$, derive the remaining two-qubit gate count exactly and show it is $O(nm)$. For $m = \lceil \log_2 n \rceil$, show the gate count becomes $O(n \log n)$.

6.12 Implement QFT and AQFT in software. Using a quantum-circuit SDK of your choice, implement

    (a) the n-qubit QFT with swaps,

    (b) the n-qubit QFT without swaps,

    (c) an AQFT with a truncation parameter m. For a periodic superposition input like Example 3 (§ 6.1.3.4), numerically compare the measurement distributions produced by (a) and (c) for several values of m.

6.13 Endianness and qubit order. Many software frameworks and papers adopt different conventions for which qubit is "most significant." Choose one $n = 3$ example state $|x_1 x_2 x_3\rangle$ and explain how a difference in endianness changes the apparent QFT circuit diagram.

7. Quantum Phase Estimation (QPE)

Contents

Quantum Phase Estimation (QPE), introduced in early quantum algorithmic work on eigenvalue estimation and order finding [18, 19], generalizes the Quantum Fourier Transform (QFT) into a universal quantum measurement technique. It is a fundamental primitive in quantum computing, forming the backbone of algorithms such as Shor's factoring algorithm [20], many Hamiltonian simulation and spectral-estimation methods, quantum chemistry routines, and a wide range of eigenvalue and eigenphase estimation procedures.

Representative Applications of QPE

Algorithm / Field	QPE's Role

Shor's algorithm (Chapter 8)	Period finding interpreted as eigenphase estimation of the modular exponentiation operator.
Quantum simulation (Chapter 10)	Estimation of energy eigenvalues of a Hamiltonian, typically using $U = e^{-iHt}$.
Quantum chemistry (Part III)	Determination of ground-state energies corresponding to eigenvalues of molecular Hamiltonians.
Linear systems (HHL) (Chapter 12)	Estimation of eigenvalues of a Hermitian matrix A in order to solve $A\boldsymbol{x} = \boldsymbol{b}$.
Amplitude estimation (Chapter 9)	Special case of QPE where the unitary U encodes success amplitudes for quantum-enhanced Monte Carlo and optimization.

We begin by examining the standard circuit implementation of QPE and analyzing its error bounds and circuit complexity. Later sections present iterative and adaptive extensions, as well as practical variations relevant to quantum simulation, linear systems, and quantum chemistry.

7.1 The Standard QPE Algorithm

The goal of QPE is to estimate the phase ϕ associated with an eigenvalue of a unitary operator U, given access to an eigenvector $|\psi\rangle$ satisfying

$$U |\psi\rangle = e^{2\pi i \phi} |\psi\rangle. \tag{7.1}$$

Because all eigenvalues of a unitary lie on the complex unit circle, we express them in exponential form. By periodicity of the exponential, we may restrict $0 \leq \phi < 1$ without loss of generality.

7.1.1 Conceptual Structure of QPE

A high-level conceptual view of QPE is shown in Fig. 7.1. The algorithm uses two registers: an n-qubit *control register* and an m-qubit *target register*. In Fig. 7.1(a), these are initialized to $|k\rangle$ and $|\psi\rangle$, respectively.

A controlled unitary U_f is defined to act jointly on both registers. It applies U^k to the target register conditional on the integer k encoded in the control register:

$$U_f |k\rangle |\psi\rangle = |k\rangle\, U^k |\psi\rangle. \tag{7.2}$$

Since $|\psi\rangle$ is an eigenvector of U, we have

$$U_f |k\rangle |\psi\rangle = e^{2\pi i k \phi} |k\rangle |\psi\rangle. \tag{7.3}$$

The control register acquires phase information about the eigenvalue of U, while the target register stores the corresponding eigenvector. Although these registers might be called the *eigenphase* and *eigenvector* registers, we use the standard terms "control" and "target" registers.

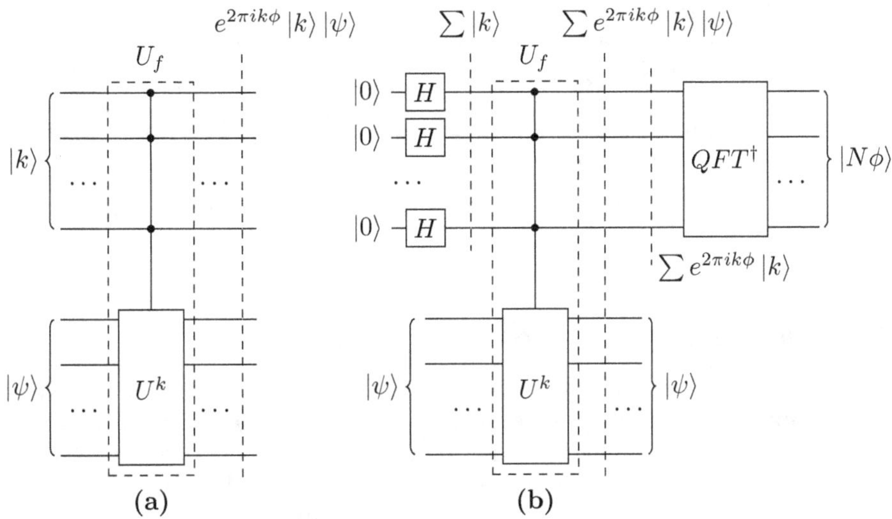

Figure 7.1: High-Level Conceptual View of QPE

Applying Hadamard gates to the control register produces the uniform superposition

$$\frac{1}{\sqrt{N}} \sum_{k=0}^{N-1} |k\rangle, \qquad N = 2^n.$$

Using linearity of U_f, Fig. 7.1(b) yields the transformed state

$$U_f \left(\frac{1}{\sqrt{N}} \sum_{k=0}^{N-1} |k\rangle |\psi\rangle \right) = \frac{1}{\sqrt{N}} \sum_{k=0}^{N-1} e^{2\pi i k\phi} |k\rangle |\psi\rangle.$$

At this stage, the two registers separate into a product state

$$\left(\frac{1}{\sqrt{N}} \sum_{k=0}^{N-1} e^{2\pi i k\phi} |k\rangle \right) \otimes |\psi\rangle,$$

since the k-dependence appears only in a phase factor that can be associated with the control register.

The resulting control-register state

$$\frac{1}{\sqrt{N}} \sum_{k=0}^{N-1} e^{2\pi i k\phi} |k\rangle$$

is a pure harmonic whose frequency is the eigenphase ϕ. Applying the inverse QFT therefore extracts ϕ (see § 6.1.3).

 We use QFT† rather than QFT because the control register accumulates phases of the form $e^{+2\pi i k\phi}$, matching the inverse-transform convention.

Key Takeaways

1. QPE estimates ϕ in an eigenvalue equation $U \left| \psi \right\rangle = e^{2\pi i \phi} \left| \psi \right\rangle$.

2. The output is an n-bit approximation $\widehat{\phi}$, i.e., a phase estimate with resolution $\sim 2^{-n}$.

3. The hard part in practice is not the QFT; it is implementing controlled powers U^{2^j} efficiently.

7.1.2 The Controlled-Unitary Subcircuit

The central component of the QPE circuit is the controlled-U subcircuit U_f, which applies U^k to the target register conditional on the control register state $\left| k \right\rangle$. For example, since the binary string 10101010 represents 170, a control state $\left| 10101010 \right\rangle$ results in a transformation of the target register by U^{170}. This subcircuit is a primitive in many quantum algorithms and finds application beyond QPE.

1 Single-Qubit Controlled-Unitary

A single-qubit controlled-U gate acts as follows:

$$
\begin{array}{l}
\left| k \right\rangle \;\text{———•———}\; \left| k \right\rangle \\[4pt]
\left| \psi \right\rangle \;\boxed{U}\; U^k \left| \psi \right\rangle
\end{array}
$$

This implements

$$\left| k \right\rangle \left| \psi \right\rangle \mapsto \left| k \right\rangle U^k \left| \psi \right\rangle, \tag{7.4}$$

with $k \in \{0, 1\}$.

2 Multi-Qubit Controlled-Unitary

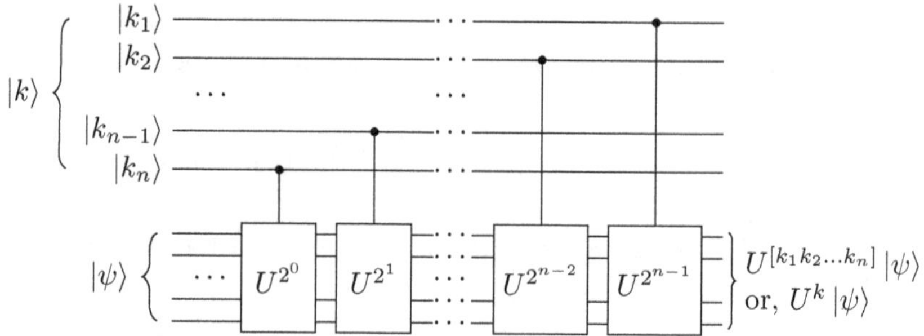

Figure 7.2: The Controlled-Unitary Subcircuit

Extending this to an n-qubit control register, Fig. 7.2 applies powers of U controlled on the bits $k_1, \ldots, k_n$ of the binary string $k_1 k_2 \ldots k_n$. The target undergoes the unitary

$$U^{2^{n-1}k_1} U^{2^{n-2}k_2} \ldots U^{2k_{n-1}} U^{k_n} = U^{2^{n-1}k_1 + 2^{n-2}k_2 + \ldots + 2k_{n-1} + k_n}$$

$$= U^{[k_1 k_2 \ldots k_n]},$$

where $[k_1 k_2 \cdots k_n]$ is the integer encoded by the binary string.

Thus,

$$|k_1 k_2 \cdots k_n\rangle |\psi\rangle \;\mapsto\; |k_1 k_2 \cdots k_n\rangle \, U^{[k_1 k_2 \cdots k_n]} |\psi\rangle .$$

Writing $k = [k_1 k_2 \cdots k_n]$ in decimal gives

$$|k\rangle |\psi\rangle \;\mapsto\; |k\rangle \, U^k |\psi\rangle . \tag{7.5}$$

Throughout this text we use the *big-endian* (i.e., big end first) convention: in a register $|k_1 k_2 \cdots k_n\rangle$, the leftmost bit k_1 is the most significant bit (MSB) and the rightmost bit k_n is the least significant bit (LSB).

It is important to note that some software frameworks—most notably `Qiskit`—use a little-endian convention in which the lowest-index qubit corresponds to the least significant bit. When comparing circuits or measurement outcomes across conventions, care must be taken to translate between qubit indices, bitstring order, and the corresponding integer values.

So far we have not assumed any property of U beyond its unitarity, so the above construction of U_f is completely general. Particular to QPE, since $|\psi\rangle$ is an eigenstate of U, this circuit realizes Eq. 7.3.

Exercise 7.1 Consider $U = X$ (the Pauli-X gate).

(a) Verify that Fig. 7.2 implements Eq. 7.5 when the target is initialized in $|+\rangle$ and $|-\rangle$.

(b) Determine the output of the control register when the target is initialized in $|0\rangle$ or $|1\rangle$.

7.1.3 The Full QPE Circuit

Substituting the controlled-unitary circuit of Fig. 7.2 into the high-level block diagram of Fig. 7.1(b) yields the complete QPE circuit shown in Fig. 7.3. The main stages of the standard QPE algorithm are then:

1. Prepare a uniform superposition in the control register and an eigenstate of U in the target register.

2. Apply the controlled-unitary subcircuit to imprint the eigenphase onto the control register (phase kickback).

3. Apply the inverse QFT on the control register and measure.

1 Superposition Initial State

The initial state is

$$|\Psi_1\rangle = \frac{1}{\sqrt{N}} \sum_{k=0}^{N-1} |k\rangle \otimes |\psi\rangle .$$

In Dirac notation, tensor products such as $|k\rangle \otimes |\psi\rangle$ are often written simply as $|k\rangle |\psi\rangle$. The explicit $\otimes$ is often retained when helpful to emphasize register separation.

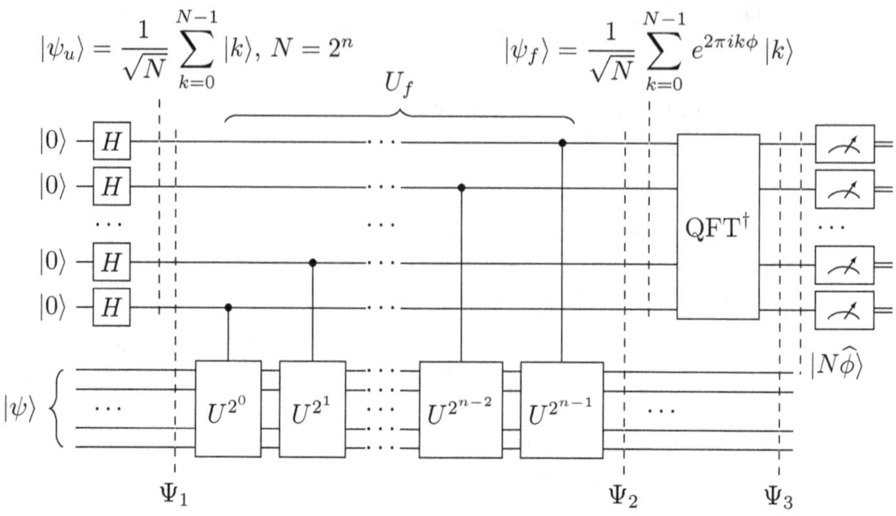

Figure 7.3: Quantum Phase Estimation (QPE) Circuit

2   Phase Kickback through the Controlled Unitary

Applying the controlled-unitary subcircuit gives

$$|\Psi_2\rangle = \frac{1}{\sqrt{N}} \sum_{k=0}^{N-1} |k\rangle \otimes U^k |\psi\rangle$$

$$= \left(\frac{1}{\sqrt{N}} \sum_{k=0}^{N-1} e^{2\pi i k \phi} |k\rangle \right) \otimes |\psi\rangle .$$

Thus, the control register becomes

$$|\psi_f\rangle = \frac{1}{\sqrt{N}} \sum_{k=0}^{N-1} e^{2\pi i k \phi} |k\rangle . \tag{7.6}$$

This transfer of phase information from the target eigenstate back onto the control register is known as *phase kickback*. Because $|\psi\rangle$ is an eigenstate of U, the target register factors out unchanged, and the eigenphase information is fully encoded in the amplitudes of the control register.

3   Inverse QFT and Measurement

Assume the phase ϕ is well approximated by the n-bit binary fraction

$$\widehat{\phi} = [0.\phi_1\phi_2\ldots\phi_n] = \sum_{j=1}^{n} \phi_j 2^{-j} .$$

Then $N\widehat{\phi}$ is an integer between 0 and $N-1$, with $N = 2^n$. Equation 7.6 becomes

$$|\psi_f\rangle = \frac{1}{\sqrt{N}} \sum_{k=0}^{N-1} e^{\frac{2\pi i k (N\widehat{\phi})}{N}} |k\rangle . \tag{7.7}$$

This is a discrete complex exponential with integer frequency index $N\phi$. Therefore, $\text{QFT}^\dagger$ maps it to the corresponding computational-basis state $|N\widehat{\phi}\rangle$ (compare with § 6.1.3.2, noting the opposite sign convention):

$$|N\widehat{\phi}\rangle = \text{QFT}^\dagger |\psi_f\rangle.$$

Here $\text{QFT}^\dagger$, rather than QFT, is used because the phase factor in Eq. 7.7 has the form $e^{+2\pi i k(\cdot)/N}$. The final composite state is

$$|\Psi_3\rangle = |N\widehat{\phi}\rangle \otimes |\psi\rangle.$$

Measuring the control register yields $N\widehat{\phi}$, corresponding to the bit string $\phi_1\phi_2\ldots\phi_n$, giving an estimate of ϕ with precision $1/N = 2^{-n}$.

> **Exercise 7.2** Let $U = X$ where X is the Pauli-X gate.
>
> (a) Verify that the circuit in Fig. 7.3 identifies the eigenvalues of X when $|\psi\rangle = |+\rangle$ or $|-\rangle$.
>
> (b) Analyze the measurement outcomes when the target register is initialized in $|0\rangle$.

Key Takeaway

If the control register is in a pure exponential state $\frac{1}{\sqrt{N}} \sum_{k=0}^{N-1} e^{2\pi i k\widehat{\phi}} |k\rangle$, then $\text{QFT}^\dagger$ maps it to the computational-basis state $|N\widehat{\phi}\rangle$. In exact-grid cases ($\phi = r/N$), QPE returns r with certainty; off-grid phases produce a distribution sharply peaked near $N\phi$.

7.1.4 QPE with Non-Eigenstate Inputs: Spectral Estimation

The standard formulation of QPE assumes that the target input $|\psi\rangle$ is an eigenvector of U as in Eq. 7.1. When $|\psi\rangle$ is not an eigenvector, QPE acts as a *spectral measurement* of U: it samples an eigenphase according to the input state's overlap with the corresponding eigenvector.

Since U is unitary, it admits an orthonormal eigenbasis $\{|\psi_j\rangle\}$ with eigenvalues $e^{2\pi i \phi_j}$. Any input state can be expanded as

$$|\psi\rangle = \sum_{j=1}^{m} c_j |\psi_j\rangle, \qquad \sum_j |c_j|^2 = 1.$$

Running QPE on this input yields an entangled final state of the form

$$|\Psi_3\rangle \approx \sum_{j=1}^{m} c_j |N\widehat{\phi}_j\rangle |\psi_j\rangle,$$

where $\widehat{\phi}_j$ is the n-bit digitization of ϕ_j. Measuring the control register outputs an estimate associated with ϕ_j with probability approximately $|c_j|^2$, and the target register collapses to the corresponding eigenvector $|\psi_j\rangle$.

Thus, QPE effectively implements a projective measurement in the eigenbasis of U, sampling eigenphases from a distribution weighted by the input state's overlaps $|c_j|^2$. This property is used in Shor's algorithm, where the target register is prepared as a superposition of eigenvectors of a modular multiplication operator, and repeated QPE runs yield phase samples that are classically post-processed to recover the period (see § 8.3).

Key Takeaways

---

1. If $|\psi\rangle = \sum_j c_j |\psi_j\rangle$ is a superposition of eigenstates of U, then QPE samples an eigenphase ϕ_j with probability $|c_j|^2$.

2. Measuring the control register also collapses the target register to the corresponding eigenstate (up to approximation error).

3. In this sense, QPE implements a spectral measurement of U, not just a numerical estimation routine.

7.2 ✳ Error and Resource Analysis

7.2.1 Error Analysis

If we relax the assumption that $N\phi$ is an integer, then $\text{QFT}^\dagger |\psi_f\rangle$ is no longer a computational-basis state but a superposition peaked near $N\phi$. Writing the post-inverse-QFT control state as $|\Phi(\phi)\rangle := \text{QFT}^\dagger |\psi_f\rangle$, we obtain the explicit expansion

$$|\Phi(\phi)\rangle = \frac{1}{N} \sum_{j=0}^{N-1} \left(\sum_{k=0}^{N-1} e^{2\pi i k \left(\phi - \frac{j}{N} \right)} \right) |j\rangle .$$

For a given phase ϕ, the probability of measuring the outcome $\widehat{\phi} \equiv [0.\phi_1\phi_2\ldots\phi_n]$ (i.e., $j = N\widehat{\phi}$) is

$$P(\widehat{\phi} \mid \phi) = \left| \langle N\widehat{\phi} \mid \Phi(\phi)\rangle \right|^2 = \left| \frac{1}{N} \sum_{k=0}^{N-1} e^{2\pi i k(\phi - \widehat{\phi})} \right|^2 = \frac{1}{N^2} \frac{\sin^2\left(\pi N (\phi - \widehat{\phi})\right)}{\sin^2\left(\pi (\phi - \widehat{\phi})\right)} .$$

Figure 7.4 illustrates $P(\widehat{\phi} \mid \phi)$ for $n = 4$ ($N = 16$). Each curve corresponds to a fixed grid value $\widehat{\phi} \in \{0, 1/N, \ldots, (N-1)/N\}$, and shows how likely that outcome is as the true phase ϕ varies. When ϕ lies exactly on the grid, the measurement outcome equals ϕ with probability 1.

When ϕ lies between grid points, the distribution spreads across nearby outcomes. In the worst-aligned case, when ϕ lies exactly halfway between two adjacent grid values, the two nearest outcomes occur with probability $\left(\frac{2}{\pi}\right)^2 \approx 0.405$ each. More generally, one can show that the probability of obtaining an estimate within error $|\widehat{\phi} - \phi| \leq 1/N$ is at least $\frac{8}{\pi^2} \approx 0.81$, independent of ϕ.

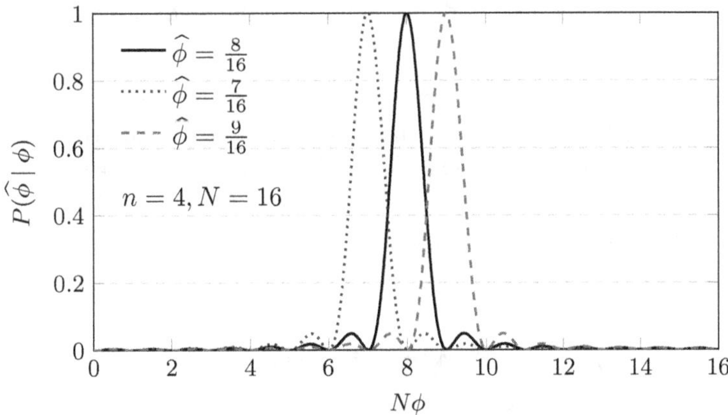

Figure 7.4: Measurement Probability Distribution in QPE

Key Takeaways

1. Off-grid phases produce a peaked distribution over integers $j \in \{0, \ldots, N-1\}$ centered near $j \approx N\phi$.

2. The peak width is $O(1)$ in the integer index j, corresponding to phase error $O(1/N)$.

3. Rounding the measured outcome j to the nearest integer $N\phi$ succeeds with probability > 0.4; increasing n (so $N = 2^n$) improves resolution exponentially.

7.2.2 Register Size Determination

The control (eigenphase) register size n is determined by the desired precision of the phase estimate (2^{-n}) and is independent of the size of the target (eigenvector) register. Specifically, since $\varepsilon \sim 2^{-n}$, we have $n \sim O(\log(1/\varepsilon))$. Thus, a larger control register allows for greater precision; for example, choosing $n = 10$ gives precision $2^{-10} \approx 10^{-3}$.

The size of the target register m is dictated by the dimension of the Hilbert space on which U acts. In many quantum algorithms, such as Shor's algorithm, the target register represents the quantum system whose eigenstate $|\psi\rangle$ encodes the solution to the problem. Consequently, the target register size is problem-dependent.

7.2.3 Circuit Complexity

The complexity of QPE hinges on whether the circuit family U^{2^k} ($k = 0, 1, \ldots, n-1$) can be implemented efficiently, and in particular whether the *controlled* versions of these powers can be realized without an exponential blow-up in resources.

If each controlled-U^{2^k} can be implemented with cost polynomial in the problem size (for example, because U has structure enabling fast exponentiation, or because U is provided as an efficiently implementable oracle for controlled powers), then the overall cost is dominated by the inverse QFT, which requires $O(n^2)$ two-qubit gates (or $O(n \log n)$ with standard approximations).

By contrast, if U^{2^k} is implemented naively as 2^k sequential applications of U, then the largest power $U^{2^{n-1}}$ dominates, and the total gate count grows exponentially in n. Thus, the practical feasibility of QPE depends on whether controlled powers of U can be executed efficiently in the intended application.

We will discuss efficient implementations of powers of unitary operators in Chapters 10 and 11.

7.3 $*$Iterative QPE Algorithms

The standard QPE circuit shown in Fig. 7.3 is powerful but resource-intensive, typically requiring many qubits and deep controlled-unitary sequences. As a result, it is best suited for fault-tolerant (fully error-corrected) quantum devices.

Iterative Quantum Phase Estimation (IQPE) is a family of QPE variants that improve resource efficiency by extracting information about ϕ through repeated short circuits, often estimating bits of ϕ sequentially. Instead of implementing an inverse Quantum Fourier Transform (QFT), IQPE relies on repeated measurements and classical post-processing, substantially reducing the number of qubits. This makes iterative approaches more compatible with near-term Noisy Intermediate-Scale Quantum (NISQ) hardware.

Overview

1. Iterative QPE reduces qubit count by replacing the inverse QFT with repeated short circuits and classical post-processing.

2. Kitaev-style IQPE extracts bits of ϕ using controlled powers U^{2^j} plus phase feedback β.

3. Bayesian and information-theoretic variants trade deeper circuits for more measurements and heavier classical inference.

7.3.1 The Hadamard Test Circuit

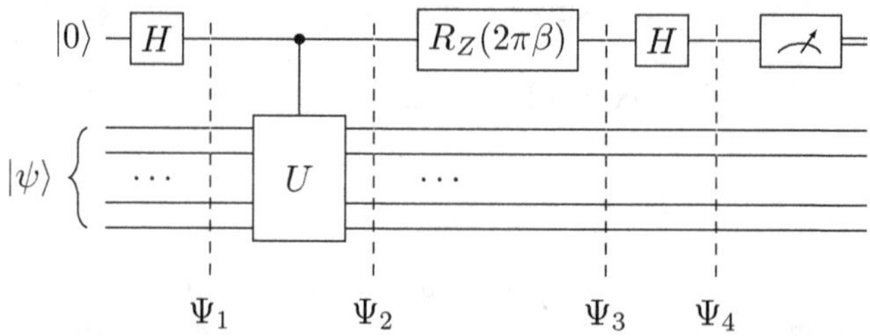

Figure 7.5: Hadamard Test Circuit for Iterative Quantum Phase Estimation

At the core of many iterative QPE methods is a Hadamard test circuit with an additional phase shift $2\pi\beta$, as shown in Fig. 7.5. The Hadamard test uses an ancilla qubit (initialized in $|0\rangle$) together with a controlled application of a unitary operator U on the target register.

The phase-shift gate is defined as

$$R_Z(2\pi\beta) = \begin{bmatrix} 1 & 0 \\ 0 & e^{2\pi i\beta} \end{bmatrix}.$$

Since $R_Z(2\pi\beta)$ acts only on the ancilla and the controlled-U acts jointly on the ancilla and target, these operations commute, so their order in the circuit can be interchanged.

1 Key Result

A key result (derivation deferred) is that the probability of measuring $|0\rangle$ on the ancilla is

$$P_X(0) = \frac{1 + \mathrm{Re}\big(e^{2\pi i\beta}\,\langle\psi|\,U\,|\psi\rangle\big)}{2}. \tag{7.8}$$

This gives access to the *real part* of the expectation value $\langle\psi|\,U\,|\psi\rangle$. To obtain the *imaginary part*, we add $\frac{1}{4}$ to β, or equivalently, apply an S gate before the final Hadamard on the ancilla. This yields

$$P_Y(0) = \frac{1 + \mathrm{Im}\big(e^{2\pi i\beta}\,\langle\psi|\,U\,|\psi\rangle\big)}{2}. \tag{7.9}$$

The two measurements correspond to projections of the ancilla state onto the X and Y axes of the Bloch sphere.

2 Application to IQPE

In iterative phase estimation, the Hadamard test is applied repeatedly with U replaced by U^k, where k is an integer. For an eigenstate of U satisfying

$$U\,|\psi\rangle = e^{2\pi i\phi}\,|\psi\rangle, \quad \phi \in [0, 1),$$

we have $\langle\psi|\,U^k\,|\psi\rangle = e^{2\pi i k\phi}$, and thus

$$P_X(0) = \frac{1 + \cos 2\pi(\beta + k\phi)}{2}, \tag{7.10a}$$

$$P_Y(0) = \frac{1 + \sin 2\pi(\beta + k\phi)}{2}. \tag{7.10b}$$

Combining them allows the fractional part of $k\phi$ to be determined from

$$\mathrm{frac}(k\phi) = \frac{\mathrm{atan2}(2P_Y(0) - 1,\, 2P_X(0) - 1)}{2\pi}, \tag{7.11}$$

where $\mathrm{atan2}(y, x)$ is the extended atan function that returns the polar angle of (x, y) uniquely in $[0, 2\pi)$.

3 Extracting a Binary Bit of ϕ

Assume ϕ is represented as a binary fraction

$$\phi = [0.\phi_1\phi_2\ldots] = \sum_{j=1}^{\infty} \phi_j 2^{-j}. \tag{7.12}$$

Choose $k = 2^{n-1}$, yielding

$$2^{n-1}\phi = [\phi_1\phi_2 \ldots \phi_{n-1}.\phi_n\phi_{n+1}\ldots].$$

Applying Eq. 7.11 gives

$$\text{frac}(2^{n-1}\phi) = [0.\phi_n\phi_{n+1}\ldots]. \tag{7.13}$$

Rounding this fractional value to the nearest multiple of 2^{-1} provides a good estimate of ϕ_n. Thus, with $k = 2^{n-1}$, we can extract the n-th binary bit of ϕ.

Exercise 7.3 Suppose $\phi = [0.\phi_1\phi_2 \ldots]$ is an infinite binary fraction.

  (a) Prove Eq. 7.13 by interpreting multiplication by 2^{n-1} as a binary shift.

  (b) Explain why rounding $\text{frac}(2^{n-1}\phi)$ to the nearest value in $\{0, \frac{1}{2}\}$ recovers ϕ_n when ϕ is not near a binary boundary.

  (c) Give an explicit example where ϕ is near a binary boundary and describe what can go wrong.

4   ✳ Derivation of Eq. 7.8

Finally, we provide a derivation of Eq. 7.8, which busy readers may skip.

The system starts in the state $|0\rangle \otimes |\psi\rangle$, where $|\psi\rangle$ is the initial state of the target register. Applying a Hadamard gate to the ancilla transforms the state to

$$|\Psi_1\rangle = \frac{1}{\sqrt{2}}\left(|0\rangle + |1\rangle\right) \otimes |\psi\rangle.$$

Next, we apply the controlled unitary U. If the control qubit is in state $|0\rangle$, U is not applied; if it is in state $|1\rangle$, U is applied to $|\psi\rangle$. The resulting state is

$$|\Psi_2\rangle = \frac{1}{\sqrt{2}}\left(|0\rangle \otimes |\psi\rangle + |1\rangle \otimes U\,|\psi\rangle\right).$$

We then apply the phase rotation gate $R_Z(2\pi\beta)$ to the ancilla, yielding

$$|\Psi_3\rangle = \frac{1}{\sqrt{2}}\left(|0\rangle \otimes |\psi\rangle + e^{2\pi i\beta}\,|1\rangle \otimes U\,|\psi\rangle\right).$$

Applying another Hadamard gate to the ancilla gives

$$|\Psi_4\rangle = |0\rangle \otimes \frac{|\psi\rangle + e^{2\pi i\beta}U\,|\psi\rangle}{2} + |1\rangle \otimes \frac{|\psi\rangle - e^{2\pi i\beta}U\,|\psi\rangle}{2}.$$

Measuring the ancilla in the computational basis, the probability of obtaining outcome $|0\rangle$ is

$$P(0) = \langle\Psi_4|\left(|0\rangle\langle 0| \otimes I\right)|\Psi_4\rangle.$$

Expanding $|\Psi_4\rangle$ gives

$$P(0) = \left(\frac{\langle\psi| + e^{-2\pi i\beta}\,\langle\psi|\,U^\dagger}{2}\right)\left(\frac{|\psi\rangle + e^{2\pi i\beta}U\,|\psi\rangle}{2}\right)$$

$$= \frac{1}{4}\left(\langle\psi|\psi\rangle + e^{2\pi i\beta}\,\langle\psi|\,U\,|\psi\rangle + e^{-2\pi i\beta}\,\langle\psi|\,U^\dagger\,|\psi\rangle + \langle\psi|\,U^\dagger U\,|\psi\rangle\right).$$

Since U is unitary, $U^\dagger U = I$, so $\langle\psi|\,U^\dagger U\,|\psi\rangle = \langle\psi|\psi\rangle = 1$, which simplifies the expression to Eq. 7.8.

> **Exercise 7.4** Derive Eq. 7.9.

7.3.2 Kitaev's IQPE Algorithm

In principle, one can fix $k = 1$ in Eq. 7.11 and estimate ϕ to high accuracy by repeating the Hadamard test many times. However, this brute-force approach is inefficient because statistical error decreases slowly with the number of circuit executions. If we perform M independent repetitions, the estimation error typically scales as $\varepsilon \sim 1/\sqrt{M}$ (often called *standard-quantum-limit* scaling).

 See QC Math [1], Chapter 14: *Fundamentals of Probability*, on Sampling Error of the Mean.

In contrast, Kitaev's IQPE algorithm [18] achieves an exponentially precise estimate of ϕ in the number of extracted bits. More precisely, n iterations determine n bits of ϕ, giving a phase resolution on the order of 2^{-n}.

1 Setup

Kitaev's IQPE executes the Hadamard test circuit in Fig. 7.5 over n iterations, replacing U in each iteration with

$$U^{2^{n-1}}, \, U^{2^{n-2}}, \, \ldots, \, U^2, \, U.$$

2 First Iteration

In the first iteration, we set $\beta = 0$ and $k = 2^{n-1}$. According to Eq. 7.13, this estimates $\mathrm{frac}(2^{n-1}\phi) = [0.\phi_n\phi_{n+1}\ldots]$, yielding an approximation of the least significant extracted bit ϕ_n.

3 Second Iteration

In the second iteration, we set $k = 2^{n-2}$ and choose

$$\beta = -\frac{1}{2}\,\mathrm{frac}(2^{n-1}\phi) = -[0.0\,\phi_n\phi_{n+1}\ldots].$$

This choice of β cancels the contribution from the previously extracted bits and isolates

$$\mathrm{frac}\big(2^{n-2}\phi + \beta\big) = [0.\phi_{n-1}].$$

Thus, we can estimate ϕ_{n-1} (typically using multiple circuit executions to reduce statistical error).

4 General Procedure

In the l-th iteration, we set $k = 2^{n-l}$. Given that we have already determined $\phi_{n-l+2}, \phi_{n-l+3}, \ldots$, we define

$$\beta = -\frac{1}{2}\,\mathrm{frac}(2^{n-l+1}\phi) = -[0.0\,\phi_{n-l+2}\phi_{n-l+3}\cdots].$$

Then the fractional part of $2^{n-l}\phi + \beta$ simplifies to

$$\mathrm{frac}(2^{n-l}\phi + \beta) = [0.\phi_{n-l+1}],$$

so we can extract ϕ_{n-l+1}. This process continues until the n-th round, where we determine ϕ_1, and we then reconstruct ϕ from the extracted bits.

Exercise 7.5 For $\phi = [0.\phi_1\phi_2\phi_3\phi_4 \ldots]$, find the values of β and k that isolate ϕ_3.

Exercise 7.6 Let

$$U = \begin{bmatrix} 1 & 0 \\ 0 & e^{2\pi i \frac{5}{16}} \end{bmatrix}.$$

Perform Kitaev's IQPE step-by-step to verify its correct operation.

5 Scaling Behavior

In Kitaev's IQPE, n iterations extract n bits of ϕ, corresponding to a phase resolution on the order of 2^{-n}. Consequently, the number of iterations scales as $n \sim O(\log(1/\varepsilon))$.

Each iteration estimates a Bernoulli probability of observing $|0\rangle$ or $|1\rangle$ on the ancilla, so it is repeated M times to reduce the probability of deciding the wrong bit. Standard concentration bounds (e.g., Hoeffding or Chernoff) imply that the bit-error probability decreases as e^{-cM} for some constant $c > 0$ under idealized independence assumptions. To keep the overall failure probability bounded as n grows, it suffices to set the error probability per bit to $O(1/n)$, which yields $M \sim O(\log n)$ repetitions per bit.

Thus, IQPE requires n sequential iterations, each repeated $O(\log n)$ times, giving a total number of circuit executions scaling as

$$Mn \sim O(n \log n) \sim O(\log(1/\varepsilon) \log \log(1/\varepsilon)).$$

Although IQPE eliminates the inverse QFT and relies instead on classical post-processing, its circuit depth can still be substantial because each iteration requires a controlled-U^{2^k} operation. In the worst case, if U^{2^k} is implemented by repeating U exactly 2^k times, then the total depth scales as $O(2^n) = O(1/\varepsilon)$, which is prohibitive at high precision. In many structured applications, however, controlled powers can be implemented more efficiently, and the main advantage of IQPE is the reduced qubit requirement.

Exercise 7.7 Assume each IQPE round estimates a Bernoulli probability P by M independent samples.

(a) State a standard concentration bound (e.g., Hoeffding) that upper-bounds $\Pr(|\widehat{P} - P| \geq \eta)$.

(b) Use your bound to choose M so that the probability of making a wrong bit decision is at most $1/n$.

(c) Briefly explain why a per-bit failure probability of $O(1/n)$ is sufficient to keep the overall failure probability bounded as n grows.

6 ✳ Iteration Order

The order in which phase bits are extracted in IQPE depends on the chosen implementation.

Kitaev's original formulation [18] follows a most significant bit (MSB)-first strategy. In this version, the first iteration estimates the most significant bit ϕ_1, and the algorithm begins with the smallest power of the unitary, U, proceeding through $U^{2^1}, U^{2^2}, \ldots, U^{2^{n-1}}$, extracting successively less significant bits.

By contrast, many modern implementations reverse this order and employ a least significant bit (LSB)-first approach. Here, the algorithm begins with the largest power, $U^{2^{n-1}}$, and works downward to U^{2^0}, determining the smallest binary fractions of ϕ first.

While the MSB-first ordering is conceptually elegant and allows for immediate phase correction after each measurement, it can be more sensitive to early rounding errors. If the true phase ϕ lies near a binary boundary, an incorrect decision of the MSB can shift all subsequent bits and yield a substantially incorrect phase estimate. For example, if $\phi = [0.01110\ldots]$, an incorrect first-bit decision to $[0.1]$ misaligns all lower bits.

The LSB-first approach mitigates this issue by refining the least significant bits first. Because early bits contribute only small corrections to the overall phase, small estimation errors have a reduced effect on later rounds. Modern extensions build upon this framework by incorporating fractional-bit refinement and Bayesian phase estimation techniques, which combine data from multiple measurements to update a phase posterior iteratively.

> **Exercise 7.8** Formulate Kitaev's IQPE algorithm starting with the most significant bit and progressing towards the least significant bit. Clearly outline the phase correction steps applied at each iteration.

7.3.3   ∗ Information-Theoretic QPE

Kitaev's approach to iterative quantum phase estimation relies on structured sequences of controlled-unitary operations and phase correction to refine an estimate of ϕ. An alternative approach, often called *Information-Theoretic QPE* (also known as the Svore–Hastings–Freedman method), uses randomized measurements combined with classical inference to extract phase information [21].

1   Measurement Procedure

Information-Theoretic QPE is also based on the Hadamard test circuit in Fig. 7.5. We consider a randomized strategy where the unitary is applied with a randomly chosen power k_i, and an additional phase shift β_i is introduced.

The probability of measuring $|0\rangle$ is (see Eq. 7.10)

$$P(0 \mid \phi) = \frac{1 + \cos\left(2\pi(\beta_i + k_i\phi)\right)}{2}. \tag{7.14}$$

Similarly,

$$P(1 \mid \phi) = 1 - P(0 \mid \phi). \tag{7.15}$$

Each measurement produces an outcome $v_i \in \{0, 1\}$. Given ϕ, the probability of observing a sequence $\{v_1, \ldots, v_s\}$ is described by the likelihood function

$$P(v_1, v_2, \ldots, v_s \mid \phi) = \prod_{i=1}^{s} P(v_i \mid \phi),$$

where $P(v_i \mid \phi)$ is determined by Eqs. 7.14 and 7.15.

2   Maximum Likelihood Estimation (MLE) for the Phase

To estimate ϕ, we seek to find the value of ϕ that maximizes the likelihood function. This is known as the maximum likelihood estimation (MLE) method. The principle behind MLE is that the parameter ϕ that makes the observed data most probable is the best estimate. As the number of measurements s increases, the likelihood function sharpens, and the estimate of ϕ converges to its true value.

This maximization is carried out classically using numerical optimization methods, which are often iterative. In the simplest (batch) form, one collects a set of measurement outcomes and then performs the MLE once using the full dataset; alternatively, the MLE can be updated sequentially as new measurements are acquired. By shifting complexity to classical post-processing, Information-Theoretic QPE avoids deep quantum circuits such as an inverse QFT, at the cost of more circuit executions.

3   Scaling Behavior

A key advantage of this approach is that it shifts the complexity of phase estimation to classical post-processing, eliminating the need for controlled-U^{2^k} gates. However, this comes at the cost of requiring significantly more quantum measurements.

The number of measurements required to achieve accuracy ε scales as $s = O(1/\varepsilon)$, which is exponentially larger than the $O(\log(1/\varepsilon))$ scaling of Kitaev's IQPE. This trade-off allows for shallower quantum circuits at the expense of a longer measurement process.

Additionally, the classical post-processing is exponentially expensive in the number of bits of precision required. Specifically, the inference step involves evaluating the likelihood function over a large space of possible ϕ values, resulting in a computational cost of $O(2^n)$, which limits the practicality of this method for high-precision applications.

Despite its computational cost, Information Theory QPE provides a valuable alternative approach to phase estimation, particularly in scenarios where quantum resources are limited but classical computational power is available.

7.3.4   ✳ Bayesian Quantum Phase Estimation

Bayesian Quantum Phase Estimation (Bayesian QPE) replaces fixed bit-by-bit phase correction rules with Bayesian inference. Instead of prescribing a deterministic schedule, Bayesian QPE selects parameters dynamically and updates a probability distribution over ϕ as measurement data arrive [22].

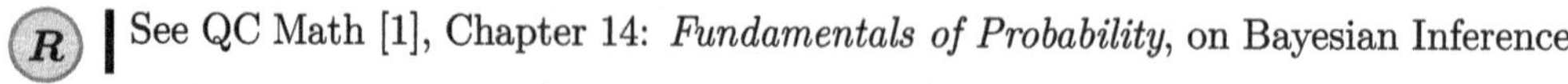

See QC Math [1], Chapter 14: *Fundamentals of Probability*, on Bayesian Inference.

1   Measurement Procedure

Bayesian QPE uses the Hadamard test circuit in Fig. 7.5, with a unitary evolution U^k and an additional phase shift $2\pi\beta$ applied through a rotation gate. The probability of obtaining outcome $v \in \{0, 1\}$ can be written as (see Eq. 7.10)

$$P(v \mid \phi, k, \beta) = \frac{1 + \cos\left(2\pi\left(\beta + k\phi - \frac{v}{2}\right)\right)}{2}.$$

This serves as the likelihood function for Bayesian updates.

2 Bayesian Inference and Posterior Updates

Bayesian QPE maintains a prior distribution $P(\phi)$, which is updated iteratively. Given outcomes $\{v_1, \ldots, v_s\}$, the posterior distribution is

$$P(\phi \mid v_1, \ldots, v_s) \propto P(\phi) \prod_{i=1}^{s} P(v_i \mid \phi, k_i, \beta_i).$$

Each measurement refines the estimate by concentrating the posterior around likely phase values.

3 Parameter Selection for k and β

A central practical issue is choosing k and β adaptively. The choice of k controls the "resolution" of the phase interrogation, while β shifts the measurement fringe to improve informativeness. Existing methods often use heuristics such as

$$k_r = \left\lceil \frac{1.25}{\sqrt{\mathrm{Var_H}(\phi)}} \right\rceil, \quad \beta_r \sim \mathcal{U}(0, 1),$$

where $\mathrm{Var_H}(\phi)$ is the Holevo variance, a circular measure of phase uncertainty. More advanced strategies choose (k, β) to minimize the expected posterior variance after the next measurement.

4 Noise-Aware Bayesian Updates

On real hardware, decoherence and gate errors perturb the ideal likelihood. A common phenomenological model replaces the ideal fringe contrast by a reduced contrast factor. For example, one may use

$$P(v \mid \phi, k, \beta) = \frac{1 + (1 - q) \cos\left(2\pi\left(\beta + k\phi - \frac{v}{2}\right)\right)}{2},$$

where $q \in [0, 1]$ is an effective noise parameter that captures loss of coherence or depolarizing effects.

5 Scaling Behavior

Bayesian QPE can achieve high precision with shallow quantum circuits by increasing k while keeping the number of rounds modest. In many idealized analyses, the number of measurement rounds needed to reach resolution ε scales as $O(\log(1/\varepsilon))$, while the *total* evolution time (informally, the sum of the applied powers k) scales as $O(1/\varepsilon)$, reflecting a Heisenberg-type trade-off between precision and interrogation time.

Compared to Kitaev's IQPE and Information-Theoretic QPE, Bayesian QPE offers:

- Robustness to noise through explicit noise-aware likelihoods.

- Adaptive learning via dynamic selection of k and β.

- Reduced quantum-circuit depth per round at the expense of heavier classical post-processing.

Despite its classical complexity, Bayesian QPE provides a powerful framework for phase estimation, particularly for near-term quantum devices where quantum coherence time is limited.

7.3.5 Other QPE Techniques

Several other variations of QPE have been developed to optimize performance under different constraints. A comparative introduction is given in Table 7.2.

QPE Variant	Quantum Measurements	Classical Complexity	Noise Robustness	Key Trade-offs
Kitaev's IQPE	$O(\log(1/\varepsilon))$	Low	Moderate	Requires controlled-U^{2^k} gates
Bayesian QPE	$O(\log(1/\varepsilon))$	High	High	Requires Bayesian updates
Robust Phase Estimation	Moderate	Low	Very High	Best for noisy gates
Machine Learning QPE	Optimized	High	Adaptive	Requires classical ML
Adaptive QPE	Fewer than Kitaev's	Medium	High	Requires real-time feedback
Information-Theoretic QPE	$O(1/\varepsilon)$	Exponential	Moderate	No deep circuits, but high classical cost

Table 7.2: Comparison of Quantum Phase Estimation (QPE) Variants

Chapter Takeaways

1. QPE estimates an eigenphase ϕ of a unitary U using controlled powers U^{2^k} and an inverse QFT, yielding an n-bit estimate with controllable precision.

2. QPE is practical only when controlled powers of U can be implemented efficiently; otherwise the cost grows exponentially in the number of output bits.

3. With a non-eigenstate input, QPE samples eigenphases according to

the input state's overlap on the eigenbasis, making it a spectral-probing primitive (not just an "eigenvalue calculator").

4. Iterative QPE (IQPE) replaces the inverse QFT by repeated short circuits and classical postprocessing; Kitaev-style and Bayesian/information-theoretic variants trade fewer qubits for more repetitions and classical inference.

Problem Set 7

7.1 Let $U = \begin{bmatrix} 1 & 0 \\ 0 & e^{2\pi i\phi} \end{bmatrix}$ and suppose the target is initialized in $|1\rangle$.

(a) Write the control-register state *before* the inverse QFT in closed form.

(b) Show that if $\phi = \frac{a}{2^n}$ for an integer a, then standard QPE returns a with probability 1.

(c) For $\phi = \frac{a}{2^n} + \delta$, derive the measurement distribution $P(\widehat{\phi} \mid \phi)$ and identify the most likely outcomes.

7.2 Assume U has eigenpairs $\{(e^{2\pi i\phi_j}, |\psi_j\rangle)\}$ and the target input is $|\psi\rangle = \sum_j c_j |\psi_j\rangle$.

(a) Derive the post-QPE joint state (before measuring the control register) in terms of $\{c_j\}$.

(b) Prove that measuring the control register samples eigenphases with probabilities $|c_j|^2$ in the idealized "exact-grid" case.

(c) Explain how this property can be used to estimate spectral features (e.g., dominant eigenphases) when $|\psi\rangle$ has unknown decomposition.

7.3 Using your preferred quantum SDK (e.g., Qiskit), implement standard QPE for $U = R_Z(2\pi\phi)$ acting on a single-qubit target.

(a) Verify the "exact-grid" case $\phi = \frac{a}{2^n}$ for several choices of a and n.

(b) Explore an off-grid case (e.g., $\phi = 0.3$) and empirically plot the histogram of outcomes.

(c) Compare the empirical histogram with the analytic distribution from the error analysis section.

7.4 Suppose U is given as a gate sequence of depth D.

(a) Estimate the depth of standard QPE if U^{2^j} is implemented by repeating U exactly 2^j times.

(b) Describe one structured setting (e.g., $U = e^{-iHt}$ for local H) where U^{2^j} can be implemented more efficiently than repetition.

(c) Give a short "engineering" discussion of why controlled-U^{2^j} can be harder than uncontrolled U^{2^j} on real devices.

7.5 Many frameworks (e.g., `Qiskit`) use a little-endian convention. Reformulate Kitaev's IQPE under the little-endian convention, making explicit how the powers U^{2^j} and the reconstructed binary fraction $[0.\phi_1 \ldots \phi_n]$ must be interpreted.

7.6 Let H be a Hermitian operator with eigenpairs $(E_j, |E_j\rangle)$ and consider $U = e^{-iHt}$.

(a) Show that $|E_j\rangle$ is an eigenstate of U and express the corresponding phase in terms of E_j and t.

(b) Discuss how the choice of t affects (i) energy resolution and (ii) phase wrap-around modulo 2π.

(c) For a two-level Hamiltonian $H = \frac{\Delta}{2} Z$, work out the phases explicitly and explain how QPE could recover Δ.

7.7 Model each estimated probability as $\widehat{P} = P + \xi$, where ξ is an additive error capturing SPAM and finite sampling.

(a) Explain qualitatively why early bit errors can corrupt all subsequent bits in MSB-first IQPE.

(b) For LSB-first IQPE, argue why early errors tend to have smaller impact on the final $\widehat{\phi}$.

(c) Propose a simple redundancy scheme (e.g., majority vote or repeated rounds) to reduce the probability of catastrophic failure.

7.8 ✳ (Information-Theoretic QPE as Classical Inference.) Consider measurements indexed by (k_i, β_i) producing outcomes $v_i \in \{0, 1\}$ with likelihood $P(v_i \mid \phi, k_i, \beta_i)$.

(a) Write the log-likelihood $\ell(\phi) = \log L(\phi)$.

(b) Derive an expression for $\ell'(\phi)$ and describe how it could be used in a gradient-based optimizer.

(c) Implement a simple grid-search MLE on $\phi \in [0, 1)$ and discuss how the grid resolution affects runtime and accuracy.

7.9 ✳ (Bayesian QPE with a Discrete Prior.) Assume a uniform prior on $\Phi = \{0, \frac{1}{2^n}, \ldots, \frac{2^n - 1}{2^n}\}$.

(a) Implement the Bayesian update rule over Φ for a sequence of measurements (k_i, β_i, v_i).

(b) Compare MAP and circular-mean estimators on the same dataset.

(c) Experiment with an adaptive choice of k_i based on the current posterior width and report the effect on convergence.

7.10 ✳ You are given a unitary U and access to an approximate eigenstate $|\psi\rangle$ whose overlap with the true eigenstate satisfies $|\langle\psi|\psi_\star\rangle|^2 \geq 0.2$. You have a limited coherence budget that restricts you to controlled-U^k with $k \leq k_{\max}$.

(a) Propose a phase-estimation strategy (standard QPE, IQPE, Bayesian, information-theoretic, or a hybrid).

(b) Justify your choice in terms of qubits, depth, and classical post-processing.

(c) Explain how you would diagnose whether the observed outcomes are dominated by the desired eigenphase or by other spectral components.

8. Shor's Algorithm and Period Finding

Contents

Shor's algorithm is one of the most celebrated results in quantum computing. It provides an exponential speedup (in the standard complexity-theoretic sense) for integer factorization, a problem with major cryptographic significance through schemes such as RSA. Shor's 1994 discovery [20] was a turning point: it offered concrete evidence that quantum computation can outperform classical computation on a practically motivated task, and it helped clarify the boundary between efficient and intractable computation in the quantum setting.

The unifying idea behind Shor's algorithm is *period finding*. We first show how factoring reduces to finding the period of modular exponentiation, and then present two closely related quantum routes to extracting that period: a QFT-based approach using partial measurement and a formulation in terms of Quantum Phase Estimation (QPE). We conclude by placing period finding in a broader framework via the Hidden Subgroup Problem, which explains why Shor's method works so well in the Abelian setting and what challenges arise in non-Abelian generalizations.

8.1 ＊Background: RSA and Integer Factorization

This section provides the classical background underlying Shor's algorithm. It draws on elementary number theory and the classical principles behind RSA cryptography. None of this material is required to understand the *quantum* component of Shor's algorithm; therefore, new readers may rely on the following Overview and skip the remainder of this section. Readers interested in the number-theoretic foundations may find this section a useful refresher.

Overview

1. Certain cryptosystems such as RSA rely on the presumed difficulty of factoring large composite integers.

2. Integer factoring can be reduced to period finding of a modular exponential function.

3. Period finding can be efficiently computed with Shor's algorithm.

8.1.1 The RSA Cryptosystem

RSA (Rivest–Shamir–Adleman, 1977) is a foundational public-key cryptosystem. It is built on the presumed difficulty of factoring large composite integers. This section introduces RSA, explains how Euler's theorem ensures correctness, and clarifies how the difficulty of factoring supports its security.

RSA uses *asymmetric cryptography*, where encryption and decryption use different keys. The public key enables anyone to encrypt a message, but only the private key—derived from secret prime factors of a large modulus—can decrypt it.

1 Key Generation

To generate an RSA key pair:

1. Choose two large, random, distinct primes p and q of similar bit length.

2. Compute the modulus $N = pq$.

3. Evaluate Euler's totient $\varphi(N) = (p-1)(q-1)$.

4. Choose a public exponent e such that $1 < e < \varphi(N)$ and $\gcd(e, \varphi(N)) = 1$. A common choice is $e = 65537$.

5. Compute the private exponent d as the modular inverse of e modulo $\varphi(N)$:

$$ed \equiv 1 \pmod{\varphi(N)}.$$

The public key is (N, e), and the private key is (N, d). The primes p and q are securely discarded after key generation.

2   Encryption and Decryption

For a plaintext message m satisfying $0 \leq m < N$, encryption computes

$$c \equiv m^e \pmod{N}.$$

Decryption recovers

$$m \equiv c^d \pmod{N}.$$

3   Correctness via Euler's Theorem

Euler's theorem states that if $\gcd(m, N) = 1$, then

$$m^{\varphi(N)} \equiv 1 \pmod{N}.$$

Since $ed \equiv 1 \pmod{\varphi(N)}$, there exists an integer k such that $ed = 1 + k\varphi(N)$. Thus,

$$c^d = m^{ed} = m^{1+k\varphi(N)} = m\,(m^{\varphi(N)})^k \equiv m \cdot 1^k \equiv m \pmod{N}.$$

Therefore, decryption inverts encryption. In practice, secure padding schemes help ensure $\gcd(m, N) = 1$ and prevent structural attacks.

■ **Example 8.1** To illustrate, consider a small numerical example (for pedagogical purposes only; real RSA uses much larger primes).

1. Select $p = 5$, $q = 11$, giving $N = 55$.

2. Compute $\varphi(N) = (p-1)(q-1) = 40$.

3. Choose $e = 3$, since $\gcd(3, 40) = 1$.

4. Determine d as the modular inverse of e modulo 40: $3d \equiv 1 \pmod{40}$, yielding $d = 27$.

5. The public key is $(N, e) = (55, 3)$; the private key is $(N, d) = (55, 27)$.

6. Encrypt $m = 9$:

$$c \equiv 9^3 \pmod{55} \equiv 729 \pmod{55} = 14.$$

7. Decrypt:

$$m' \equiv 14^{27} \pmod{55}.$$

8. Verification of $m' = m$: Since $ed = 81 = 2 \times 40 + 1$, Euler's theorem gives

$$m' \equiv 9^{ed} \equiv 9^{2 \times 40 + 1} \equiv (9^{40})^2 \cdot 9 \equiv 1^2 \cdot 9 \equiv 9 \pmod{55}.$$

■

8.1.2   Security and the Role of Factoring

The security of RSA rests on the assumption that factoring a large composite modulus $N = pq$ is computationally hard. Knowing p and q allows one to compute $\varphi(N)$ and derive the private key:

$$\text{Factoring } N \implies (p, q) \implies \varphi(N) \implies d.$$

Currently, the best-known classical factoring algorithms, such as the General Number Field Sieve, have sub-exponential time complexity:

$$T_{\text{classical}}(N) = \exp\left(O\left((\log N)^{1/3}(\log\log N)^{2/3}\right)\right),$$

making 2048-bit RSA keys effectively secure against classical attacks.

In contrast, Shor's algorithm factors N in polynomial time,

$$T_{\text{quantum}}(N) = O((\log N)^3),$$

which would efficiently recover p and q and thus the private key.

From a complexity-theoretic perspective, factoring is a function problem rather than a decision problem, but its associated decision version ("Does N have a nontrivial factor at most k?") lies in NP, since a proposed factor can be verified in polynomial time. Factoring is not known to be NP-complete, and is widely believed to be hard for classical computers.

8.1.3 Factoring via Period Finding

The central insight of Shor's algorithm is that factoring a composite integer N reduces to *finding the period* of a modular exponential function. This transforms a hard number-theoretic problem into one that can be efficiently solved on a quantum computer using the Quantum Fourier Transform (QFT) or, equivalently, Quantum Phase Estimation (QPE).

1 The Reduction Idea

Suppose N is an odd composite number that is not a prime power. To factor N:

1. Randomly select a such that $1 < a < N$ and $\gcd(a, N) = 1$. If $\gcd(a, N) \neq 1$, then $\gcd(a, N)$ is already a nontrivial factor of N.

2. Consider
$$f(x) = a^x \bmod N.$$
Since there are finitely many residues modulo N, the values must repeat. Let r be the smallest positive integer such that
$$a^r \equiv 1 \pmod{N}, \quad \text{equivalently} \quad a^{x+r} \equiv a^x \pmod{N}.$$
This r is called the *period* (or *order*) of a modulo N.

Once r is known, one can often compute a nontrivial factor of N.

2 Recovering Factors from the Period

From
$$a^r \equiv 1 \pmod{N} \quad \Longrightarrow \quad a^r - 1 \equiv 0 \pmod{N},$$
and
$$a^r - 1 = (a^{r/2} - 1)(a^{r/2} + 1),$$
if r is even and $a^{r/2} \not\equiv -1 \pmod{N}$, then
$$p = \gcd(a^{r/2} - 1, N), \qquad q = \gcd(a^{r/2} + 1, N)$$
yield nontrivial factors of N, with at least one being nontrivial.

If r is odd or $a^{r/2} \equiv -1 \pmod{N}$, the attempt fails and a new a must be chosen. For a random valid a, the success probability is at least $1/2$, so only a few repetitions are typically needed.

■ **Example 8.2** Let $N = 55$ (from Example 8.1) and choose $a = 2$ with $\gcd(2, 55) = 1$. Consider

$$f(x) = 2^x \bmod 55.$$

The successive powers of 2 modulo 55 are:

x	1	2	3	4	5	6	7	8	9	10
$2^x \bmod 55$	2	4	8	16	32	9	18	36	17	34

x	11	12	13	14	15	16	17	18	19	20
$2^x \bmod 55$	13	26	52	49	43	31	7	14	28	1

Since $2^{20} \equiv 1 \pmod{55}$, the period (order) of 2 modulo 55 is $r = 20$.

Because r is even and

$$2^{r/2} = 2^{10} \equiv 34 \not\equiv -1 \pmod{55},$$

we extract factors via

$$\gcd(2^{r/2} - 1, 55) = \gcd(33, 55) = 11, \quad \gcd(2^{r/2} + 1, 55) = \gcd(35, 55) = 5.$$

Thus $55 = 5 \times 11$ is recovered from the period of $f(x)$.

Below are the periods for more values of a where $\gcd(a, 55) = 1$:

a	3	4	6	12	16	21	34	36	53	54
$r \pmod{55}$	20	10	10	4	5	2	2	5	20	2

These r values are all divisors of $\varphi(55) = 40$. ∎

Exercise 8.1 Work out the factors of 55 with $a = 6$, 12, and 21.

Exercise 8.2 Let $N = 21$ and choose $a = 2$ with $\gcd(a, N) = 1$.

(a) Compute $f(x) = 2^x \bmod 21$ for $x = 0, 1, 2, \ldots$ until the values repeat, and determine the period r.

(b) Check whether r is even and whether $2^{r/2} \not\equiv -1 \pmod{21}$.

(c) Use $\gcd(2^{r/2} \pm 1, 21)$ to recover a nontrivial factor of 21.

3 Period Finding as the Classical Bottleneck

The reduction above shows that factoring can be converted into the task of finding the period r of

$$f(x) = a^x \bmod N.$$

Classically, this period-finding step appears to require exponential time in $n = \log N$, since one may need to evaluate successive powers of a until the sequence repeats.

The quantum advantage is that the QFT, or equivalently QPE, can recover information about r in polynomial time. This is achieved by using superposition to encode many powers of a coherently and using interference to extract the underlying periodic structure.

The next section introduces Shor's quantum circuit, showing how the Quantum Fourier Transform enables efficient period finding and, consequently, efficient integer factorization.

Key Takeaways

---

1. Choose a with $1 < a < N$ and $\gcd(a, N) = 1$; define the modular exponential function $f(x) = a^x \bmod N$.

2. The goal is to find the order r, the smallest positive integer with $a^r \equiv 1 \pmod{N}$.

3. If r is even and $a^{r/2} \not\equiv -1 \pmod{N}$, then $\gcd(a^{r/2} \pm 1, N)$ yields a nontrivial factor of N.

8.2 Period Finding via Partial Measurement and QFT

As detailed in § 8.1, Shor's key insight is that factoring a composite integer N can be reduced to a *period-finding* problem, a task that quantum computation can solve efficiently using interference and the Quantum Fourier Transform (QFT).

We choose an integer a that is coprime to N, i.e., $\gcd(a, N) = 1$. The modular exponential function

$$f(x) = a^x \bmod N \tag{8.1}$$

is periodic with some integer period r, meaning

$$f(x + jr) = f(x),$$

for all non-negative integers x and j. Once r is known, classical post-processing can recover the factors of N. Thus, the quantum part of Shor's algorithm reduces to finding the period r of $f(x)$.

8.2.1 Conceptual Structure of Shor's Algorithm

We begin with the partial measurement model of § A.3. The modular exponential function Eq. 8.1 is implemented as a unitary operation U_f acting across two registers. The resulting circuit is shown in Fig. 8.1.

Let $n = \lceil \log_2 N \rceil$, so the target register has n qubits and can represent all residues modulo N. If $N < 2^n$, then U_f acts nontrivially only on the first N computational basis states of the target register and leaves the remainder unchanged. The initial state of the target register, $|\psi\rangle$, can be chosen so that all relevant eigenphases can occur upon measurement. A precise explanation of this point is deferred to § 8.3.

We choose the control register to have $2n$ qubits for a reliable recovery of r, so its dimension is $Q = 2^{2n}$. We prepare a uniform superposition in the control register:

$$|\Psi_1\rangle = \frac{1}{\sqrt{Q}} \sum_{x=0}^{Q-1} |x\rangle \, |\psi\rangle.$$

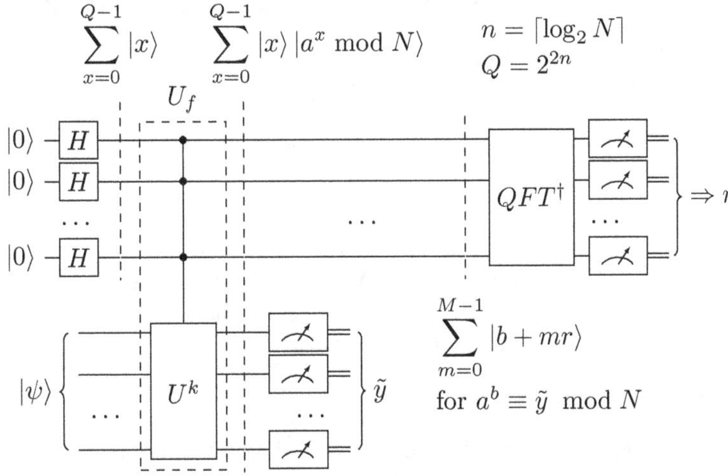

Figure 8.1: High-Level View of Shor's Algorithm

The composite unitary U_f transforms $|\Psi_1\rangle$ into

$$|\Psi_2\rangle = \frac{1}{\sqrt{Q}} \sum_{x=0}^{Q-1} |x\rangle \, |f(x)\rangle = \frac{1}{\sqrt{Q}} \sum_{x=0}^{Q-1} |x\rangle \, |a^x \bmod N\rangle .$$

The notation $|a^x \bmod N\rangle$ denotes the computational basis state $|j\rangle$ where $j \equiv a^x$ (mod N).

Suppose we measure the second register and observe a value $\tilde{y}$. The measurement collapses the control register to a superposition of all x satisfying $f(x) = \tilde{y}$:

$$|\tilde{x}\rangle = \frac{1}{\sqrt{M}} \sum_{m=0}^{M-1} |b + mr\rangle ,$$

where b is the smallest non-negative integer satisfying $a^b \equiv \tilde{y}$ (mod N), and

$$M = \left\lfloor \frac{Q-1-b}{r} \right\rfloor + 1$$

is the number of integers $x \in \{0, 1, \ldots, Q-1\}$ for which $f(x) = \tilde{y}$.

This state is precisely the periodic superposition examined in § 6.1.3 (Extracting Periodicity via QFT). Applying the QFT to the control register produces sharp peaks near integer multiples of Q/r, from which r can be recovered using continued fractions. The technical details of the success probability and the continued-fraction recovery are deferred to § 8.3.

Shor's algorithm is efficient because both the QFT and the unitary U_f can be implemented efficiently, as we now discuss.

8.2.2   Efficient Implementation of the Controlled Unitary

The controlled unitary U_f can be implemented using the same construction as in QPE (see § 7.1.2 and Fig. 7.2). Here we choose U such that

$$U \,|j\rangle = |aj \bmod N\rangle, \tag{8.2}$$

for $j = 0, 1, \ldots, N - 1$.

Since $\gcd(a, N) = 1$, multiplication by a modulo N is a permutation on $\{0, 1, \ldots, N - 1\}$, and thus U is a permutation matrix in the computational basis. Any permutation matrix is unitary because its columns form an orthonormal set. Therefore, U is a valid unitary operator.

1   Implementation of U

In practice, the unitary U is implemented as a reversible circuit computing modular multiplication. Its action can be expressed as

$$|j\rangle \, |0\rangle \;\mapsto\; |j\rangle \, |aj \bmod N\rangle .$$

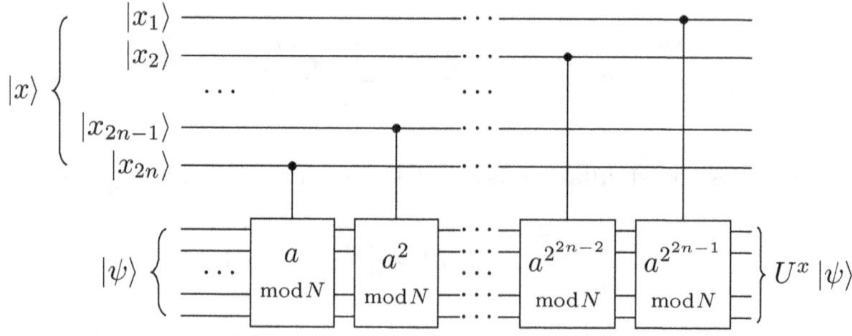

To make this operation reversible, additional ancilla qubits are introduced, and the intermediate workspace is cleared using uncomputation (see Fig. 4.2). With these techniques, U can be implemented with at most polynomial overhead in $\log N$ (see § 4.3.5.1).

2   Implementation of U_f

Figure 8.2: Implementation of the Controlled Unitary in Shor's Algorithm

For the purposes of U_f (introduced in Fig. 7.2), which must apply U^x conditioned on the integer x encoded in the control register, we require powers of U. Because modular multiplication satisfies

$$a(b \bmod N) \bmod N = ab \bmod N,$$

repeated application of U corresponds to taking higher powers of a modulo N:

$$U^k \,|j\rangle = |a^k j \bmod N\rangle . \tag{8.3}$$

Thus, implementing controlled-U^k amounts to replacing a with a^k in Eq. 8.2. The resulting controlled unitary U_f is shown in Fig. 8.2.

The efficiency of Shor's algorithm depends critically on implementing U^k without performing k sequential applications of U. Fortunately, $a^k \bmod N$ can be computed classically in time $O((\log k)(\log N)^2)$ using repeated squaring. This allows us to precompute

$$a^{2^0} \bmod N, \quad a^{2^1} \bmod N, \quad a^{2^2} \bmod N, \quad a^{2^3} \bmod N, \ \ldots$$

and implement each controlled-U^{2^j} using only a polynomial-size quantum circuit.

■ **Example 8.3** Consider $N = 55$ and $a = 2$. To implement U^{2^5}, we precompute:

$$2^2 \equiv 4, \quad 2^4 \equiv 16, \quad 2^8 \equiv 16^2 \equiv 36, \quad 2^{16} \equiv 36^2 \equiv 31, \quad 2^{32} \equiv 31^2 \equiv 26 \quad (\bmod\ 55),$$

with only 5 squarings instead of 32. We can also precompute transformations like

$$U^{2^5} |31\rangle = |26 \cdot 31 \bmod 55\rangle = |36\rangle.$$

■

Shor's algorithm benefits from the fact that modular exponentiation is efficiently computable classically; hence the corresponding quantum unitaries can also be implemented efficiently.

8.2.3 Quantum Circuit for Shor's Algorithm

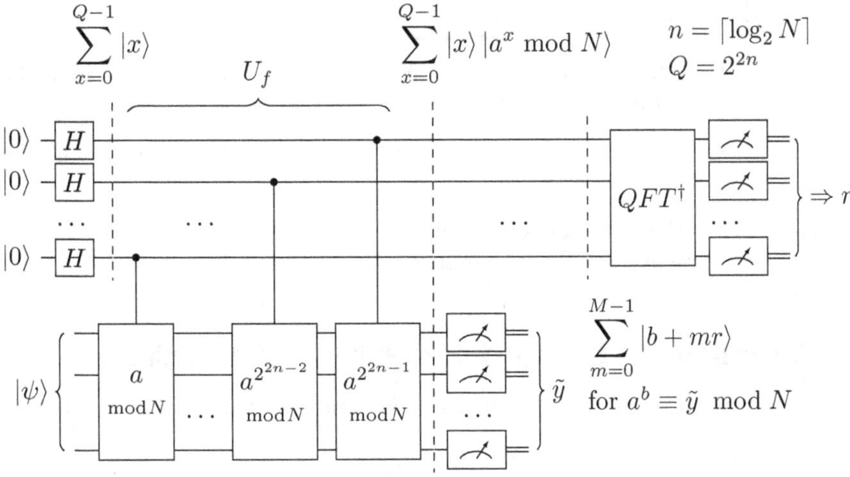

Figure 8.3: Quantum Circuit for Shor's Algorithm

Combining the controlled-unitary circuit for U_f in Fig. 8.2 with the block diagram of Fig. 8.1, we arrive at the complete quantum circuit for Shor's algorithm shown in Fig. 8.3. Its resemblance to the QPE circuit in Fig. 7.3 is no accident, as we will discuss next.

Key Takeaways

1. One oracle call correlates $|x\rangle$ with $|f(x)\rangle$ in superposition.

2. Measuring the target register collapses the control register to a periodic superposition $\sum_m |b + mr\rangle$.

3. Applying $\text{QFT}^\dagger$ maps periodicity in x to peaks near multiples of Q/r, enabling recovery of r by continued fractions.

8.3   Period Finding via Quantum Phase Estimation

While Shor's algorithm can be understood intuitively through the QFT-based formulation—emphasizing measurement, interference, and Fourier analysis—Quantum Phase Estimation (QPE, Chapter 7) provides a rigorous and conceptually clean framework for period finding.

As discussed in § 8.1, factoring a composite integer N reduces to finding the period r of the modular exponential function $f(x) = a^x \bmod N$, where $1 < a < N$ is chosen such that $\gcd(a, N) = 1$. The central problem is to determine the smallest positive integer r satisfying $a^r \equiv 1 \pmod{N}$.

The key idea of the QPE-based approach is that $f(x)$ can be encoded into a unitary operator U whose eigenvalues contain the phase information from which r can be recovered. Thus, period finding becomes an instance of eigenphase estimation. Applying the QPE circuit of Fig. 7.3 to this unitary yields precisely the same circuit as Fig. 8.3, which we obtained earlier from the partial-measurement formulation.

Overview

1. Define $U|j\rangle = |aj \bmod N\rangle$; since $\gcd(a, N) = 1$, U is a permutation and hence unitary.

2. On the r-cycle, U has eigenvalues $e^{2\pi i k/r}$, so the eigenphases encode the rationals k/r.

3. QPE estimates an eigenphase k/r, and classical continued fractions recovers the denominator r (the order).

8.3.1   Properties of the Modular-Multiplication Unitary

As introduced in Eq. 8.2, the modular-multiplication unitary U is defined as

$$U|j\rangle = |aj \bmod N\rangle, \qquad 0 \le j \le N - 1. \tag{8.4}$$

This operator acts on $n = \lceil \log_2 N \rceil$ qubits. Since multiplication by a modulo N is a permutation of $\{0, 1, \ldots, N - 1\}$, the operator U is unitary.

1   Eigenvalues and Eigenvectors

The eigenvectors of U supported on the r-cycle

$$1, \; a \bmod N, \; a^2 \bmod N, \; \ldots, \; a^{r-1} \bmod N$$

are

$$|u_k\rangle = \frac{1}{\sqrt{r}} \sum_{l=0}^{r-1} e^{-2\pi i l k/r} |a^l \bmod N\rangle, \qquad k = 0, 1, \ldots, r-1, \tag{8.5}$$

with eigenvalues $\lambda_k = e^{2\pi i k/r}$. Indeed,

$$U|u_k\rangle = e^{2\pi i k/r} |u_k\rangle. \tag{8.6}$$

Thus, the eigenphases of U encode the rational numbers k/r, from which the period r can be recovered.

Proof. Using Eq. 8.5, together with $U|j\rangle = |aj \bmod N\rangle$ and $a^r \equiv 1 \pmod{N}$, we have

$$U|u_k\rangle = \frac{1}{\sqrt{r}} \sum_{l=0}^{r-1} e^{-2\pi i l k/r} |a^{l+1} \bmod N\rangle.$$

Let $m = l + 1$. As l runs from 0 to $r - 1$, m runs from 1 to r, and since $a^r \equiv 1 \pmod{N}$ we identify $|a^r \bmod N\rangle = |a^0 \bmod N\rangle$. Thus

$$U|u_k\rangle = \frac{1}{\sqrt{r}} \sum_{m=0}^{r-1} e^{-2\pi i (m-1)k/r} |a^m \bmod N\rangle$$

$$= e^{2\pi i k/r} \frac{1}{\sqrt{r}} \sum_{m=0}^{r-1} e^{-2\pi i m k/r} |a^m \bmod N\rangle.$$

The sum is precisely $|u_k\rangle$, so Eq. 8.6 follows. $\qquad\square$

2 The r-Dimensional Eigen-Subspace

The vectors in Eq. 8.5 span the r-dimensional subspace generated by

$$|a^0 \bmod N\rangle, \ |a^1 \bmod N\rangle, \ \ldots, \ |a^{r-1} \bmod N\rangle.$$

More generally, U decomposes the computational basis into disjoint cycles, and each cycle admits a similar Fourier eigenbasis. Shor's algorithm focuses on the cycle generated by 1, whose length is exactly the order r of a modulo N.

If r is small, the eigenstructure simplifies correspondingly. For example, when $a \equiv -1 \pmod{N}$, we have $r = 2$ and

$$|u_0\rangle = \tfrac{1}{\sqrt{2}}(|1\rangle + |a \bmod N\rangle), \qquad |u_1\rangle = \tfrac{1}{\sqrt{2}}(|1\rangle - |a \bmod N\rangle),$$

with eigenvalues $+1$ and -1.

3 Uniform Superposition of the Eigenvectors

In standard QPE the target register is initialized in an eigenstate of U. In Shor's algorithm, however, we use the computational basis state $|1\rangle$, which has the expansion

$$|1\rangle = \frac{1}{\sqrt{r}} \sum_{k=0}^{r-1} |u_k\rangle. \tag{8.7}$$

Proof. Expand the right-hand side using Eq. 8.5:

$$\frac{1}{\sqrt{r}}\sum_{k=0}^{r-1}|u_k\rangle = \frac{1}{\sqrt{r}}\sum_{k=0}^{r-1}\left(\frac{1}{\sqrt{r}}\sum_{l=0}^{r-1}e^{-2\pi ilk/r}|a^l \bmod N\rangle\right)$$

$$= \frac{1}{r}\sum_{l=0}^{r-1}\left(\sum_{k=0}^{r-1}e^{-2\pi ilk/r}\right)|a^l \bmod N\rangle.$$

The inner sum is a geometric series over the r-th roots of unity:

$$\sum_{k=0}^{r-1}e^{-2\pi ilk/r} = \begin{cases} r, & l \equiv 0 \pmod r, \\ 0, & \text{otherwise.} \end{cases}$$

Thus only the $l = 0$ term survives, and

$$\frac{1}{\sqrt{r}}\sum_{k=0}^{r-1}|u_k\rangle = |a^0 \bmod N\rangle = |1\rangle.$$

$\square$

Thus $|1\rangle$ is a uniform superposition of all eigenvectors supported on the r-cycle, allowing all eigenphases k/r to occur with equal probability in the phase-estimation procedure (see § 7.1.4).

8.3.2   Extracting the Period via QPE

We now outline how QPE applied to U yields the period r.

1   QPE Circuit for Modular Multiplication

Let $n = \lceil \log_2 N \rceil$ and choose the control register to have $t = 2n$ qubits, so its dimension is $Q = 2^t = 2^{2n}$. The QPE circuit in Fig. 8.3 proceeds as follows:

1. Initialize the control register in a uniform superposition and the target register in $|1\rangle$:

$$|\Psi_0\rangle = \frac{1}{\sqrt{Q}}\sum_{x=0}^{Q-1}|x\rangle \otimes |1\rangle.$$

2. Apply controlled-U^{2^j} gates for $j = 0, 1, \ldots, t-1$. Equivalently, the circuit applies U^x conditioned on $|x\rangle$, producing

$$|\Psi_1\rangle = \frac{1}{\sqrt{Q}}\sum_{x=0}^{Q-1}|x\rangle \otimes U^x|1\rangle.$$

Using Eqs. 8.6 and 8.7 to expand $|1\rangle$ in the eigenbasis of U, we obtain

$$|\Psi_1\rangle = \frac{1}{\sqrt{r}}\sum_{k=0}^{r-1}\left(\frac{1}{\sqrt{Q}}\sum_{x=0}^{Q-1}e^{2\pi ixk/r}|x\rangle\right) \otimes |u_k\rangle.$$

Thus, conditioned on eigenvector $|u_k\rangle$, the control register carries the eigenphase $\phi_k = k/r$.

3. Apply the inverse QFT to the control register. For each fixed k, the state

$$\frac{1}{\sqrt{Q}} \sum_{x=0}^{Q-1} e^{2\pi i x k / r} |x\rangle$$

is mapped close to a computational basis state peaked near Qk/r.

4. Measure the control register to obtain an integer $y \in \{0, 1, \ldots, Q-1\}$ satisfying

$$\left| \frac{y}{Q} - \frac{k}{r} \right| \leq \frac{1}{2Q}$$

with high probability.

Exercise 8.3 Let $Q = 2^t$ and suppose f has period r on $\{0, 1, \ldots, Q-1\}$.

(a) Assume first that $r \mid Q$. Explain why the QFT of a coset superposition has support only on indices y that are integer multiples of Q/r.

(b) Now consider the case $r \nmid Q$. Describe qualitatively what changes in the measurement distribution on y.

2 ✳ Recovering the Period by Continued Fractions

The measurement yields a rational approximation y/Q to $\phi_k = k/r$. Since $r < N \leq 2^n$ and $Q = 2^{2n} \geq N^2$, we have

$$\left| \frac{y}{Q} - \frac{k}{r} \right| \leq \frac{1}{2Q} \leq \frac{1}{2N^2} < \frac{1}{2r^2}.$$

By the standard continued-fraction guarantee, k/r is then the unique fraction with denominator at most N lying within $1/(2r^2)$ of y/Q, so the continued-fraction expansion of y/Q recovers k/r and hence r.

■ **Example 8.4** Let $N = 55$ and $a = 2$. The period is $r = 20$. Here $n = \lceil \log_2 55 \rceil = 6$, so we take $t = 2n = 12$ control qubits and $Q = 2^{12} = 4096$. Suppose a measurement yields $y = 205$. Then

$$\frac{y}{Q} = \frac{205}{4096} \approx 0.05005,$$

which is close to the phase $\phi_1 = \frac{1}{20} = 0.05$.

We expand $\frac{205}{4096}$ as a continued fraction:

$$\frac{205}{4096} = [0; 19, 1, 50, 4].$$

The convergents are

$$\frac{0}{1}, \quad \frac{1}{19}, \quad \frac{1}{20}, \quad \frac{51}{1019}, \quad \frac{205}{4096}.$$

Considering denominators less than $N = 55$, the nontrivial candidates are $r' = 19$ and $r' = 20$. We test each candidate by checking whether $2^{r'} \equiv 1 \pmod{55}$:

$$2^{19} \equiv 28 \not\equiv 1 \pmod{55}, \qquad 2^{20} \equiv 1 \pmod{55}.$$

Thus $r = 20$ is recovered. We can then compute

$$\gcd\left(2^{r/2} \pm 1, \, 55\right) = \gcd(2^{10} \pm 1, \, 55)$$

to obtain the nontrivial factors 5 and 11 of 55. ∎

The randomness in the choice of a and in the eigenphase k/r makes Shor's algorithm probabilistic. However, the probability of obtaining a useful period is bounded below by a constant, and a few repetitions suffice to factor N with high success probability.

Exercise 8.4 Let $Q = 2^t$ and suppose a QPE/QFT-based routine produces an outcome y such that

$$\left| \frac{y}{Q} - \frac{k}{r} \right| \le \frac{1}{2Q}.$$

Explain why continued fractions is a natural classical tool to recover a candidate denominator r from y/Q.

Exercise 8.5 In the example above we chose $t = 2n$ control qubits. Explore what can go wrong if we use only $t = n$ control qubits.

(a) For $N = 55$, we have $n = \lceil \log_2 55 \rceil = 6$ and hence $Q = 2^n = 64$. If the true phase is $\phi_1 = \frac{1}{20}$, compute $Q\phi_1$ and identify the nearest integers y.

(b) Compute the continued fraction of the closer approximation (for example, $\frac{3}{64}$) and list its convergents with denominators less than $N = 55$. What candidate denominators would the continued-fraction method suggest?

Key Takeaways

1. The QPE/QFT measurement produces y so that y/Q approximates k/r for some k.

2. Choosing $Q = 2^{2n}$ with $n = \lceil \log_2 N \rceil$ typically gives enough precision that k/r is the unique fraction with denominator $< N$ close to y/Q.

3. Continued fractions recovers k/r, and testing candidate denominators verifies the correct r.

8.3.3   Resource Estimation

1   A Rough Resource Estimate for Breaking RSA-2048

Let n denote the bit length of the integer N to be factored. In the standard order-finding formulation of Shor's algorithm, one uses a work register of size about n qubits together with a control register of size about $2n$ qubits, plus ancillary workspace for reversible modular arithmetic. Thus, even at an elementary level, the circuit width is naturally expected to scale linearly with n, with a rough footprint of a few multiples of n logical qubits. A concrete low-width implementation due to Häner, Roetteler, and Svore uses only $2n + 2$ qubits, illustrating that linear-in-n width is indeed achievable in explicit circuit constructions [23].

The dominant cost comes from modular exponentiation. Using repeated squaring, one performs $O(n)$ controlled modular multiplications. With straightforward reversible arithmetic, a modular multiplication costs about $O(n^2)$ gate layers up to implementation-dependent constants, so the overall depth of the period-finding circuit is about $O(n^3)$. In comparison, the inverse QFT contributes only $O(n^2)$ gates and is therefore subdominant. This rough scaling is consistent with explicit Toffoli-based constructions, where the circuit depth is $O(n^3)$ and the overall gate count is $O(n^3 \log n)$ [23].

For RSA-2048, where $n = 2048$, this elementary estimate suggests a circuit width on the order of several thousand logical qubits and a logical gate count or logical depth on the order of $n^3 \approx 8.6 \times 10^9$, again up to substantial constant-factor variation depending on the chosen arithmetic circuits and the degree of parallelization. This is only a rough estimate, but it already explains why Shor's algorithm, despite being polynomial-time, still lies firmly in the fault-tolerant regime for cryptographically relevant key sizes.

2   ✳ From Elementary Scaling to Modern Resource Estimates

Modern resource estimates refine this picture by optimizing arithmetic circuits, exploiting parallelism, and incorporating explicit assumptions about fault-tolerant architecture and error correction. A widely cited estimate by Gidney and Ekerå concluded that factoring an RSA-2048 integer could be achieved in about 8 hours using roughly 2×10^7 noisy physical qubits under surface-code-based assumptions [24]. In their corresponding abstract logical circuit model, factoring an n-bit RSA integer uses about $3n + 0.002\, n \log n$ logical qubits and about $0.3 n^3 + 0.0005\, n^3 \log n$ Toffoli gates [24]. For $n = 2048$, this is roughly 6.2×10^3 logical qubits and 2.6×10^9 Toffoli gates.

More recently, Cain *et al.* argued that cryptographically relevant Shor-type attacks may be possible with far fewer physical qubits if one assumes reconfigurable neutral-atom connectivity, high-rate quantum error-correcting codes, and efficient logical instruction sets [25]. Their study reports RSA-2048 runtime estimates ranging from about 10^3 days with roughly $(1.1–1.3) \times 10^4$ physical qubits to about 97 days with roughly 1.0×10^5 physical qubits, depending on the adopted space–time tradeoff [25]. These results are architecture-specific theoretical estimates rather than demonstrations, but they show that the practical resource outlook for Shor's algorithm continues to evolve rapidly.

The main lesson is therefore twofold: the textbook scaling for Shor's algorithm is already enough to show why RSA-2048 is beyond near-term devices, yet serious fault-tolerant estimates depend strongly on circuit design, parallelization strategy, and the underlying hardware and code architecture [24, 25].

8.4   ✳ Generalization: The Hidden Subgroup Problem

Shor's factoring algorithm, Simon's algorithm, the discrete logarithm algorithm, and several number-theoretic procedures share a common structure: they can all be formulated as instances of the Hidden Subgroup Problem (HSP). The HSP provides a unifying abstraction for many quantum algorithms that exhibit exponential speedups over known classical methods.

This section introduces the HSP and its standard quantum solution strategy, highlights representative Abelian examples, and explains why the non-Abelian case is substantially more difficult.

 | This section builds on QC Math [1], Chapter 4: *Sets, Groups, and Functions.*

8.4.1 Groups, Subgroups, and Cosets

Let G be a finite group written additively. A subgroup $H \leq G$ is a subset closed under the group operation, containing the identity 0, and containing the inverse $-h$ of every $h \in H$.

For any $x \in G$, the (left) coset of H through x is

$$x + H = \{x + h : h \in H\}.$$

Cosets satisfy:

1. All cosets have the same size $|H|$.

2. Any two cosets are either identical or disjoint.

3. The cosets of H partition G.

> **Definition 8.1 — H-Periodic Function.** A function $f : G \to S$ is H-*periodic* if
>
> $$f(x) = f(y) \quad \Longleftrightarrow \quad x - y \in H.$$
>
> Equivalently, f is constant on each coset $x + H$ and takes distinct values on distinct cosets.

■ **Example 8.5 — Cosets and H-Periodic Functions.** Let $G = \mathbb{Z}_{12}$ with addition modulo 12, and let H be the cyclic subgroup generated by 4:

$$H = \langle 4 \rangle = \{0, 4, 8\} \leq \mathbb{Z}_{12}.$$

Equivalently, $H = \{k \cdot 4 \bmod 12 : k \in \mathbb{Z}\}$. The cosets of H are

$$\begin{aligned}
0 + H &= \{0, 4, 8\}, \\
1 + H &= \{1, 5, 9\}, \\
2 + H &= \{2, 6, 10\}, \\
3 + H &= \{3, 7, 11\}.
\end{aligned}$$

These four cosets are disjoint and together partition $\mathbb{Z}_{12}$, and each has size $|H| = 3$.

Now define $f : \mathbb{Z}_{12} \to S$ by assigning one distinct label to each coset, for example

$$f(x) = \begin{cases}
h_0, & x \in 0 + H, \\
h_1, & x \in 1 + H, \\
h_2, & x \in 2 + H, \\
h_3, & x \in 3 + H,
\end{cases}$$

where h_0, h_1, h_2, h_3 represent four distinct symbols (e.g., $0, 1, 2, 3$). Then

$$f(x) = f(y) \quad \Longleftrightarrow \quad x - y \in H,$$

because $f(x) = f(y)$ holds exactly when x and y lie in the same coset, which is equivalent to saying that $x - y$ is a multiple of 4 modulo 12. ■

8.4.2 The Hidden Subgroup Problem (HSP)

> **Definition 8.2 — Hidden Subgroup Problem (HSP).** Given a finite group G and oracle access to a function
>
> $$f : G \to S$$
>
> that is promised to be constant on cosets of some subgroup $H \leq G$ and distinct on different cosets, determine H.

In the quantum setting, we are given oracle access to f (we can query x and obtain $f(x)$), and the goal is to recover the hidden subgroup H from the resulting pattern of repeated values. In most applications, "determine H" means finding generators for H.

■ Example 8.6 — Period Finding as an Instance of the HSP. Consider the additive group $G = \mathbb{Z}_N$. Suppose we are given oracle access to a function $f : \mathbb{Z}_N \to S$ that is promised to have *period* r, meaning

$$f(x) = f(x + r) \quad \text{for all } x \in \mathbb{Z}_N,$$

and that r is the smallest positive integer with this property.

Assume for simplicity that $r \mid N$. Define the subgroup

$$H = \langle r \rangle = \{0, r, 2r, \dots, N - r\} \leq \mathbb{Z}_N,$$

which has size $|H| = N/r$. Then for any $x \in \mathbb{Z}_N$, the coset $x + H$ is

$$x + H = \{x, x + r, x + 2r, \dots, x + (N/r - 1)r\}.$$

By construction, f is constant on each such coset. The promise that r is minimal implies that different cosets give different function values. Therefore,

$$f(x) = f(y) \quad \Longleftrightarrow \quad x - y \in \langle r \rangle = H,$$

so f is H-periodic in exactly the sense required by Def. 8.2. In other words, period finding over $\mathbb{Z}_N$ is precisely the HSP for the subgroup $H = \langle r \rangle$.

In Shor's order-finding subroutine, one considers the function

$$f(x) = a^x \bmod N,$$

defined over a suitable domain (often $\mathbb{Z}_Q$ for a chosen Q), and the unknown period r is the multiplicative order of a modulo N, i.e., the smallest positive r such that $a^r \equiv 1 \pmod{N}$. Determining the hidden subgroup $H = \langle r \rangle$ is therefore equivalent to determining the period r.

■

Key Takeaways

1. The HSP: given f that is constant on cosets of a hidden subgroup $H \leq G$, determine H (typically via generators).

2. One query plus partial measurement produces a random coset state $|x_0 + H\rangle$.

3. Applying QFT_G and measuring yields information about H; repeating and post-processing reconstructs H efficiently for finite Abelian groups.

8.4.3   Quantum Algorithm for the HSP

At a high level, quantum algorithms for the HSP exploit three structural features:

1. the ability to prepare a uniform superposition over G,

2. the ability to convert oracle access to f into coset states,

3. the ability to reveal information about H by applying a quantum Fourier transform (QFT) over G and measuring.

The overall pattern is: create a superposition over all inputs, query the oracle once, measure to obtain a random coset state, Fourier transform and measure, and repeat enough times that the resulting samples allow classical post-processing to reconstruct H.

1   Step 1: Create a Uniform Superposition

Prepare the state

$$|\Psi_0\rangle = \frac{1}{\sqrt{|G|}} \sum_{x \in G} |x\rangle \, |0\rangle \, .$$

The first register stores a group element $x \in G$ (the input register), while the second register is a workspace (the output register) that will store the function value $f(x)$.

2   Step 2: Query the Oracle

Assume oracle access to f via a unitary

$$U_f : |x\rangle \, |y\rangle \mapsto |x\rangle \, |y \oplus f(x)\rangle \, ,$$

where $\oplus$ denotes an appropriate reversible encoding of the range (e.g., bitwise XOR if S is encoded in binary). Applying U_f to $|\Psi_0\rangle$ yields

$$|\Psi_1\rangle = \frac{1}{\sqrt{|G|}} \sum_{x \in G} |x\rangle \, |f(x)\rangle \, .$$

At this point, the input register is coherently correlated with the function values.

3   Step 3: Measure the Output Register

Measure the second register in the computational basis, obtaining some outcome $s \in S$. Because f is constant on cosets of H and distinct on different cosets, the event $f(x) = s$ identifies exactly one coset $x_0 + H$ (for some representative x_0 satisfying $f(x_0) = s$). Conditioned on this outcome, the post-measurement state of the first register becomes the uniform superposition over that coset:

$$|x_0 + H\rangle = \frac{1}{\sqrt{|H|}} \sum_{h \in H} |x_0 + h\rangle \, .$$

Thus a single oracle query and one measurement produce a *random coset state*. The representative x_0 is random and typically unimportant, while the subgroup H is common to all coset states and is what we wish to learn.

4 Step 4: Apply the QFT over G

To extract information about H, we apply the quantum Fourier transform over G, denoted QFT_G. In full generality, the Fourier transform is described using irreducible representations of G. Let $\widehat{G}$ denote a complete set of inequivalent irreducible representations (irreps) of G, and let d_ρ be the dimension of an irrep $\rho \in \widehat{G}$. One standard definition of QFT_G is the unitary map

$$|x\rangle \longmapsto \frac{1}{\sqrt{|G|}} \sum_{\rho \in \widehat{G}} \sqrt{d_\rho} \sum_{i,j=1}^{d_\rho} \rho(x)_{i,j} \, |\rho, i, j\rangle,$$

where $|\rho, i, j\rangle$ labels the irrep ρ together with matrix indices i, j. This is the non-Abelian analogue of the familiar QFT on $\mathbb{Z}_N$, where the output basis is labeled by a single frequency k.

Connection to the usual QFT on $\mathbb{Z}_N$.

If $G = \mathbb{Z}_N$ is Abelian, every irrep is one-dimensional $(d_\rho = 1)$, so each $\rho(x)$ is a complex phase, i.e., a *character*. Writing characters as $\chi_k(x) = e^{2\pi i k x/N}$, the above reduces to

$$|x\rangle \longmapsto \frac{1}{\sqrt{N}} \sum_{k=0}^{N-1} e^{2\pi i k x/N} |k\rangle,$$

which is the standard QFT on $\mathbb{Z}_N$.

✳ Why the QFT reveals information about H (Abelian case).

For finite Abelian G, the Fourier basis is labeled by characters $\chi \in \widehat{G}$. Define the set of characters that are *trivial on H* by

$$\mathrm{Ann}(H) := \{\chi \in \widehat{G} : \chi(h) = 1 \text{ for all } h \in H\}.$$

The key fact is that the Fourier transform of a coset state has support only on $\mathrm{Ann}(H)$. More precisely, for any coset state $|x_0 + H\rangle$, measuring $\mathrm{QFT}_G |x_0 + H\rangle$ yields a uniformly random character $\chi \in \mathrm{Ann}(H)$. The coset representative x_0 affects only an overall phase $\chi(x_0)$, which does not change measurement probabilities. (Many texts denote $\mathrm{Ann}(H)$ by $H^\perp$.)

✳ Non-Abelian intuition.

A non-Abelian group is one whose operation is not commutative. For non-Abelian G, irreps are matrix-valued. The analogous statement is that only those representations ρ that contain an H-invariant component can appear with nonzero probability after applying QFT_G to a coset state. This condition is captured by the averaging operator

$$\frac{1}{|H|} \sum_{h \in H} \rho(h),$$

which acts as a projector onto the subspace invariant under $\rho(h)$ for all $h \in H$.

5 Step 5 and Step 6: Measure and Classically Recover H

After applying QFT_G to the coset state, we measure in the Fourier basis. In the finite Abelian case, each measurement returns a random character $\chi \in \mathrm{Ann}(H)$. Repeating the procedure yields independent samples from $\mathrm{Ann}(H)$, and classical post-processing reconstructs H from these samples (often by reducing the problem to solving linear constraints determined by the sampled characters). For all finite Abelian groups, this reconstruction can be done efficiently.

In contrast, for many non-Abelian groups, efficiently extracting enough information from the measured representation labels (and their internal indices) to reconstruct H remains difficult, and the HSP over certain non-Abelian families is a major open problem in quantum algorithms.

■ **Example 8.7 — Shor's Algorithm Viewed as an HSP Algorithm.** Fix integers $N > 1$ and a coprime to N, and define the function

$$f(x) = a^x \bmod N.$$

Let r be the order of a modulo N, i.e., the smallest positive integer such that $a^r \equiv 1 \pmod N$. Then f is periodic with period r: for all integers x, one has $f(x+r) = f(x)$. Choose a modulus Q (typically a power of two with $N^2 \le Q < 2N^2$) and consider $G = \mathbb{Z}_Q$ with addition modulo Q. Under the idealized promise that $r \mid Q$, define the hidden subgroup

$$H = \langle r \rangle = \{0, r, 2r, \ldots, Q - r\} \le \mathbb{Z}_Q.$$

Then f is constant on cosets of H and distinct on different cosets, so order finding fits Def. 8.2.

Now match the general HSP steps with the circuit in Fig. 8.3:

1. Uniform superposition (Step 1). The Hadamard gates on the first register prepare a uniform superposition over $\mathbb{Z}_Q$,

$$\frac{1}{\sqrt{Q}} \sum_{x=0}^{Q-1} |x\rangle.$$

2. Oracle query (Step 2). The subcircuit labeled U_f coherently computes $f(x) = a^x \bmod N$, producing a state of the form $\sum_x |x\rangle |f(x)\rangle$.

3. Coset state via partial measurement (Step 3). Measuring the second register yields an outcome $\tilde{y}$ and collapses the first register to a uniform superposition over one coset $b + \langle r \rangle$.

4. Fourier transform over G (Step 4). The block labeled $\text{QFT}^\dagger$ applies the (inverse) QFT on the first register, i.e., the Fourier transform over $G = \mathbb{Z}_Q$.

5. Measurement in the Fourier basis (Step 5). Measuring the first register after $\text{QFT}^\dagger$ produces a sample whose distribution encodes the period r.

6. Classical reconstruction of H (Step 6). Classical post-processing (typically via continued fractions) recovers $H = \langle r \rangle$, equivalently the period r.

Thus Shor's order-finding routine is an Abelian HSP algorithm on $G = \mathbb{Z}_Q$, where the hidden subgroup is generated by the period r. ■

8.4.4 Abelian Hidden Subgroup Problems

A key reason the HSP is well understood in the Abelian setting is that finite Abelian groups have a simple canonical structure. In particular, every finite Abelian group G is isomorphic to a direct product of cyclic groups,

$$G \cong \mathbb{Z}_{N_1} \times \cdots \times \mathbb{Z}_{N_m}.$$

A *cyclic group* is a group generated by a single element g, meaning every element of the group can be written as kg for some integer k (additive notation), or as g^k (multiplicative notation).

Concretely, up to a relabeling of elements, every $x \in G$ can be represented as an m-tuple

$$x = (x_1, \ldots, x_m), \qquad x_j \in \mathbb{Z}_{N_j},$$

and the group operation becomes componentwise addition modulo each N_j. One can choose the decomposition so that $N_1 \mid N_2 \mid \cdots \mid N_m$.

For such a decomposition, the QFT over G factors into independent QFTs on each cyclic component:

$$\mathrm{QFT}_G \;=\; \mathrm{QFT}_{\mathbb{Z}_{N_1}} \otimes \cdots \otimes \mathrm{QFT}_{\mathbb{Z}_{N_m}}.$$

Consequently, the HSP admits an efficient quantum solution for all finite Abelian groups. Below are several representative Abelian examples.

1   Shor's Algorithm

Shor's algorithm can be phrased as an Abelian HSP over a cyclic group, as illustrated in Example 8.7. Here is a brief summary. Consider the function

$$f(x) = a^x \bmod N,$$

which has (unknown) period r, meaning $f(x) = f(x + r)$ for all x. In the idealized finite setting, take $G = \mathbb{Z}_Q$ for a chosen Q and the hidden subgroup

$$H = \langle r \rangle \leq \mathbb{Z}_Q, \qquad H = \{kr \bmod Q : k \in \mathbb{Z}\},$$

(assuming $r \mid Q$ for simplicity). Fourier sampling over $\mathbb{Z}_Q$ yields values close to integer multiples of Q/r, corresponding to rationals of the form k/r. The tensor-product structure does not appear explicitly here because order finding uses the special case $m = 1$, namely $G = \mathbb{Z}_Q$. Continued fractions then recover r classically, which in turn enables a factorization of N.

2   Bernstein–Vazirani Problem

The Bernstein–Vazirani problem fits the Abelian HSP template over

$$G = (\mathbb{Z}_2)^n, \qquad f(x) = s \cdot x \pmod 2,$$

where $s \in (\mathbb{Z}_2)^n$ is the hidden bit string and $s \cdot x$ is the binary dot product. The hidden subgroup is the hyperplane

$$H = \{h \in (\mathbb{Z}_2)^n : s \cdot h = 0\},$$

since $f(x) = f(y)$ holds exactly when $x \oplus y \in H$. Fourier sampling over $(\mathbb{Z}_2)^n$ returns s directly (up to the trivial ambiguity when $s = 0$).

3   Simon's Problem

In Simon's problem, $G = (\mathbb{Z}_2)^n$, and we are given oracle access to a function f promised to satisfy

$$f(x) = f(x \oplus L) \quad \text{for all } x,$$

for a hidden $L \neq 0$, and to be distinct on different pairs $\{x, x \oplus L\}$. The hidden subgroup is

$$H = \{0, L\}.$$

Fourier sampling produces a random $y \in (\mathbb{Z}_2)^n$ satisfying

$$y \cdot L = 0 \quad (\text{mod } 2).$$

After $O(n)$ samples, these linear constraints determine L with high probability.

4   Discrete Logarithms

The discrete logarithm problem is the following inverse problem in modular arithmetic: given a group element h known to be a power of a generator g, recover the exponent. Concretely, in the multiplicative group $\mathbb{Z}_p^{\times}$ (with p prime), given g and h, find an integer k such that $h \equiv g^k \pmod{p}$. This problem underlies classic public-key schemes such as Diffie–Hellman and ElGamal, much as integer factoring underlies RSA. In this sense, discrete logarithms and factoring are parallel "hard problems" in classical cryptography, and both admit efficient quantum algorithms via the Abelian HSP framework.

The discrete logarithm problem can be phrased as an Abelian HSP as follows. Let p be prime and g a generator of the multiplicative group $\mathbb{Z}_p^{\times}$. Given $h = g^k \bmod p$, define

$$G = \mathbb{Z}_n \times \mathbb{Z}_n, \qquad f(x, y) = g^x h^y \bmod p,$$

where $n = p - 1$. Then

$$f(x, y) = f(x', y') \quad \Longleftrightarrow \quad (x - x', y - y') \in H,$$

where the hidden subgroup is the one-dimensional subgroup

$$H = \langle (k, -1) \rangle = \{(tk, -t) : t \in \mathbb{Z}_n\} \leq \mathbb{Z}_n \times \mathbb{Z}_n.$$

Recovering H therefore recovers k, the discrete logarithm.

8.4.5 Non-Abelian Hidden Subgroup Problems

The Abelian HSP succeeds because Fourier sampling produces outcomes (frequencies) that can be interpreted as linear constraints on H, and these constraints can be combined efficiently by classical post-processing.

When G is non-Abelian, the situation changes in two ways:

1. The Fourier transform becomes representation-valued. For Abelian groups, all irreps are one-dimensional, so the Fourier basis is labeled by characters. For non-Abelian groups, irreps can have dimension $d_\rho > 1$, so the QFT basis is labeled by a representation name ρ together with internal indices.

2. Measurement outcomes are richer, but harder to interpret. After creating a coset state and applying QFT_G, a measurement typically returns an irrep label ρ and (depending on the chosen measurement basis) additional data within the d_ρ-dimensional representation space. Although many non-Abelian QFTs can be implemented in polynomial time for important families of groups, efficiently extracting H from these outcomes is often the true bottleneck.

In short, in the non-Abelian setting, implementing QFT_G is frequently not the main difficulty; rather, the challenge is to design measurements and classical post-processing that recover H from the distribution over irreps and their internal degrees of freedom.

1 The Dihedral Group and Approximate SVP

The dihedral group

$$D_N = \langle r, s : r^N = s^2 = 1,\ srs = r^{-1} \rangle$$

is the symmetry group of a regular N-gon. Its elements can be written as r^k (rotations) and sr^k (reflections), where $k \in \{0, 1, \ldots, N-1\}$.

A common HSP instance on D_N hides a reflection subgroup of size 2,

$$H = \{1, sr^t\},$$

for an unknown shift t. While Fourier sampling over D_N is straightforward to implement, turning the samples into the shift t (and hence into generators for H) has resisted a polynomial-time solution. This case is central because an efficient quantum algorithm for the dihedral HSP would imply strong algorithms for certain lattice problems, including versions of the approximate Shortest Vector Problem (SVP). Such a breakthrough would undermine many lattice-based post-quantum cryptographic schemes.

2 The Symmetric Group and Graph Isomorphism

Let $G = S_n$, the group of all permutations of n labeled vertices. A permutation $\pi \in S_n$ acts on a labeled graph by relabeling its vertices. If G_1 is a graph on $\{1, 2, \ldots, n\}$, write $\pi(G_1)$ for the relabeled graph.

Define a function

$$f(\pi) = \text{the adjacency matrix of } \pi(G_1).$$

Then

$$f(\pi) = f(\sigma) \quad \Longleftrightarrow \quad \pi(G_1) = \sigma(G_1) \quad \Longleftrightarrow \quad \pi^{-1}\sigma \in \text{Aut}(G_1),$$

where

$$\text{Aut}(G_1) = \{\gamma \in S_n : \gamma(G_1) = G_1\}$$

is the automorphism group of G_1. Thus f is constant on the left cosets of $\text{Aut}(G_1)$ and distinct on different cosets, so this is an HSP instance with hidden subgroup $\text{Aut}(G_1)$.

To connect with graph isomorphism, suppose we are given two graphs G_1 and G_2. If they are isomorphic, there exists a permutation τ such that $G_2 = \tau(G_1)$. In that case, the set of all isomorphisms from G_1 to G_2 is exactly the coset

$$\tau\,\text{Aut}(G_1).$$

so finding an isomorphism amounts to finding a particular coset element.

The non-Abelian HSP over S_n is therefore closely linked to graph isomorphism. Although the QFT over S_n can be described and (in principle) implemented, designing measurements and efficient classical post-processing that recover $\text{Aut}(G_1)$ (or an isomorphism coset) from Fourier samples remains an open problem.

Chapter Takeaways

1. Shor-style algorithms reduce number-theoretic problems (factoring, discrete logarithm, related tasks) to *period/order finding*, where the period is extracted from Fourier/phase information.

2. The quantum core prepares a superposition over inputs, queries an oracle that hides periodic structure, and applies a Fourier-type transform so that measurement yields information about the hidden period.

3. Continued fractions (and simple verification tests) convert the measured rational approximation into the true period/order with high probability.

4. The unifying abstraction is the Hidden Subgroup Problem (HSP): many exponential quantum speedups arise from identifying a hidden subgroup $H \leq G$ from an H-periodic function.

5. Abelian HSP admits a standard efficient solution strategy via Fourier sampling, while the non-Abelian case is substantially harder and remains a major barrier for broader generalizations.

Problem Set 8

8.1 Estimate the circuit width and depth needed to use Shor's algorithm to factor a 512, 1024, and 2048 bit number. Clearly state which cost model you are using (e.g., logical Toffoli count, T-count, or depth in a particular gate set), and what assumptions you make about modular multiplication.

8.2 Factor 55 on a QC platform or simulator, using Shor's algorithm. Report (i) the value(s) of a tried, (ii) the period estimate(s) recovered from the measurement data, and (iii) the final factors. (If you use a compiled/demo implementation, explain what has been "compiled away" compared with the textbook circuit.)

8.3 Let $N = 35$ and choose $a = 6$ (note $\gcd(6, 35) = 1$).

    (a) Find the order r of 6 modulo 35.

    (b) Decide whether the period-to-factors step succeeds. If it fails, explain exactly which condition fails.

    (c) Try a different a (still coprime to 35) for which the step succeeds.

8.4 Let $N = 55$, $a = 2$, and $n = \lceil \log_2 N \rceil = 6$. Compare the two choices $t = n$ and $t = 2n$ for the number of control qubits, with $Q = 2^t$.

    (a) For $t = n$, search for an outcome $y \in \{0, 1, \ldots, Q - 1\}$ such that $\frac{y}{Q}$ is a good approximation to some $\frac{k}{r}$ with $r \leq N$, but the continued-fraction reconstruction (restricted to denominators $< N$) is ambiguous or returns an incorrect candidate first.

    (b) Repeat for $t = 2n$ and explain why the ambiguity is much less likely.

8.5 Let N be odd and composite, and let a satisfy $\gcd(a, N) = 1$. Suppose the order of a modulo N is r.

(a) Prove that if r is even and $a^{r/2} \not\equiv -1 \pmod{N}$, then at least one of $\gcd(a^{r/2} - 1, N)$ or $\gcd(a^{r/2} + 1, N)$ is a nontrivial factor of N.

(b) Give a concrete example (small N and a) where r is odd and the method fails.

(c) Give a concrete example where r is even but $a^{r/2} \equiv -1 \pmod{N}$, and the method fails.

8.6 Let $U\,|j\rangle = |aj \bmod N\rangle$ for $0 \le j \le N - 1$ with $\gcd(a, N) = 1$.

(a) Show that U is a permutation matrix in the computational basis.

(b) Conclude that U is unitary.

(c) For $N = 15$ and $a = 2$, write the cycle decomposition of this permutation.

8.7 Assume $U\,|u_k\rangle = e^{2\pi i k/r}\,|u_k\rangle$ and that the control register has dimension Q.

(a) Starting from $\frac{1}{\sqrt{Q}} \sum_{x=0}^{Q-1} |x\rangle\,|u_k\rangle$, show that controlled-$U^x$ produces

$$\frac{1}{\sqrt{Q}} \sum_{x=0}^{Q-1} e^{2\pi i x k/r}\,|x\rangle\,|u_k\rangle.$$

(b) Explain why applying QFT† to the control register concentrates probability near integers close to Qk/r.

8.8 (Filling the Gap: $r \nmid Q$) In Shor's algorithm one typically chooses Q as a power of two with $N^2 \le Q < 2N^2$, so r need not divide Q.

(a) Explain why the "ideal" HSP promise (distinct values on distinct cosets) may fail when working modulo Q rather than over an infinite domain.

(b) Nevertheless, the QFT/QPE measurement still yields an approximation to k/r. Give an intuitive explanation (one paragraph) of why the peaks remain informative.

8.9 ✳ (A Toy Discrete Logarithm Instance) Let $p = 29$, $g = 2$, and $h = 18$ in $\mathbb{Z}_{29}^{\times}$.

(a) Find the discrete logarithm k such that $h \equiv g^k \pmod{29}$.

(b) Let $n = p - 1 = 28$ and define $f(x, y) = g^x h^y \bmod p$ over $G = \mathbb{Z}_n \times \mathbb{Z}_n$. Show that

$$f(x, y) = f(x', y') \quad \Longleftrightarrow \quad (x - x', y - y') \in \langle (k, -1) \rangle.$$

(c) Explain how Fourier sampling over G could, in principle, recover k.

8.10 ✳ (A Toy Non-Abelian Symmetry Problem) Let D_8 be the dihedral group of the square. Consider the hidden subgroup

$$H = \{e, sr^t\}$$

for an unknown $t \in \{0, 1, 2, 3\}$ (using the standard presentation $D_8 = \langle r, s : r^4 = s^2 = e, \ srs = r^{-1} \rangle$).

(a) List the four possible hidden subgroups H (one for each t).

(b) Explain (at a high level) how a coset state of H could be produced with one oracle query.

(c) Compute the left cosets of H explicitly for one chosen t, and show that they partition D_8.

8.11 $*$ (Fourier Transform of a Coset State) Let $G = \mathbb{Z}_N$ (additive notation), and assume r is a positive divisor of N. Define the subgroup

$$H = \langle r \rangle = \{0, r, 2r, \ldots, N - r\} \le \mathbb{Z}_N,$$

so that $|H| = N/r$. For a coset representative $x \in \mathbb{Z}_N$, define the (normalized) coset state

$$|x + H\rangle = \frac{1}{\sqrt{|H|}} \sum_{h \in H} |x + h\rangle .$$

Show that the quantum Fourier transform on $\mathbb{Z}_N$ satisfies

$$\mathrm{QFT}_{\mathbb{Z}_N} |x + H\rangle = \sqrt{\frac{r}{N}} \sum_{k=0}^{r-1} e^{2\pi i k x / r} |k \cdot (N/r)\rangle .$$

Conclude that the Fourier-transformed state has support only on basis states $|j\rangle$ where j is an integer multiple of N/r.

Hint: Evaluate the geometric series $\sum_{t=0}^{|H|-1} e^{2\pi i j (x+tr)/N}$ and determine when it vanishes.

8.12 $*$ (Bernstein–Vazirani as HSP) Let $G = (\mathbb{Z}_2)^n$ with group operation $\oplus$ (bitwise XOR). Fix a secret $s \in (\mathbb{Z}_2)^n$ and define

$$f(x) = s \cdot x \pmod 2,$$

where $s \cdot x = \sum_{i=1}^n s_i x_i \pmod 2$ is the binary dot product. Define the hyperplane

$$H = \{h \in (\mathbb{Z}_2)^n : s \cdot h = 0\}.$$

(a) Prove the HSP promise explicitly by showing

$$f(x) = f(y) \quad \Longleftrightarrow \quad x \oplus y \in H.$$

Equivalently, prove that f is constant on each coset $x \oplus H$ and takes distinct values on distinct cosets.

(b) Show that the orthogonal subgroup (annihilator)

$$H^\perp := \{z \in (\mathbb{Z}_2)^n : z \cdot h = 0 \text{ for all } h \in H\}$$

is exactly $H^\perp = \{0, s\}$.

Hint: For (b), use the fact that $\dim(H) = n - 1$ when $s \ne 0$, hence $\dim(H^\perp) = 1$.

8.13 $*$ (Simon as HSP, Explicitly) Show that Simon's problem is an instance of the HSP over $G = (\mathbb{Z}_2)^n$ with hidden subgroup $H = \{0, L\}$. Then explain why repeated Fourier sampling yields linear constraints of the form $y \cdot L = 0 \pmod 2$.

9. Amplitude Amplification and Estimation

Contents

Many quantum algorithms are *probabilistic*: after running a circuit and measuring, we may or may not obtain a desired "success" outcome. We model success by a projector Π onto a "good" subspace. If the final state has amplitude $\sqrt{p}$ in the good subspace, then measurement succeeds with probability p. The techniques in this chapter provide coherent control over this success probability. Rather than repeating the circuit many times, we use interference to amplify the good component or to estimate p accurately, often with far fewer applications of the underlying circuit.

Historically, the core rotation mechanism underlying these methods was popularized by Grover's 1996 search algorithm [26], and was soon abstracted into the general framework of amplitude amplification and amplitude estimation by Brassard *et al.* [27]. Amplitude estimation includes quantum counting as an influential early specialization to Grover's setting [28]. We also cover two widely used variants developed to address practical algorithm-design needs: fixed-point amplification, which avoids over-rotation when p is not known precisely [29], and oblivious amplitude amplification, which applies amplification in ancilla-postselection settings and plays a central role in linear-combination-of-unitaries and block-encoding constructions [30, 31].

Projectors and Reflections: a Refresher

An orthogonal projector onto a subspace $S \subset \mathcal{H}$ is a linear operator $\Pi : \mathcal{H} \to \mathcal{H}$ satisfying

$$\Pi^2 = \Pi, \qquad \Pi^\dagger = \Pi.$$

Equivalently, if $\{|s_k\rangle\}$ is an orthonormal basis of S, then

$$\Pi = \sum_k |s_k\rangle\langle s_k|.$$

An orthogonal projector has eigenvalues 0 and 1, corresponding to the orthogonal decomposition

$$\mathcal{H} = S \oplus S^\perp, \qquad \Pi|\psi\rangle = \begin{cases} |\psi\rangle, & |\psi\rangle \in S, \\ 0, & |\psi\rangle \in S^\perp. \end{cases}$$

Given an orthogonal projector Π, the associated reflection operator is defined as

$$R = 2\Pi - I.$$

This operator is Hermitian and unitary ($R^\dagger = R$ and $R^2 = I$). It reflects states about the subspace S and has eigenvalues ± 1:

$$R|\psi\rangle = \begin{cases} |\psi\rangle, & |\psi\rangle \in S, \\ -|\psi\rangle, & |\psi\rangle \in S^\perp. \end{cases}$$

A particularly important special case is the rank-one projector

$$\Pi_\psi = |\psi\rangle\langle\psi|,$$

which yields the reflection about the vector $|\psi\rangle$

$$R_\psi = 2|\psi\rangle\langle\psi| - I.$$

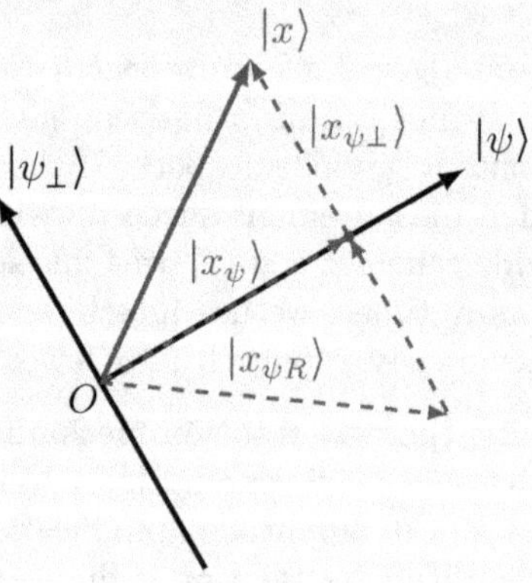

$$|x_\psi\rangle = \Pi_\psi |x\rangle, \qquad |x_{\psi\perp}\rangle = (I - \Pi_\psi)|x\rangle, \qquad |x_{\psi R}\rangle = R_\psi |x\rangle$$

Products of reflections generate rotations. In particular, if a two-dimensional subspace $\mathcal{K} \subset \mathcal{H}$ is invariant under two reflections $R_1 = 2\Pi_1 - I$ and $R_2 = 2\Pi_2 - I$, then $R_2 R_1$ maps $\mathcal{K}$ to itself and acts as a 2×2 unitary on $\mathcal{K}$; in a real plane, this action is a rotation.

For more details, see QC Math [1], Section 11.2: *Vector-Valued Functions*.

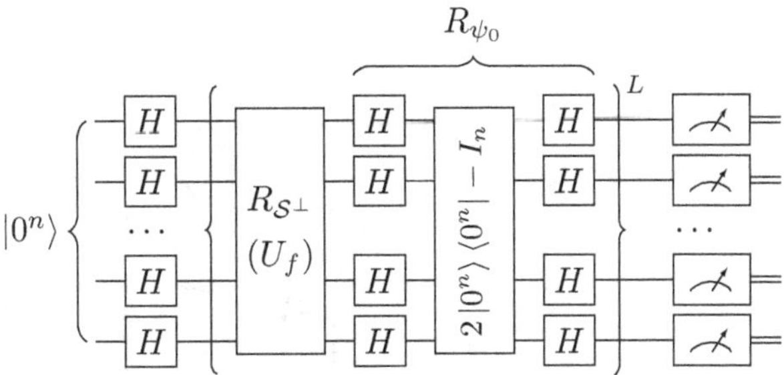

Figure 9.1: Grover-Search Iterate Circuit

9.1 Amplitude Amplification

Amplitude amplification is a general quantum procedure for boosting the success probability of a probabilistic algorithm by coherently increasing the amplitude on its "good" outcomes. The core mechanism is a two-reflection rotation in a two-dimensional invariant subspace. It is best understood by first examining a concrete setting: Grover's unstructured search problem.

9.1.1 Grover Search as a Motivating Example

Consider the unstructured search problem, where we are given oracle access to a Boolean function $f : \{0,1\}^n \to \{0,1\}$ that marks a subset of M "good" items among $N = 2^n$ possibilities, with $N \gg M$. The task is to find an x such that $f(x) = 1$ using as few oracle queries as possible. In the classical setting, this requires $\Theta(N/M)$ oracle queries to find a marked item with constant probability, whereas Grover's algorithm reduces this number to $\Theta(\sqrt{N/M})$ queries.

1 Good Subspace and Two-Dimensional Reduction

Define the *solution (good) subspace* $\mathcal{S} = \mathrm{span}\{|x\rangle : f(x) = 1\}$ and its orthogonal complement $\mathcal{S}^\perp$. Let $\Pi_{\mathcal{S}}$ be the orthogonal projector onto $\mathcal{S}$, namely $\Pi_{\mathcal{S}} = \sum_{f(x)=1} |x\rangle\langle x|$. Starting from $|0^n\rangle$, apply $H^{\otimes n}$ to prepare the uniform superposition (see Fig. 9.1)

$$|\psi_0\rangle = H^{\otimes n} |0^n\rangle = \frac{1}{\sqrt{N}} \sum_{x\in\{0,1\}^n} |x\rangle .$$

Introduce the normalized "uniform" directions in the good and bad subspaces

$$|\psi_{\mathcal{S}}\rangle = \frac{1}{\sqrt{M}} \sum_{f(x)=1} |x\rangle , \qquad |\psi_{\mathcal{S}^\perp}\rangle = \frac{1}{\sqrt{N-M}} \sum_{f(x)=0} |x\rangle .$$

Then $|\psi_0\rangle$ lies in the two-dimensional subspace $\mathcal{K} = \mathrm{span}\{|\psi_{\mathcal{S}}\rangle, |\psi_{\mathcal{S}^\perp}\rangle\}$ and can be written as

$$|\psi_0\rangle = \sin\theta_0 |\psi_{\mathcal{S}}\rangle + \cos\theta_0 |\psi_{\mathcal{S}^\perp}\rangle , \qquad \sin\theta_0 = \sqrt{\frac{M}{N}}, \quad \cos\theta_0 = \sqrt{\frac{N-M}{N}}.$$

Although $\Pi_{\mathcal{S}}$ is generally high-rank (when $M > 1$), its action on the particular plane $\mathcal{K}$ is effectively rank-1: within $\mathcal{K}$, $\Pi_{\mathcal{S}}$ acts as $|\psi_{\mathcal{S}}\rangle\langle\psi_{\mathcal{S}}|$ because $\mathcal{K} \cap \mathcal{S} = \mathrm{span}\{|\psi_{\mathcal{S}}\rangle\}$.

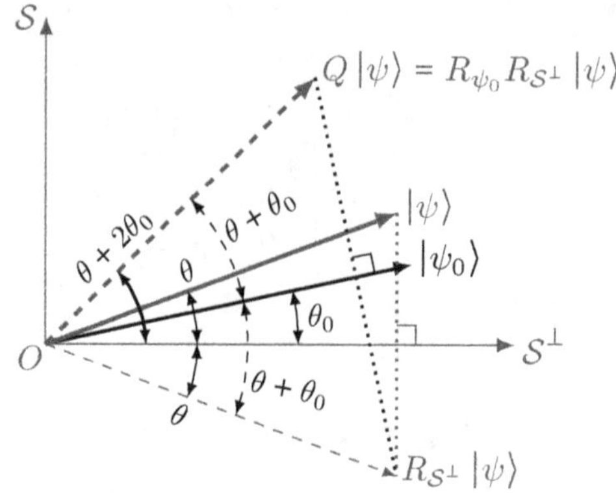

Figure 9.2: Two-Reflection Rotation Geometry in Grover's Algorithm

2　Oracle Reflection

The phase oracle is the unitary

$$U_f\,|x\rangle = (-1)^{f(x)}\,|x\rangle = \begin{cases} -\,|x\rangle, & \text{if } f(x) = 1 \text{ (marked, good)}, \\ |x\rangle, & \text{if } f(x) = 0. \end{cases}$$

Thus, U_f flips the sign of the component in $\mathcal{S}$ while leaving $\mathcal{S}^\perp$ unchanged. Geometrically, it is a reflection about the axis $\mathcal{S}^\perp$, so it is natural to view U_f as $R_{\mathcal{S}^\perp}$. Equivalently,

$$U_f = I - 2\Pi_{\mathcal{S}} = R_{\mathcal{S}^\perp}.$$

3　Diffusion Reflection

The diffusion step is the reflection about the initial direction $|\psi_0\rangle$:

$$R_{\psi_0} = 2\,|\psi_0\rangle\langle\psi_0| - I.$$

A standard circuit identity (used in Fig. 9.1) is

$$R_{\psi_0} = H^{\otimes n}\left(2\,|0^n\rangle\langle 0^n| - I\right)H^{\otimes n}.$$

Here the middle operation $2\,|0^n\rangle\langle 0^n| - I$ applies a phase flip to all computational basis states except $|0^n\rangle$.

4　One Grover Step as a Rotation

Define one Grover iterate by

$$Q = R_{\psi_0} U_f.$$

Both U_f and R_{ψ_0} preserve the plane $\mathcal{K} = \text{span}\{|\psi_{\mathcal{S}}\rangle, |\psi_{\mathcal{S}^\perp}\rangle\}$, so Q acts within $\mathcal{K}$. In this plane, U_f is a reflection about the $\mathcal{S}^\perp$ axis, and R_{ψ_0} is a reflection about the axis spanned by $|\psi_0\rangle$. The composition of two reflections is a rotation by twice the angle between the reflection axes, hence Q acts as a rotation by $2\theta_0$ toward $\mathcal{S}$, as illustrated in Fig. 9.2. Concretely, if

$$|\psi(\theta)\rangle = \sin\theta\,|\psi_{\mathcal{S}}\rangle + \cos\theta\,|\psi_{\mathcal{S}^\perp}\rangle,$$

then

$$Q\,|\psi(\theta)\rangle = |\psi(\theta + 2\theta_0)\rangle.$$

5   After L Iterations

Starting from $|\psi_0\rangle = |\psi(\theta_0)\rangle$, after L Grover iterations we obtain

$$|\psi_L\rangle = Q^L |\psi_0\rangle = \sin\big((2L+1)\theta_0\big) |\psi_S\rangle + \cos\big((2L+1)\theta_0\big) |\psi_{S\perp}\rangle. \tag{9.1}$$

Therefore, the success probability (measuring a marked item) is

$$p_L = \|\Pi_S |\psi_L\rangle\|^2 = \sin^2\big((2L+1)\theta_0\big). \tag{9.2}$$

This probability increases as the state approaches the S axis, but it decreases again if we iterate too long and overshoot the optimum.

6   Choosing the Iteration Count

To maximize p_L, choose $(2L+1)\theta_0 \approx \pi/2$, which yields

$$L \approx \left\lfloor \frac{\pi}{4\theta_0} - \frac{1}{2} \right\rfloor. \tag{9.3}$$

When $M \ll N$, we have $\theta_0 \approx \sqrt{M/N}$, so $L = \Theta(\sqrt{N/M})$. In particular, for a unique marked item ($M = 1$), Grover search uses $\Theta(\sqrt{N})$ oracle calls, giving a quadratic speedup over classical exhaustive search.

Exercise 9.1 Work in the two-dimensional subspace $\mathcal{K} = \mathrm{span}\{|\psi_S\rangle, |\psi_{S\perp}\rangle\}$ for Grover search.

(a) Show that U_f acts as -1 on $|\psi_S\rangle$ and as $+1$ on $|\psi_{S\perp}\rangle$.

(b) Conclude that the restriction of U_f to $\mathcal{K}$ is $\mathrm{diag}(-1, 1)$ in the ordered basis $\{|\psi_S\rangle, |\psi_{S\perp}\rangle\}$.

Exercise 9.2 Let $\theta_0 \in (0, \pi/2)$ and define $p_L = \sin^2((2L+1)\theta_0)$.

(a) Show that the maximum of p_L over integers $L \geq 0$ is achieved by choosing L so that $(2L+1)\theta_0$ is closest to $\pi/2$.

(b) Prove the bound $p_L \geq \cos^2(\theta_0)$ for this nearest-choice rule.

Key Takeaways

1. Grover search evolves in a two-dimensional plane $\mathcal{K} = \mathrm{span}\{|\psi_S\rangle, |\psi_{S\perp}\rangle\}$, even though the full space may be large.

2. The Grover iterate $Q = R_{\psi_0} R_{S\perp}$ is the product of two reflections, so it acts as a rotation on $\mathcal{K}$.

3. After L steps, the success probability is $p_L = \sin^2((2L+1)\theta_0)$, where $p_0 = \sin^2\theta_0 = M/N$.

4. Choosing $L \approx \left\lfloor \frac{\pi}{4\theta_0} - \frac{1}{2} \right\rfloor$ uses $\Theta(\sqrt{N/M})$ oracle calls and makes the success probability close to 1 when $N \gg M$.

9.1.2 From Grover Search to Amplitude Amplification

Grover's search is governed by two reflections: a reflection that marks the good subspace and a reflection about the prepared initial state. Their product acts as a rotation in a two-dimensional invariant subspace, steadily increasing the weight on good outcomes until over-rotation occurs. Amplitude amplification abstracts this same mechanism into a general-purpose primitive: given any state-preparation routine with nonzero success probability, it coherently boosts that success probability using the same two-reflection rotation.

1   Registers and Notation

In many applications, the algorithm operates on two registers: an ancilla register of n qubits and a system register of n qubits. The total Hilbert space is

$$\mathcal{H} = \mathcal{H}_{\mathrm{anc}} \otimes \mathcal{H}_{\mathrm{sys}}, \qquad \dim(\mathcal{H}_{\mathrm{anc}}) = 2^n, \quad \dim(\mathcal{H}_{\mathrm{sys}}) = 2^n.$$

We typically start from the all-zero state on both registers. For compactness, throughout this subsection we write

$$|0\rangle \equiv |0^n\rangle |0^n\rangle, \qquad I \equiv I_{2n} = I_{\mathrm{anc}} \otimes I_{\mathrm{sys}},$$

with the understanding that all operators act on the full $(2n)$-qubit register unless stated otherwise.

2   Good and Bad Subspaces

In Grover search, $\mathcal{S}$ is the span of marked computational basis states. In amplitude amplification, $\mathcal{S} \subseteq \mathcal{H}$ is still the *good subspace*, but it may be defined by an arbitrary success criterion (for example, a designated flag qubit being $|1\rangle$, or a verifier accepting). We write $\mathcal{S}^\perp$ for the orthogonal complement (the *bad subspace*), and we let $\Pi_{\mathcal{S}}$ denote the orthogonal projector onto $\mathcal{S}$.

Given any state $|\psi\rangle \in \mathcal{H}$, its success probability is

$$p(\psi) = \|\Pi_{\mathcal{S}} |\psi\rangle\|^2 = \langle\psi| \Pi_{\mathcal{S}} |\psi\rangle.$$

In particular, for the prepared initial state $|\psi_0\rangle$ defined below, we write

$$p_0 = \langle\psi_0| \Pi_{\mathcal{S}} |\psi_0\rangle = \sin^2 \theta_0,$$

which defines an angle $\theta_0 \in [0, \pi/2]$ consistent with the geometry in Fig. 9.2.

3   General State Preparation

Let A be a state-preparation unitary acting on the full register (system and any ancillas), and define

$$|\psi_0\rangle = A |0\rangle.$$

In Grover search, $A = H^{\otimes n}$ and there is no separate ancilla register, but in general A may include ancillas and may prepare an entangled state across the two registers. The goal of amplitude amplification is to transform $|\psi_0\rangle$ into a state with substantially larger overlap with $\mathcal{S}$ using as few calls to A, $A^\dagger$, and the marking operation as possible.

To make the two-dimensional rotation explicit, define the normalized projections of $|\psi_0\rangle$ onto $\mathcal{S}$ and $\mathcal{S}^\perp$:

$$|\psi_{\mathcal{S}}\rangle = \frac{\Pi_{\mathcal{S}} |\psi_0\rangle}{\|\Pi_{\mathcal{S}} |\psi_0\rangle\|}, \qquad |\psi_{\mathcal{S}^\perp}\rangle = \frac{(I - \Pi_{\mathcal{S}}) |\psi_0\rangle}{\|(I - \Pi_{\mathcal{S}}) |\psi_0\rangle\|}.$$

Then

$$|\psi_0\rangle = \sin\theta_0\,|\psi_S\rangle + \cos\theta_0\,|\psi_{S\perp}\rangle, \qquad \sin^2\theta_0 = p_0.$$

Exercise 9.3 Consider the vectors $|\psi_S\rangle$ and $|\psi_{S\perp}\rangle$ in amplitude amplification.

(a) Identify precisely when each of these normalized vectors is *not* defined.

(b) Interpret those exceptional cases in terms of the success probability p_0 and explain what amplitude amplification should do in each case.

4 The Two Reflections

Amplitude amplification uses the same two-reflection structure as Grover search.

Marking Reflection

The success oracle (or marking operation) is the phase flip on the good subspace,

$$R_{S\perp} \equiv U_f = I - 2\Pi_S.$$

This operator acts as $-I$ on S and as $+I$ on $S^\perp$. Geometrically, it is a reflection about the axis $S^\perp$.

State Reflection

The second reflection is about the prepared initial state:

$$R_{\psi_0} = 2\,|\psi_0\rangle\langle\psi_0| - I.$$

In circuits, R_{ψ_0} is implemented by conjugating the $|0\rangle$-reflection with A:

$$R_{\psi_0} = A\,(2\,|0\rangle\langle0| - I)\,A^\dagger.$$

This is the structure indicated by the brace labeled R_{ψ_0} in the circuit of Fig. 9.3.

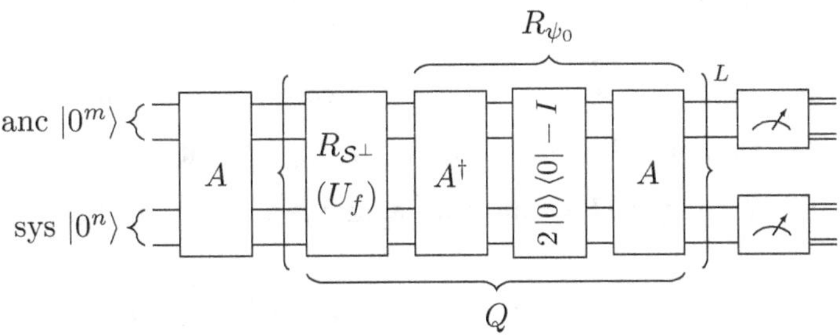

Figure 9.3: Circuit for Quantum Amplitude Amplification

5 Two-Dimensional Invariance

Although S and $S^\perp$ can be high-dimensional, both $R_{S\perp}$ and R_{ψ_0} preserve the two-dimensional subspace

$$\mathcal{K} = \operatorname{span}\{|\psi_S\rangle, |\psi_{S\perp}\rangle\}.$$

Indeed, $R_{S\perp}$ acts as a sign flip on $|\psi_S\rangle$ and fixes $|\psi_{S\perp}\rangle$, while R_{ψ_0} is a reflection about the axis spanned by $|\psi_0\rangle \in \mathcal{K}$. Therefore, the amplification dynamics reduce to planar rotations in $\mathcal{K}$, exactly as depicted in Fig. 9.2.

6 Amplification Iterate and Its Action

Define the amplitude amplification iterate by the same two-reflection product used in Grover search:

$$Q = R_{\psi_0} R_{\mathcal{S}\perp}. \tag{9.4}$$

As illustrated in Fig. 9.2, the composition of these reflections acts as a rotation by $2\theta_0$ within the invariant plane $\mathcal{K}$.

Therefore, all of the rotation formulas derived for Grover's algorithm carry over directly: the expression for the state after L iterations $|\psi_L\rangle$ (Eq. 9.1), the corresponding success probability p_L (Eq. 9.2), and the near-optimal choice of L (Eq. 9.3) remain valid without change, with θ_0 determined by the initial overlap of $|\psi_0\rangle$ with $\mathcal{S}$.

In the literature, the iterate Q in Eq. 9.4 is sometimes called a *walk operator*. Repeated applications Q^L generate a discrete-time unitary dynamics (a sequence of "steps") on an invariant subspace, analogous to a quantum walk.

7 Circuit Implementation

A circuit implementation of amplitude amplification is shown in Fig. 9.3. It generalizes the Grover iterate in Fig. 9.1 by replacing the Hadamard state preparation with an arbitrary unitary A.

The amplitude amplification procedure, following the blocks in Fig. 9.3, is as follows.

1. Initialize the joint register in $|0\rangle$ and apply A to prepare $|\psi_0\rangle$.

2. Repeat L times: apply the marking reflection $R_{\mathcal{S}\perp}$, then apply the state reflection R_{ψ_0} (equivalently apply $Q = R_{\psi_0} R_{\mathcal{S}\perp}$).

3. Measure; with high probability the measurement outcome is "good" (as specified by $\Pi_{\mathcal{S}}$).

Exercise 9.4 Let $|\psi_0\rangle = \sin\theta_0 \, |\psi_{\mathcal{S}}\rangle + \cos\theta_0 \, |\psi_{\mathcal{S}\perp}\rangle$ with $|\psi_{\mathcal{S}}\rangle \in \mathcal{S}$ and $|\psi_{\mathcal{S}\perp}\rangle \in \mathcal{S}^\perp$.

(a) Compute $R_{\psi_0} |\psi_{\mathcal{S}}\rangle$ and $R_{\psi_0} |\psi_{\mathcal{S}\perp}\rangle$ explicitly.

(b) Verify that R_{ψ_0} preserves $\mathcal{K}$ and is a reflection (i.e., Hermitian and unitary) on $\mathcal{K}$.

(c) Show that $R_{\mathcal{S}\perp}$ leaves $|\psi_{\mathcal{S}\perp}\rangle$ fixed and flips the sign of $|\psi_{\mathcal{S}}\rangle$.

(d) Conclude that $Q = R_{\psi_0} R_{\mathcal{S}\perp}$ preserves $\mathcal{K}$.

Key Takeaways

1. Amplitude amplification applies to any state preparation A with initial success probability $p_0 = \|\Pi_{\mathcal{S}} A |0\rangle\|^2$.

2. The same two reflections are used:

$$R_{\mathcal{S}\perp} = I - 2\Pi_{\mathcal{S}}, \qquad R_{\psi_0} = 2|\psi_0\rangle\langle\psi_0| - I, \quad |\psi_0\rangle = A|0\rangle.$$

3. The iterate $Q = R_{\psi_0} R_{\mathcal{S}^\perp}$ preserves the plane spanned by the good and bad components of $|\psi_0\rangle$ and rotates by $2\theta_0$, where $p_0 = \sin^2 \theta_0$.

4. It boosts the success probability from p_0 to near 1 using $\Theta(1/\sqrt{p_0})$ applications of the marking operation (query-optimal in the black-box model).

9.1.3 Optimality and Limitations of the Quadratic Speedup

1 Lower Bound and Optimality in the Black-Box Model

In the black-box (oracle) model, the input is not given explicitly; instead, the only way to learn which items are marked is to query an oracle such as U_f (or equivalently f). The cost of an algorithm is measured primarily by the number of oracle queries it makes, while the remaining computation is treated as negligible in comparison.

Within this model, one can prove a lower bound: any quantum algorithm that finds a marked item with constant success probability must use $\Omega(\sqrt{N/M})$ oracle queries in the worst case [32, 33]. In particular, there is no algorithm with $o(\sqrt{N/M})$ queries that succeeds uniformly over all oracles with M marked items. Grover's algorithm matches this lower bound up to constant factors, so it is query-optimal. (See § 1.2.2 for asymptotic notations such as $\Omega(\cdot)$ and $o(\cdot)$.)

This also clarifies the limit of "black-box boosting": viewing amplitude amplification as a generic procedure that raises a success probability p_0 to a constant using $\Theta(1/\sqrt{p_0})$ coherent iterations, the same lower-bound perspective indicates that the $\Theta(1/\sqrt{p_0})$ scaling cannot be improved in the black-box setting without additional assumptions.

2 Beyond the Black-Box Model

It is equally important to understand what this optimality does not say. The lower bound is specific to the black-box model, where the oracle hides all structure and the algorithm is not allowed any shortcut other than querying it. Many quantum speedups arise precisely because the input is not an arbitrary black box: the problem instance has algebraic or analytic structure that can be exploited. For example, period finding (and hence factoring and discrete logarithms) leverages structured function evaluation together with the quantum Fourier transform to reveal global periodic information, which is fundamentally different from unstructured search.

3 Implications for BQP and NP

Finally, Grover optimality should not be over-interpreted as evidence about NP vs. BQP. The statement "black-box search cannot be faster than quadratic" does not imply that NP-complete problems are outside BQP. What it does imply is narrower: if one treats an NP-complete problem instance as an unstructured black box, so that the only way to test candidate solutions is by querying a verifier oracle, then quantum computation cannot improve over classical exhaustive search by more than a quadratic factor. The actual relationship between BQP and NP remains an open problem in complexity theory, and it is common (though unproven) to believe that $\mathrm{NP} \not\subseteq \mathrm{BQP}$.

9.2 Amplitude Estimation and Quantum Counting

Amplitude estimation addresses a basic task that appears throughout quantum algorithms: given a quantum procedure that succeeds with some (unknown) probability p, estimate p more efficiently than by repeated sampling. This capability is useful on its own (e.g., estimating expectation values), and it also supports amplitude amplification by providing the information needed to choose an iteration count without over-rotation.

9.2.1 From Amplitude Amplification to Amplitude Estimation

1   High-Level View

Amplitude estimation is obtained by running quantum phase estimation (QPE) on the amplitude-amplification iterate, as illustrated in Fig. 9.4. Amplitude amplification converts the unknown success probability p into a rotation angle θ of a fixed unitary Q, and QPE can learn that angle with controlled applications of Q.

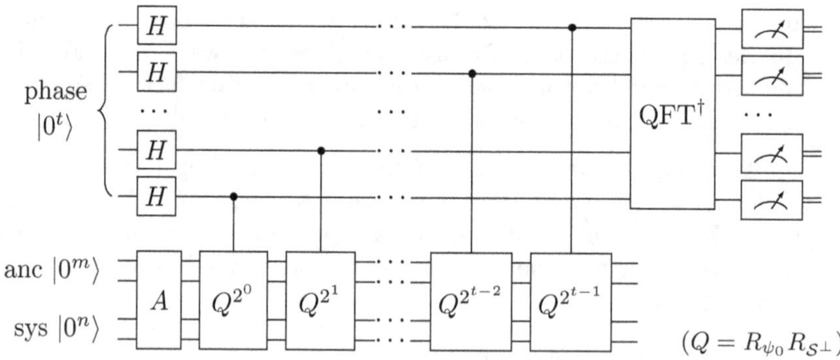

Figure 9.4: Quantum Circuit for Amplitude Estimation

Concretely, suppose a unitary A prepares

$$A\,|0\rangle = |\psi_0\rangle = \sin\theta\,|\psi_\mathcal{S}\rangle + \cos\theta\,|\psi_{\mathcal{S}\perp}\rangle,$$

where $|\psi_\mathcal{S}\rangle \in \mathcal{S}$ and $|\psi_{\mathcal{S}\perp}\rangle \in \mathcal{S}^\perp$ are normalized. The success probability is

$$p = \langle\psi_0|\,\Pi_\mathcal{S}\,|\psi_0\rangle = \sin^2\theta.$$

Define the amplification iterate $Q = R_{\psi_0} R_{\mathcal{S}\perp}$. On the invariant subspace $\mathcal{K} = \mathrm{span}\{|\psi_\mathcal{S}\rangle, |\psi_{\mathcal{S}\perp}\rangle\}$, the operator Q acts as a rotation by angle 2θ. Equivalently, Q has eigenvalues $e^{\pm i\, 2\theta}$ on $\mathcal{K}$.

Since $|\psi_0\rangle$ has support on both eigenvectors, applying QPE to Q with input $|\psi_0\rangle$ produces an estimate of an eigenphase $\pm 2\theta$, from which we can recover θ and hence $p = \sin^2\theta$.

Exercise 9.5 Let Q be the amplification iterate (see Eq. 9.4) and suppose $\widetilde{Q} = -Q$.

(a) Show that $Q^L\,|\psi_0\rangle$ and $\widetilde{Q}^L\,|\psi_0\rangle$ differ by only a global phase.

(b) Conclude that replacing a reflection by its negative (e.g., R_{ψ_0} by $-R_{\psi_0}$) cannot change any measurement probabilities, but *can* change the eigenphases reported by quantum phase estimation by an additive shift of π.

2 Resolving the Sign Ambiguity

Because $|\psi_0\rangle$ is typically a superposition of the two eigenvectors of Q, QPE may output either eigenphase branch. This is not a problem: the two eigenphases correspond to rotations by $+2\theta$ and -2θ, and the success probability $p = \sin^2 \theta$ depends only on $|\theta|$.

If we represent the eigenphases in $[0, 2\pi)$, then

$$\phi_+ = 2\theta, \qquad \phi_- = 2\pi - 2\theta,$$

since -2θ modulo 2π equals $2\pi - 2\theta$. QPE is usually described as outputting an estimate $\widetilde{\varphi} \in [0, 1)$ of the phase fraction $\varphi = \phi/(2\pi)$, so the two possible phase fractions are

$$\varphi_+ = \frac{\phi_+}{2\pi} = \frac{\theta}{\pi}, \qquad \varphi_- = \frac{\phi_-}{2\pi} = 1 - \frac{\theta}{\pi}.$$

Given an outcome $\widetilde{\varphi}$, we map it to an estimate $\widetilde{\theta} \in [0, \pi/2]$ via

$$\widetilde{\theta} = \pi \min\{\widetilde{\varphi}, 1 - \widetilde{\varphi}\},$$

and output

$$\widetilde{p} = \sin^2 \widetilde{\theta}.$$

Exercise 9.6 Suppose the postprocessing incorrectly uses $\widetilde{\theta} = \pi\widetilde{\varphi}$ instead of $\widetilde{\theta} = \pi \min\{\widetilde{\varphi}, 1 - \widetilde{\varphi}\}$. Construct a concrete numerical example (choose θ) where this produces a wrong estimate of p.

3 Complexity Advantage

Classically, estimating p to additive error ε by independent sampling requires $\Theta(1/\varepsilon^2)$ trials in general. In contrast, QPE estimates the eigenphase 2θ to additive error $O(\varepsilon)$ using $O(1/\varepsilon)$ controlled applications of Q, so amplitude estimation achieves additive error ε with query complexity $O(1/\varepsilon)$ up to logarithmic factors in the failure probability [27]. This quadratic improvement in precision is the central advantage of amplitude estimation.

Exercise 9.7 Let $p = \sin^2 \theta$ and $\widetilde{p} = \sin^2 \widetilde{\theta}$ with $\theta, \widetilde{\theta} \in [0, \pi/2]$.

(a) Show that $|\widetilde{p} - p| \leq 2|\widetilde{\theta} - \theta|$.

(b) Use this to translate an additive error guarantee $|\widetilde{\theta} - \theta| \leq \varepsilon$ into an additive error bound on $\widetilde{p}$.

4 Variants Beyond Standard QPE

The standard formulation of amplitude estimation uses quantum phase estimation (QPE) explicitly [27]. Two broad families of variants are common. First, one may replace the fully parallel QPE subroutine with *iterative* or *semiclassical* phase-estimation implementations (see § 7.3), which reduce ancilla requirements at the cost of longer runtime and more adaptivity. Second, there are "QPE-free" approaches that still rely on the same amplification operator Q, but avoid phase estimation altogether by estimating success probabilities at a sequence of amplification depths and then performing classical postprocessing (for example, maximum-likelihood estimation). These methods typically trade circuit depth for additional repetitions while retaining the characteristic $O(1/\varepsilon)$ scaling in many regimes [34, 35].

> **Exercise 9.8** Amplitude estimation can be performed either using QPE on controlled powers of Q, or using QPE-free methods that sample at multiple amplification depths.
>
> (a) Explain why QPE-based AE requires controlled-Q^{2^k} operations.
>
> (b) Explain why iterative/QPE-free AE can avoid controlled operations but may require more circuit repetitions.

9.2.2 Quantum Counting

Quantum counting is amplitude estimation specialized to Grover's unstructured search setting, where the unknown quantity is the number of marked items M (equivalently, the marked fraction M/N). Let $N = 2^n$ and let M basis states be marked by the oracle. Preparing the uniform superposition

$$|\psi_0\rangle = \frac{1}{\sqrt{N}} \sum_{x \in \{0,1\}^n} |x\rangle$$

yields success probability $p = M/N$.

Writing $p = \sin^2 \theta$ gives $\sin^2 \theta = M/N$. Applying amplitude estimation to the Grover iterate Q produces an estimate $\widetilde{\theta}$, and hence

$$\widetilde{p} = \sin^2 \widetilde{\theta}, \qquad \widetilde{M} = N\widetilde{p} = N \sin^2 \widetilde{\theta}.$$

The estimate $\widetilde{M}$ is what is typically called *quantum counting*.

With t controlled applications of Q, amplitude estimation (and thus quantum counting) can estimate p to additive error $O(1/t)$, which translates into estimating M to additive error $O(N/t)$. By comparison, to achieve additive error $O(1/t)$ classically one needs $O(t^2)$ oracle calls (independent samples), so quantum counting provides the same quadratic improvement in precision as amplitude estimation.

> **Key Takeaways**
>
> 1. Amplitude estimation (AE) is the "measurement" counterpart of amplitude amplification: it estimates $p = \|\Pi_{\mathcal{S}} |\psi_0\rangle\|^2$ rather than just boosting it.
>
> 2. The AA iterate Q has eigenphases $\pm 2\theta$ on the relevant two-dimensional subspace, where $p = \sin^2 \theta$.
>
> 3. Running quantum phase estimation on controlled powers of Q yields an estimate of θ, and hence an estimate of p.

4. AE can estimate p within additive error ε (i.e., $|\widetilde{p} - p| \leq \varepsilon$) using $O(1/\varepsilon)$ uses of Q, vs. $\Theta(1/\varepsilon^2)$ samples classically.

9.3 ∗ Extensions and Variants

Thus far we have focused on the standard form of amplitude amplification—with Grover search as its canonical example—which boosts the success probability of a given success criterion, and on amplitude estimation—with quantum counting as a special case—which estimates the success probability itself. Over the years, many extensions and refinements have been developed to address new algorithmic settings and practical constraints. Table 9.1 summarizes several commonly encountered variants; in the remainder of this section we focus on two widely used ones: fixed-point/robust amplification and oblivious amplitude amplification.

Variant / Extension	Informal Description
Grover's Search (Baseline)	The canonical instance of amplitude amplification: start from a simple initial state (often uniform), apply a phase-marking oracle for the "good" basis states, then reflect about the start state to rotate amplitude toward solutions.
Multi-Solution Grover	The same two-reflection rotation when there are multiple marked items: the "good" subspace is the span of all marked basis states, and the optimal iteration count depends on the marked fraction M/N.
Amplitude Amplification Wrapper	A general-purpose wrapper for any quantum routine A that succeeds with probability $p > 0$: alternate a phase flip on the "good" subspace with a reflection about $A\lvert 0\rangle$ (implemented using A and $A^\dagger$) to boost success probability in $\Theta(1/\sqrt{p})$ iterations.
Quantum-Walk Search	A structured-search extension that replaces Grover's diffusion step with a quantum walk tailored to the geometry of the search space (e.g., a graph or Markov chain), often improving performance when locality or structure is available.
Parallel / Batched Amplification	A resource-management variant that runs multiple amplification instances or batches candidates, useful for producing multiple solutions or for trading circuit depth against parallel resources while retaining the same amplification principle.
Oblivious Amplitude Amplification (OAA)	An input-independent amplification method used in linear combination of unitaries and block-encoding settings: the "good" event is typically an ancilla flag (e.g., measuring $\lvert 0^n\rangle$), and amplification boosts that postselection success using reflections that depend only on the circuit/ancilla structure, not on the unknown system state.

Amplitude Estimation (AE) / Quantum Counting	An estimation extension: instead of merely boosting success, combine amplification-style operators with phase estimation to estimate the success probability p (or, in Grover's setting, the number of marked items), enabling quantum speedups for Monte Carlo–type tasks.
Fixed-Point / Robust Amplification	A stability-oriented extension that avoids "overshooting" when p is unknown or when monotone convergence is desired, typically using variable-phase reflections or schedules that guarantee steady progress toward the good subspace.

Table 9.1: Common Variants and Extensions of Amplitude Amplification

---

Overview

---

1. Standard AA can *over-rotate* when p_0 (or θ_0) is not known accurately, causing success probability to decrease after it first becomes large.

2. Robust amplification replaces reflections by selective phase shifts of the form $I + (e^{i\gamma} - 1)P$. By choosing a phase schedule, one can avoid oscillation and guarantee monotone progress toward the good subspace, without knowing p_0 precisely.

3. Oblivious amplitude amplification (OAA) is AA with a "good ancilla" condition (typically $|0^n\rangle$) and reflections that do not depend on the unknown system input state.

4. OAA is the standard method to make postselected LCU/block-encoding constructions succeed with near-unit probability.

9.3.1   Fixed-Point and Robust Amplitude Amplification

1   Motivation: Avoiding Over-Rotation

Recall the standard amplitude-amplification iterate

$$Q = R_{\psi_0} R_{\mathcal{S}^\perp},$$

which acts as a rotation by $2\theta_0$ on the invariant plane $\mathcal{K} = \mathrm{span}\{|\psi_{\mathcal{S}}\rangle, |\psi_{\mathcal{S}^\perp}\rangle\}$, where $p_0 = \sin^2\theta_0$. After L applications,

$$p_L = \sin^2\big((2L + 1)\theta_0\big).$$

If θ_0 is unknown, then a near-optimal choice $L \approx \frac{\pi}{4\theta_0} - \frac{1}{2}$ cannot be made reliably. Even if one only knows a lower bound on p_0, standard amplification can still overshoot and undo earlier progress. In many algorithmic settings (especially when amplification is used as an internal subroutine), it is preferable to have a monotone, "no-regret" procedure.

Exercise 9.9 Consider standard amplitude amplification with initial success probability $p_0 = \sin^2 \theta_0$.

   (a) Show that if $p_0 = \frac{1}{2}$ (i.e., $\theta_0 = \frac{\pi}{4}$), then the success probability stays equal to $\frac{1}{2}$ for all $L \geq 0$.

   (b) Show that if $p_0 = \frac{1}{4}$ (i.e., $\theta_0 = \frac{\pi}{6}$), then a single amplification step yields success probability 1.

   (c) Find all $\theta_0 \in (0, \frac{\pi}{2}]$ for which there exists an integer $L \geq 0$ such that after L amplification steps the success probability is exactly 1.

Fixed-point and robust amplitude amplification address this issue by modifying the reflections so that the success probability increases *monotonically* toward a target value, without requiring precise knowledge of p_0. The price is typically a modest increase in the number of oracle calls (often by logarithmic factors), but the benefit is a procedure that is stable, predictable, and easier to use as a subroutine.

2   Variable-Phase Reflections

The key idea is to replace the π-phase reflections used in standard amplitude amplification with *selective phase shifts*. In our convention, the standard reflections are

$$R_{\mathcal{S}^\perp} = I - 2\Pi_{\mathcal{S}}, \qquad R_{\psi_0} = 2\,|\psi_0\rangle\langle\psi_0| - I.$$

Their phase-shifted counterparts are

$$R_{\mathcal{S}^\perp}(\phi) = I + (e^{i\phi} - 1)\Pi_{\mathcal{S}}, \qquad R_{\psi_0}(\varphi) = I + (e^{i\varphi} - 1)\,|\psi_0\rangle\langle\psi_0|. \tag{9.5}$$

When $\phi = \varphi = \pi$ these reduce to $R_{\mathcal{S}^\perp}$ and R_{ψ_0} up to a global phase. More generally, by choosing a sequence of phases

$$\phi = (\phi_1, \varphi_1, \phi_2, \varphi_2, \ldots, \phi_L, \varphi_L),$$

one obtains a generalized amplification sequence

$$Q_\phi = R_{\psi_0}(\varphi_L)\, R_{\mathcal{S}^\perp}(\phi_L) \cdots R_{\psi_0}(\varphi_1)\, R_{\mathcal{S}^\perp}(\phi_1),$$

whose net effect on $\mathcal{K}$ is still a controlled transformation of the success amplitude, but no longer a rigid rotation that necessarily oscillates.

Exercise 9.10 Verify that setting $\phi = \varphi = \pi$ recovers the standard amplitude amplification iterate up to a global phase.

Exercise 9.11 Let P be a projector ($P^2 = P$). Define $U(\gamma) = I + (e^{i\gamma} - 1)P$. Show that $U(\gamma)$ is unitary for all real γ.

3   Fixed-Point Guarantee

A fixed-point amplification procedure is designed so that once the success probability becomes large, further applications do not decrease it. It uses variable phases so that the effective rotation step shrinks as the state approaches the good subspace (i.e., as $\theta \to \frac{\pi}{2}$). Equivalently, the phase schedule is chosen so that the induced map on the success amplitude is monotone increasing on the promised range. Consequently, once the state is close to S, further applications act almost like the identity (or a small correction) on the good component, rather than rotating it away. In this sense, S becomes an *attractive fixed point* of the iteration.

Assume only a lower bound $p_0 \geq p_{\min} > 0$ is known. For any target failure probability $\delta \in (0,1)$, there exist choices of phases ϕ such that the resulting procedure outputs a state with success probability at least $1 - \delta$ for *all* inputs satisfying $p_0 \geq p_{\min}$, without requiring an estimate of p_0 itself. The required number of uses of the underlying primitives (the state preparation A and the marking oracle) scales as

$$O\left(\frac{1}{\sqrt{p_{\min}}} \log \frac{1}{\delta}\right),$$

so the cost remains essentially quadratic in $1/\sqrt{p_{\min}}$, up to a logarithmic dependence on the desired accuracy [29, 36].

Many fixed-point constructions can be interpreted as designing a polynomial transformation of the success amplitude, and can be placed within the broader framework of quantum singular value transformation (QSVT, § 11.6).

4   Circuit Implementation

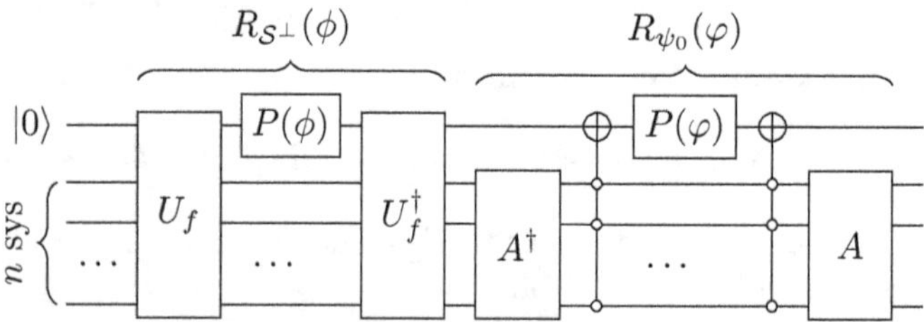

Figure 9.5: Circuit for Phase-Shifted Grover Iterate

A circuit for one phase-shifted Grover iterate

$$Q(\phi, \varphi) = R_{\psi_0}(\varphi)\, R_{S^\perp}(\phi) \tag{9.6}$$

is shown in Fig. 9.5. It differs from the standard Grover iterate only in that the two π-phase reflections are replaced by selective phase shifts with angles ϕ and φ. Throughout this discussion, we use $P(\gamma) = \operatorname{diag}(1, e^{i\gamma})$ for the single-qubit phase gate.

Selective Phase Shift on the Marked Subspace

The left block, labeled $R_{S^\perp}(\phi)$, implements a phase $e^{i\phi}$ on marked computational basis states and does nothing on unmarked states:

$$|x\rangle \mapsto e^{i\phi} |x\rangle \quad \text{for } x \in S, \qquad |x\rangle \mapsto |x\rangle \quad \text{for } x \notin S.$$

It is realized using the standard *compute–phase–uncompute* pattern. Let U_f be a Boolean oracle that computes the predicate $f(x)$ into an ancilla bit:

$$U_f |x\rangle |0\rangle = |x\rangle |f(x)\rangle, \qquad S = \{x : f(x) = 1\}.$$

Starting from $|x\rangle\,|0\rangle$, the first U_f computes $f(x)$ into the ancilla, the phase gate $P(\phi)$ applies a phase factor $e^{i\phi f(x)}$, and $U_f^\dagger$ uncomputes the ancilla back to $|0\rangle$. The net effect on the search register is therefore

$$|x\rangle \;\mapsto\; e^{i\phi f(x)}\,|x\rangle,$$

which is exactly the selective phase shift $R_{S^\perp}(\phi)$.

Selective Phase Shift on the Start State

The right block, labeled $R_{\psi_0}(\varphi)$, is the corresponding phase-shifted "state reflection." Since $|\psi_0\rangle = A\,|0^n\rangle$, we implement it by conjugation:

$$R_{\psi_0}(\varphi) = A\,R_0(\varphi)\,A^\dagger, \qquad R_0(\varphi) = I + (e^{i\varphi} - 1)\,|0^n\rangle\langle 0^n|. \tag{9.7}$$

The operator $R_0(\varphi)$ acts as

$$|0^n\rangle \mapsto e^{i\varphi}\,|0^n\rangle, \qquad |x\rangle \mapsto |x\rangle \quad \text{for all } x \neq 0^n.$$

In the circuit, one first applies $A^\dagger$, then performs a selective phase on the computational basis state $|0^n\rangle$ using an ancilla flag and a phase gate $P(\varphi)$, and finally applies A to return to the $|\psi_0\rangle$ basis. When $\varphi = \pi$, this implements R_{ψ_0} up to a global phase.

5   When to Use Robust Amplification

Fixed-point and robust variants are especially useful when:

- p_0 is unknown, but a lower bound $p_{\min}$ is available;

- overshooting is unacceptable because amplification is nested inside a larger algorithm;

- one wants a tunable failure probability δ with predictable behavior.

By contrast, when p_0 is known accurately (as in idealized analyses of Grover search), standard amplitude amplification is often simpler and has slightly smaller constant factors.

9.3.2   Oblivious Amplitude Amplification

 Readers unfamiliar with block encoding and the linear-combination-of-unitaries (LCU) method may wish to return to this topic after studying §§ 11.2 and 11.3.

Oblivious amplitude amplification (OAA) is a specialization of amplitude amplification in which the "good" subspace is defined by an *ancilla condition* (typically the ancilla being $|0^n\rangle$), and the amplification reflections are constructed so that they do not depend on the unknown input state of the system register. This is precisely the setting that arises in block-encoding and LCU constructions, where postselection on an ancilla outcome yields the desired operator action only with a small success probability (see § 11.2.4).

1   Setup and Conditions

In this context, we work with two registers: an ancilla register of n qubits and a system register. Let the total Hilbert space be $\mathcal{H} = \mathcal{H}_{\mathrm{anc}} \otimes \mathcal{H}_{\mathrm{sys}}$, where $\dim(\mathcal{H}_{\mathrm{anc}}) = 2^n$.

Define the *good* and *bad* subspaces

$$\mathcal{S} = \operatorname{span}\{|0^n\rangle\} \otimes \mathcal{H}_{\text{sys}}, \qquad \mathcal{S}^\perp = \mathcal{H} \ominus \mathcal{S},$$

and let

$$\Pi_{\mathcal{S}} = |0^n\rangle\langle 0^n| \otimes I$$

be the orthogonal projector onto $\mathcal{S}$.

The relevant state-preparation unitary is typically provided by a block encoding itself, so we denote it by U (rather than by A). We assume that for every system state $|\psi\rangle \in \mathcal{H}_{\text{sys}}$,

$$U\big(|0^n\rangle\,|\psi\rangle\big) = |0^n\rangle\, \frac{A\,|\psi\rangle}{\alpha} + |\Phi_\perp(\psi)\rangle, \qquad |\Phi_\perp(\psi)\rangle \in \mathcal{S}^\perp, \tag{9.8}$$

where A is a linear operator on $\mathcal{H}_{\text{sys}}$ and $\alpha \geq 1$ is a known normalization factor. Equation Eq. 9.8 is exactly the statement that U is a block encoding of A/α with respect to the ancilla subspace $\mathcal{S}$.

To perform OAA, we require:

- the ability to implement U and $U^\dagger$;

- the ability to implement the reflection about $\mathcal{S}$ (equivalently, a reflection about $|0^n\rangle$ on the ancilla register);

- a way to choose an iteration count. This is simplest when one has a usable bound relating α and $\|A\|$ (e.g., $\|A\| \leq 1$ with known α), while robust or fixed-point variants can avoid requiring precise knowledge of the initial success probability.

Crucially, the reflections used by OAA can be implemented without any knowledge of the input system state $|\psi\rangle$.

However, the *initial* postselection success probability can still depend on $|\psi\rangle$. For a given $|\psi\rangle$, the success probability of the "good" ancilla outcome is

$$p_0(\psi) = \|\Pi_{\mathcal{S}} U(|0^n\rangle\,|\psi\rangle)\|^2 = \left\| \frac{A\,|\psi\rangle}{\alpha} \right\|^2,$$

which may vary with $|\psi\rangle$. The term "oblivious" refers to the fact that the reflections used below depend only on $\mathcal{S}$ and on U, and hence are independent of $|\psi\rangle$, even though the amount of amplitude available to amplify may depend on $|\psi\rangle$.

2   Oblivious Reflections

As in standard amplitude amplification, OAA is built from two reflections.

The Marking Reflection

The marking reflection is the reflection about the axis $\mathcal{S}^\perp$:

$$R_{\mathcal{S}^\perp} = I - 2\Pi_{\mathcal{S}} = (I - 2|0^n\rangle\langle 0^n|) \otimes I.$$

This operator acts as $-I$ on $\mathcal{S}$ and as $+I$ on $\mathcal{S}^\perp$, i.e., it flips the phase of the "good" subspace and leaves the "bad" subspace unchanged.

The State Reflection

In standard amplitude amplification, the state reflection about $|\psi_0\rangle = A\,|0\rangle$ can be written as a conjugation of a reflection about the start state:

$$R_{\psi_0} = A\left(2\,|0\rangle\langle 0| - I\right)A^\dagger.$$

In OAA, the relevant "start space" is not a single vector but the entire subspace $\mathcal{S} = \mathrm{span}\{|0^n\rangle\} \otimes \mathcal{H}_{\mathrm{sys}}$. Conjugating the reflection about $\mathcal{S}$ by the state-preparation unitary U yields

$$R_U = U\left(2\Pi_{\mathcal{S}} - I\right)U^\dagger = 2\left(U\Pi_{\mathcal{S}}U^\dagger\right) - I,$$

which is the reflection about the prepared subspace $U(\mathcal{S})$.

The OAA Iterate

The OAA iterate is the product

$$Q = R_U R_{\mathcal{S}^\perp}. \tag{9.9}$$

For each fixed input $|\psi\rangle$, the state $U(|0^n\rangle\,|\psi\rangle)$ decomposes into its $\mathcal{S}$ and $\mathcal{S}^\perp$ components, and hence lies in the two-dimensional subspace spanned by these two projections. The operator Q preserves this plane and acts as a rotation within it, with a rotation angle determined by $p_0(\psi)$. As in standard amplitude amplification, repeated applications of Q increase the overlap with the good subspace $\mathcal{S}$.

In general, the optimal number of iterations depends on the initial value of $p_0(\psi)$, but robust or fixed-point variants of OAA [36] can achieve near-unit success probability without requiring precise knowledge of $p_0(\psi)$.

3 What OAA Achieves

Viewed operationally, OAA turns the *postselected* implementation in Eq. 9.8 into an implementation that succeeds with near-unit probability, without measuring the ancilla and without conditioning on $|\psi\rangle$. Viewed more structurally (see § 11.2.4), OAA provides a way to convert a probabilistic block-encoding construction into a near-deterministic one, enabling near-optimal query and gate complexity in LCU-based algorithms.

> **Exercise 9.12** Explain why OAA can still be called *oblivious* even though p_0 depends on $|\psi\rangle$.

Chapter Takeaways

1. Amplitude amplification (AA) is a general success-boosting wrapper: from a routine that succeeds with probability p_0, AA increases success using repeated reflections on a two-dimensional invariant subspace.

2. The Grover/AA iterate Q acts as a rotation with angle determined by $p_0 = \sin^2\theta$, explaining both the quadratic speedup and the need to avoid "over-rotation" when p_0 is unknown.

3. Amplitude estimation (AE) is the measurement counterpart: it estimates p to additive error ε using $O(1/\varepsilon)$ applications of Q, yielding a quadratic improvement over classical sampling; quantum counting is AE applied

to estimating the number of marked items.

4. Iterative / QPE-free AE variants avoid controlled-Q^{2^k} (and sometimes avoid phase estimation entirely) by sampling multiple amplification depths and using classical inference, trading deeper coherence for more repetitions.

5. Robust/fixed-point amplification and oblivious amplitude amplification modify the basic template to improve stability, composability, and compatibility with block-encoding and linear-algebraic subroutines.

Problem Set 9

9.1 Provide an algebraic proof of Eq. 9.1 by direct matrix multiplication, without relying on the geometric interpretation.

9.2 Let $\mathcal{K} = \mathrm{span}\{|\psi_S\rangle, |\psi_{S\perp}\rangle\}$ be the two-dimensional invariant subspace of amplitude amplification, and let

$$Q = R_{\psi_0} R_{S\perp}.$$

(a) Write the 2×2 matrix of Q in the ordered basis $\{|\psi_S\rangle, |\psi_{S\perp}\rangle\}$.

(b) Show that Q has eigenvalues $e^{\pm i 2\theta_0}$.

(c) Express the corresponding eigenvectors as superpositions of $|\psi_S\rangle$ and $|\psi_{S\perp}\rangle$, and verify explicitly that $|\psi_0\rangle$ has support on both.

9.3 In amplitude estimation, phase estimation on Q may output either eigenphase branch corresponding to $\phi_+ = 2\theta$ or $\phi_- = 2\pi - 2\theta$.

(a) Derive the two possible phase fractions $\varphi_\pm = \phi_\pm/(2\pi)$.

(b) Show that the mapping

$$\widetilde{\theta} = \pi \min\{\widetilde{\varphi}, 1 - \widetilde{\varphi}\}$$

recovers $\theta \in [0, \pi/2]$ (up to estimation error) regardless of which branch was observed.

(c) Explain why this resolves the sign ambiguity for estimating $p = \sin^2 \theta$.

9.4 (Quantum counting.) Let $N = 2^n$ and suppose M computational basis states are marked.

(a) Show that the uniform superposition $|\psi_0\rangle = \frac{1}{\sqrt{N}} \sum_x |x\rangle$ has success probability $p = M/N$.

(b) Write $p = \sin^2 \theta$ and express M in terms of θ and N.

(c) If amplitude estimation outputs $\widetilde{\theta}$ with additive error at most ε, derive an additive-error bound on $\widetilde{M} = N \sin^2 \widetilde{\theta}$ in terms of N and ε.

9.5 (Reflection convention and global phase.) Two common conventions for a reflection about $|\psi\rangle$ are

$$R_\psi = 2\,|\psi\rangle\langle\psi| - I \quad\text{and}\quad \widehat{R}_\psi = I - 2\,|\psi\rangle\langle\psi|.$$

(a) Explain why replacing R_ψ by $\widehat{R}_\psi$ cannot change any measurement probabilities in amplitude amplification.

(b) Describe how this replacement shifts eigenphases reported by QPE on the resulting iterate.

9.6 (Oblivious amplitude amplification.) Assume U satisfies the block form

$$U\big(|0^m\rangle\,|\psi\rangle\big) = |0^m\rangle\,\frac{A\,|\psi\rangle}{\alpha} + |\Phi_\perp(\psi)\rangle, \qquad |\Phi_\perp(\psi)\rangle \in \mathcal{S}^\perp,$$

with $\mathcal{S} = \operatorname{span}\{|0^m\rangle\} \otimes \mathcal{H}_{\mathrm{sys}}$. Define $R_0 = 2\Pi_\mathcal{S} - I$ and $R_U = U R_0 U^\dagger$, and let $Q = R_U R_0$.

(a) Show that for each fixed $|\psi\rangle$, the vector $U\big(|0^m\rangle\,|\psi\rangle\big)$ lies in the two-dimensional subspace spanned by its projections onto $\mathcal{S}$ and $\mathcal{S}^\perp$.

(b) Prove that Q preserves this two-dimensional subspace.

(c) Explain why the reflections are called "oblivious" in this setting.

9.7 (Selective phase shifts from projectors.) Let P be a projector and define $U(\gamma) = I + (e^{i\gamma} - 1)P$.

(a) Show that $U(\gamma)$ is unitary and determine its eigenvalues and eigenspaces.

(b) Specialize to $P = \Pi_\mathcal{S}$ and interpret $R_{\mathcal{S}^\perp}(\phi) = I + (e^{i\phi} - 1)\Pi_\mathcal{S}$ as a selective phase shift on the good subspace.

9.8 (Phase-shifted Grover iterate.) Let

$$Q(\phi, \varphi) = R_{\psi_0}(\varphi)\, R_{\mathcal{S}^\perp}(\phi),$$

where

$$R_{\mathcal{S}^\perp}(\phi) = I + (e^{i\phi} - 1)\Pi_\mathcal{S}, \qquad R_{\psi_0}(\varphi) = I + (e^{i\varphi} - 1)\,|\psi_0\rangle\langle\psi_0|.$$

(a) Work in the basis $\{|\psi_\mathcal{S}\rangle, |\psi_{\mathcal{S}^\perp}\rangle\}$ and derive the 2×2 matrix of $Q(\phi, \varphi)$ restricted to $\mathcal{K}$.

(b) For small deviations $\phi = \pi + \Delta_\phi$ and $\varphi = \pi + \Delta_\varphi$, give a first-order description (up to a global phase) of how the action on $\mathcal{K}$ differs from a rigid rotation.

9.9 (Simulation of Grover search and amplitude estimation.) Choose $n \in \{4, 5, 6\}$ and mark exactly one basis state $|x^\star\rangle$.

(a) Simulate Grover search and plot the success probability p_L as a function of iteration count L.

(b) Verify numerically that the near-optimal $L \approx \left\lfloor \frac{\pi}{4}\sqrt{N} - \frac{1}{2} \right\rfloor$ maximizes p_L.

(c) Implement amplitude estimation (using QPE or an iterative/QPE-free variant) to estimate $p = M/N = 1/N$. Report the estimate $\widetilde{p}$ and compare the scaling of error vs. the number of uses of Q.

9.10 (Interference as an Algorithmic Resource.) Choose one of the following "interference primitives" and write a short report (1–2 pages) with a worked example:

(a) Phase kickback (as used in Deutsch–Jozsa),

(b) Phase estimation as "interference in the Fourier basis",

(c) Amplitude amplification as repeated constructive/destructive interference.

Your report should (i) state the primitive, (ii) show a minimal circuit, and (iii) explain what changes in the probability distribution after measurement.

10. Foundations of Hamiltonian Simulation

Contents

We now turn to Hamiltonian simulation, one of the central algorithmic primitives in quantum computing. At its core, Hamiltonian simulation concerns the efficient implementation of quantum time evolution generated by a Hamiltonian, namely the unitary operator e^{-iHt}.

Hamiltonian simulation underlies a wide range of quantum algorithms and applications, including quantum chemistry and materials modeling, condensed-matter physics, adiabatic quantum computation, quantum linear systems algorithms, and eigenvalue estimation via quantum phase estimation. As a result, it plays a foundational role both in theoretical algorithm design and in practical quantum applications.

This chapter focuses on one of the earliest and most direct approaches to Hamiltonian simulation: *product-formula methods*, commonly referred to as *Trotterization*. We begin by formulating the Hamiltonian simulation problem and discussing its physical interpretation. We then introduce first-order and higher-order Trotter–Suzuki formulas, analyze their error behavior and complexity, and explain fundamental limitations such as the no-fast-forwarding theorem.

A key emphasis of the chapter is the efficient implementation of Hamiltonians expressed as sums of Pauli strings. Such *Pauli Hamiltonians* arise naturally in many physical settings and admit particularly simple circuit constructions. We show how exponentials of Pauli strings can be implemented using only single-qubit gates and CNOTs, making Trotterization practical for a wide class of problems.

More advanced Hamiltonian simulation techniques—such as methods based on linear combinations of unitaries, quantum walks and qubitization, quantum singular value transformation, and randomized simulation schemes—will be introduced in Chapter 11. Together, these chapters present a progression from intuitive, physically motivated methods to asymptotically optimal simulation algorithms.

10.1 Conceptual Framing of Hamiltonian Simulation

Hamiltonian simulation concerns how a quantum computer can reproduce the real-time evolution of systems governed by a Hamiltonian operator. This introductory section formalizes the problem, clarifies related physical notions, and highlights the central role simulation plays in phase-estimation algorithms.

10.1.1 Problem Formulation and Physical Meaning

Hamiltonian simulation refers to the problem of efficiently implementing the unitary time-evolution operator

$$U(t) = e^{-iHt}, \tag{10.1}$$

generated by a Hermitian operator H known as the *Hamiltonian*.

1   Definition

Formally, the Hamiltonian simulation problem is the following: given a Hermitian operator H that admits an efficient description (for example, as a sparse matrix, a k-local Hamiltonian, or a linear combination of efficiently implementable unitaries), a time parameter $t > 0$, and an error tolerance $\varepsilon > 0$, construct a quantum circuit U_{sim} satisfying

$$\| U_{\mathrm{sim}} - e^{-iHt} \| \le \varepsilon.$$

Unless stated otherwise, $\| \cdot \|$ denotes the operator norm. The goal is to achieve this with computational cost scaling polynomially in $\log N$, linearly or nearly linearly in $t\|H\|$, and with as mild a dependence on $1/\varepsilon$ as possible (product formulas typically scale polynomially in $1/\varepsilon$, while more advanced methods can achieve polylogarithmic dependence).

2   Hermitian Operators and Unitary Exponentials

If H is Hermitian ($H = H^\dagger$), then $U = e^{-iHt}$ is unitary. Indeed,

$$e^{-iHt}(e^{-iHt})^\dagger = e^{-iHt}e^{iH^\dagger t} = e^{-i(H-H^\dagger)t} = e^{0I} = I.$$

Conversely, every unitary operator U can be written as the exponential of a Hermitian operator H, often called a generator of U. Indeed, since all eigenvalues of U have the form $e^{-i\theta_j}$, choosing an eigenbasis $\{|\psi_j\rangle\}$ and defining

$$H = \sum_j \theta_j |\psi_j\rangle\langle\psi_j|$$

gives $e^{-iH} = U$, with each θ_j determined up to addition of integer multiples of 2π.

3   Relation to Quantum Phase Estimation

A central use of Hamiltonian simulation appears in quantum phase estimation (QPE, Chapter 7). The controlled powers U^{2^k} required by QPE can be implemented by simulating time evolution under the generator of U. If $U = e^{-iH}$, then

$$U^{2^k} = e^{-iH\,2^k}.$$

Thus, the ability to simulate Hamiltonian evolution for exponentially long times—while maintaining controlled accuracy—is exactly what enables efficient phase estimation for Hamiltonian systems and for many algorithms built upon them.

4   Physical Meaning: Hamiltonian Evolution and Hamiltonian Dynamics

Before discussing specific simulation algorithms, it is helpful to clarify several closely related concepts. Although these terms are sometimes used interchangeably in informal discussions, they refer to distinct ideas.

While Hamiltonian simulation refers to the algorithmic task of approximating the time evolution generated by a Hamiltonian, *Hamiltonian evolution* (or time evolution) is the physical, continuous-time propagation of quantum states. The operator e^{-iHt} describes the evolution of a quantum state $|\psi(t)\rangle$ according to the Schrödinger equation; throughout this chapter we use units with $\hbar = 1$, so

$$i\frac{d}{dt}\,|\psi(t)\rangle = H\,|\psi(t)\rangle, \qquad |\psi(t)\rangle = e^{-iHt}\,|\psi(0)\rangle.$$

Closely related to Hamiltonian evolution, *Hamiltonian dynamics* refers more broadly to the behavior induced by a Hamiltonian: how excitations propagate, how correlations develop, how observables change in time, and how the system responds to perturbations. Hamiltonian dynamics governs scattering amplitudes, spectral functions, transition rates, absorption and emission spectra, and multi-point correlation functions. Efficient Hamiltonian simulation enables a quantum computer to reproduce these dynamical properties and to explore regimes inaccessible to classical computation.

■ **Example 10.1 — Estimating Energy Levels of a Hamiltonian.** The energy levels of a quantum system are given by the eigenvalues of its Hamiltonian H. Let $|\psi_k\rangle$ be an eigenstate of H with eigenvalue E_k:

$$H\,|\psi_k\rangle = E_k\,|\psi_k\rangle.$$

The time-evolution operator defined in Eq. 10.1 is an analytic function of H. Thus,

$$U(t)\,|\psi_k\rangle = e^{-iE_k t}\,|\psi_k\rangle.$$

For $t = 1$, this simplifies to

$$U(1)\,|\psi_k\rangle = e^{-iE_k}\,|\psi_k\rangle,$$

so the eigenvalue E_k can be recovered (modulo 2π) from the eigenphase of $U(1)$. ■

10.1.2   Approaches to Hamiltonian Simulation

Hamiltonian simulation is a broad problem, and different physical models and access assumptions lead to different algorithmic approaches. A useful first division is between analog simulators, which implement dynamics directly, and digital methods, which approximate e^{-iHt} using quantum circuits.

1   Digital vs. Analog Quantum Simulators

Digital Hamiltonian simulation discretizes time evolution and realizes it through circuit-based methods such as Trotter–Suzuki product formulas, linear-combination-of-unitaries techniques, and quantum signal processing.

Analog quantum simulators provide a complementary approach. Rather than approximating e^{-iHt} with a sequence of discrete gates, an analog simulator physically engineers a controllable quantum system whose natural Hamiltonian closely matches the target Hamiltonian. Cold-atom lattices, trapped ions with tunable couplings, and superconducting qubit arrays with engineered interactions are common examples. These devices implement Hamiltonian evolution directly in real time, without digital decomposition or circuit synthesis.

Analog simulators can reach large system sizes and long coherence times, but typically offer less programmability and weaker error control compared to digital algorithms. In this chapter we focus on digital Hamiltonian simulation algorithms; later chapters will consider the physical applications of Hamiltonian evolution and quantum dynamics in more detail.

2   Digital Simulation Algorithms

From the circuit-model perspective, most algorithmic methods for approximating e^{-iHt} fall into a few recurring families. They differ mainly in (i) how they access H (sparse-access oracles, local terms, LCU/block-encodings), and (ii) how the gate complexity scales with $t\|H\|$ and the error tolerance ε.

Product-Formula (Trotter–Suzuki) Methods

When H is expressed as a sum of terms $H = \sum_{j=1}^{m} H_j$ for which each $e^{-iH_j \Delta t}$ can be implemented efficiently, one discretizes time and approximates e^{-iHt} using Trotter–Suzuki product formulas. This approach is conceptually simple and often practical on near-term devices. It is the main focus of this chapter.

Polynomial-Approximation Methods (QSP/QSVT)

These methods embed H (or H/α) into a larger unitary via a block-encoding (or a closely related signal-oracle model), and then apply polynomial transformations to approximate the desired function of H (here, the exponential). In many settings this yields asymptotically near-optimal scaling in $t\|H\|$ and polylogarithmic dependence on $1/\varepsilon$. We develop the required primitives in Chapter 11, and later revisit how they apply to Hamiltonian simulation in Part III.

Variational and Adiabatic Approaches (Simulation-Adjacent)

Not every physically motivated "simulation task" requires implementing e^{-iHt} directly. For instance, VQE targets ground-state properties via variational preparation and measurement, while adiabatic quantum computing follows a continuous-time Hamiltonian path whose endpoint encodes the desired state. These approaches are discussed elsewhere (e.g., Chapter 13) and are best viewed as complementary tools for approximately extracting spectral or dynamical information, rather than as direct black-box implementations of $U(t)$.

10.2 Trotterization: Product-Formula Methods

One of the earliest and most intuitive approaches to Hamiltonian simulation uses product-formula approximations, collectively known as Trotterization. Suppose the Hamiltonian admits a decomposition

$$H = \sum_{j=1}^{m} H_j,$$

where each term H_j can be exponentiated efficiently. In general,

$$e^{-iHt} \neq \prod_{j=1}^{m} e^{-iH_j t},$$

unless the terms H_j mutually commute.

10.2.1 First-Order Trotterization

The key observation is that even when the H_j do not commute, the exact evolution can be approximated by splitting the total evolution time t into r small intervals and applying the Lie–Trotter product formula:

$$e^{-iHt} \approx \left(\prod_{j=1}^{m} e^{-i\frac{H_j t}{r}} \right)^r. \tag{10.2}$$

The parameter r controls the number of time slices: larger r yields a smaller Trotter error at the expense of increased circuit depth. A single first-order Trotter step is illustrated in Fig. 10.1.

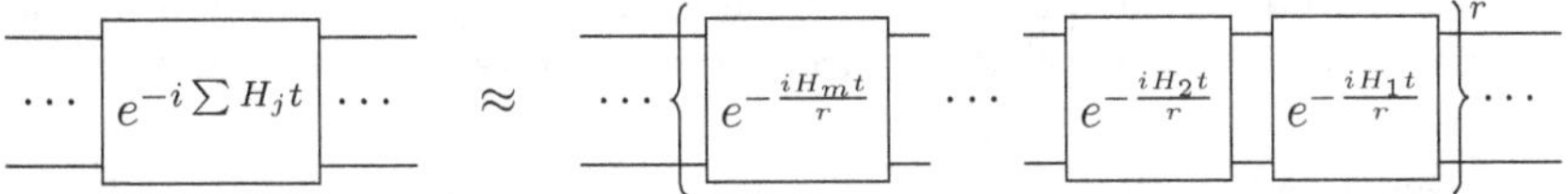

Figure 10.1: First-Order Trotterization Circuit

By convention, the gates in a circuit diagram are arranged from left to right in the temporal order in which they act on the qubits. However, the corresponding matrix multiplication follows the opposite order: a sequence of gates $U_m, \ldots, U_2, U_1$ implements the composite unitary $U = U_1 U_2 \cdots U_m$, where U_m is applied first.

Exercise 10.1 Let $H = X + Y$, where X and Y are Pauli operators. Write out the first-order Trotterization circuit in Fig. 10.1 explicitly using single-qubit rotation gates R_x and R_y.

Exercise 10.2 Let $H = H_1 + H_2$ with $[H_1, H_2] \neq 0$. Using the BCH formula to second order in Δt, show that

$$e^{-i(H_1+H_2)\Delta t} - e^{-iH_1\Delta t}e^{-iH_2\Delta t} = O\big((\Delta t)^2\big),$$

and identify the leading commutator term.

Trotterization is particularly effective in quantum chemistry and condensed-matter applications, where the Hamiltonian often has a natural decomposition into few-body interactions or sparse coupling terms. In such cases, each exponential $e^{-iH_j t/r}$ acts on only a small number of qubits and can be implemented efficiently.

10.2.2 Higher-Order Trotterization

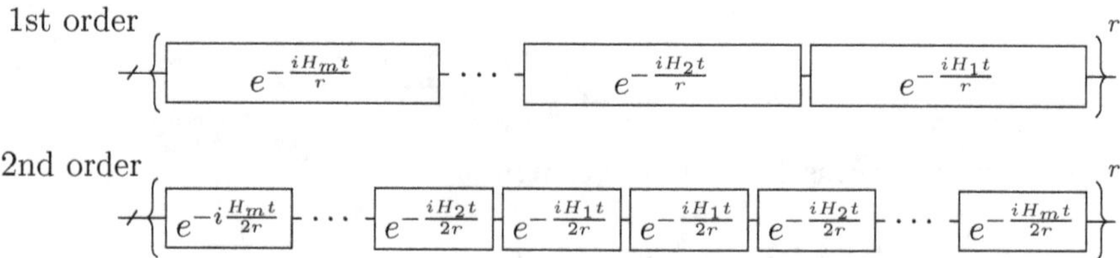

Figure 10.2: First- and Second-Order Trotterization Comparison

To improve accuracy without increasing r dramatically, we can use higher-order *Trotter–Suzuki formulas*. These formulas recursively cancel higher-order terms in the Baker–Campbell–Hausdorff (BCH) expansion and achieve substantially smaller error per time slice, at the cost of a more complex circuit structure per step.

The second-order (Strang) formula,

$$e^{-iHt} \approx \left(\prod_{j=m}^{1} e^{-i\frac{H_j t}{2r}} \prod_{j=1}^{m} e^{-i\frac{H_j t}{2r}} \right)^r, \tag{10.3}$$

can be implemented as shown in Fig. 10.2. Compared to the first-order formula, the gate sequence is applied forward and then in reverse order within each time slice.

Exercise 10.3 Show that the second-order (Strang) product formula is time-reversal symmetric:

$$S_2(\Delta t)^\dagger = S_2(-\Delta t).$$

Explain briefly why such symmetry is consistent with the improved error scaling compared to first-order Trotterization.

Exercise 10.4 Consider three terms $H = H_1 + H_2 + H_3$. Compare the two first-order steps

$$S(\Delta t) = e^{-iH_1\Delta t}e^{-iH_2\Delta t}e^{-iH_3\Delta t}, \qquad S'(\Delta t) = e^{-iH_3\Delta t}e^{-iH_2\Delta t}e^{-iH_1\Delta t}.$$

Are $S(\Delta t)$ and $S'(\Delta t)$ equal? Expand each to second order in Δt and relate the difference to commutators.

10.2.3   Complexity and the No-Fast-Forwarding Theorem

1   Trotter Error and Step Complexity

Second-order Trotterization achieves an error scaling of $O(t^3/r^2)$, a significant improvement over the first-order error of $O(t^2/r)$. Equivalently, to maintain a target error tolerance ε, the required number of steps r scales as $O(t^2/\varepsilon)$ for first-order formulas and $O(t^{3/2}/\varepsilon^{1/2})$ for second-order formulas.

More precisely, Trotter error arises from the noncommutativity of Hamiltonian terms and is governed by nested commutators, such as $[H_j, H_k]$ and $[H_j, [H_k, H_\ell]]$. When terms commute exactly, product formulas are exact for any r. Consequently, Trotterization is highly effective for *nearly commuting* Hamiltonians where these commutators have small operator norms. This is frequently the case in quantum chemistry and condensed-matter systems with weak interactions or local coupling, explaining why product-formula methods often outperform their worst-case theoretical bounds in practice.

Intuition Behind the BCH Expansion

The Baker–Campbell–Hausdorff (BCH) formula explains why the product $\exp(A)\exp(B)$ differs from $\exp(A + B)$ when A and B do not commute. At a high level, the BCH formula states that

$$e^A e^B = \exp\left(A + B + \tfrac{1}{2}[A, B] + \tfrac{1}{12}[A, [A, B]] - \tfrac{1}{12}[B, [A, B]] + \cdots\right).$$

The key idea is that noncommutativity produces additional commutator terms. Whenever $AB \neq BA$, the exponential of a sum cannot be factored into a simple product of exponentials. The extra commutator terms measure how much the operators disturb one another. Trotterization works because when we divide the time into tiny pieces, the commutator terms become very small. For example,

$$e^{(A+B)\Delta t} \approx e^{A\Delta t} e^{B\Delta t}$$

up to error of order $(\Delta t)^2$ caused by the $[A, B]$ term in the BCH expansion.

Thus, product formulas can be viewed as repeatedly applying a small-step approximation in which the BCH error terms are suppressed by powers of the step size. Higher-order Suzuki formulas systematically cancel more of these error terms.

---

2   A Fundamental Lower Bound: No Fast Forwarding

While higher-order product formulas can further suppress the dependence on ε (at the cost of increased circuit depth per step), the scaling with respect to the total evolution time t cannot generally be reduced below linear. This fundamental speed limit is formalized by the *no-fast-forwarding (NFF) theorem*. It states that for a generic or "black-box" Hamiltonian, implementing the evolution operator e^{-iHt} to constant accuracy requires $\Omega(t)$ queries.

In essence, the NFF theorem implies that there is no universal shortcut to quantum dynamics. While algorithmic refinements can reduce overhead factors— such as the number of Trotter slices needed for precision—they cannot "jump ahead" to simulate arbitrary dynamics in time $o(t)$. It is important to note that this lower bound is computational (typically proved within a black-box model) rather than a constraint of relativistic causality. Generic quantum evolution is computationally irreducible; fast-forwarding is possible only for specific classes of Hamiltonians that possess additional, exploitable algebraic structure.

3   Where Product Formulas Sit in the Algorithm Landscape

Although product formulas are conceptually simple, physically motivated, and widely used, their cost can grow quickly for long simulation times or very small error tolerance. The more advanced methods introduced in the next chapter—block-encoding, qubitization, and quantum signal processing—achieve asymptotically optimal scaling for general Hamiltonians, but Trotterization remains valuable because of its simplicity and its strong performance in many structured models.

Further mathematical background can be found in the discussion of the *BCH, Trotter, and Trotter–Suzuki Formulas* in QC Math [1], Section 11.3: *Matrix-Valued Functions.*

10.3   ✳ Pauli Hamiltonians and Their Direct Implementations

A particularly important special case of

$$H = \sum_{j=1}^{m} H_j$$

is when each H_j is proportional to a Pauli string. Concretely, we consider Hamiltonians of the form

$$H = \sum_{j=1}^{m} a_j \boldsymbol{P}_j, \qquad a_j \in \mathbb{R}, \quad \boldsymbol{P}_j \in \{I, X, Y, Z\}^{\otimes n}. \tag{10.4}$$

In the literature, these are often described as Hamiltonians expressed as *sums of Pauli strings, Pauli decompositions*, or simply *Pauli Hamiltonians*.

This representation is important for several reasons:

- Many qubit Hamiltonians in quantum chemistry and condensed matter are naturally (or after mapping) written as sums of Pauli strings and are often sparse in the sense that m is much smaller than 4^n.

- Fermionic ladder operators in second quantization (after, e.g., Jordan–Wigner or Bravyi–Kitaev mappings) become sums of Pauli strings.

- Any square matrix can be expanded in the Pauli-string basis, although the expansion can be dense in the worst case.

Most importantly for product-formula methods, each propagator $e^{-i a_j \boldsymbol{P}_j t}$ can be implemented efficiently as a quantum circuit. Therefore, Pauli Hamiltonians are excellent candidates for Trotterization. They also appear broadly (for example) as cost unitaries in the Quantum Approximate Optimization Algorithm (QAOA) and as structured unitaries in quantum signal processing constructions.

 This section builds on QC Math [1], Chapter 12: *Pauli Matrices, Strings, and Groups.*

10.3.1 Pauli String Operators

A Pauli string operator on n qubits is

$$\boldsymbol{P} = \bigotimes_{j=1}^{n} P_j = P_1 \otimes P_2 \otimes \cdots \otimes P_n, \qquad P_j \in \{I, X, Y, Z\}.$$

For brevity, we often drop the $\otimes$ symbols and write $\boldsymbol{P} = P_1 P_2 \cdots P_n$, implicitly indicating that P_j acts on qubit j.

1 Properties and Exponentiation

Pauli strings are Hermitian $(\boldsymbol{P}^\dagger = \boldsymbol{P})$, unitary $(\boldsymbol{P}^\dagger \boldsymbol{P} = I)$, and involutory $(\boldsymbol{P}^2 = I)$. Hence

$$e^{i\theta \boldsymbol{P}} = \cos\theta\, I + i\sin\theta\, \boldsymbol{P}, \tag{10.5}$$

for real θ.

If $\boldsymbol{P}\,|\psi\rangle = \lambda\,|\psi\rangle$ with $\lambda \in \{+1, -1\}$, then $|\psi\rangle$ is also an eigenvector of $e^{-i\theta \boldsymbol{P}}$ with eigenvalue $e^{-i\theta\lambda}$.

2 Diagonalization via Local Basis Changes

A convenient circuit viewpoint is to map each non-identity single-qubit Pauli to Z using a *local* basis change. A standard choice is:

$$HZH = X, \qquad (SH)\,Z\,(HS^\dagger) = Y, \qquad IZI = Z, \tag{10.6}$$

where H is the Hadamard gate and $S = \mathrm{diag}(1, i)$ is the phase gate. Equivalently, to *diagonalize* a Pauli operator, we use the inverse maps

$$H^\dagger X H = Z, \qquad (HS^\dagger)Y(SH) = Z, \qquad I^\dagger Z I = Z.$$

For a Pauli string $\boldsymbol{P} = P_1 \cdots P_n$, define a tensor-product basis change

$$\boldsymbol{V}(\boldsymbol{P}) = \bigotimes_{j=1}^{n} V(P_j), \qquad V(P_j) = \begin{cases} I, & P_j = I \text{ or } Z, \\ H, & P_j = X, \\ SH, & P_j = Y. \end{cases} \tag{10.7}$$

Then $\boldsymbol{V}(\boldsymbol{P})^\dagger \boldsymbol{P}\, \boldsymbol{V}(\boldsymbol{P})$ is a tensor product of Z's and I's (i.e., diagonal in the computational basis). This reduces implementing $e^{-i\theta \boldsymbol{P}}$ to implementing a diagonal phase conditioned on a parity of Z eigenvalues, as we will demonstrate next.

10.3.2 Direct Implementation of Pauli Exponentials

We now build the implementation pattern step by step, starting with single-qubit Paulis and then moving to multi-qubit strings. As the examples accumulate, a general recipe will emerge. Readers familiar with the basics can skip directly to § 10.3.2.6.

1 Case $P = Z$

Since $R_z(\varphi) = e^{-i\frac{\varphi}{2}Z}$, we have

$$e^{-i\theta Z} = R_z(2\theta). \tag{10.8}$$

A circuit implementation is shown below:

$$q - \boxed{R_z(2\theta)} -$$

2 Case $P = X$

Using $X = HZH$, we obtain

$$e^{-i\theta X} = H\, e^{-i\theta Z}\, H = H\, R_z(2\theta)\, H. \tag{10.9}$$

The corresponding circuit is shown below:

$$q - \boxed{H} - \boxed{R_z(2\theta)} - \boxed{H} -$$

The main intuition is that X behaves like Z after a change of basis. The Hadamard gate diagonalizes X by converting it into Z, so implementing $e^{-i\theta X}$ reduces to implementing $e^{-i\theta Z}$ in the rotated basis.

3 Case $P = Y$

Using

$$Y = (SH)\, Z\, (HS^\dagger),$$

we obtain

$$e^{-i\theta Y} = (SH)\, e^{-i\theta Z}\, (HS^\dagger) = SH\, R_z(2\theta)\, HS^\dagger. \tag{10.10}$$

The corresponding circuit is shown below:

$$q - \boxed{S^\dagger} - \boxed{H} - \boxed{R_z(2\theta)} - \boxed{H} - \boxed{S} -$$

Here the intuition is similar to the X case: Y is diagonalized into Z by a local basis change. The difference is that the basis change for Y is SH rather than H.

4 Multi-Qubit Z-Strings and the CNOT Ladder

We now move to multiple qubits. Consider a k-qubit Z-string,

$$Z_{[1:k]} = Z_1 Z_2 \cdots Z_k.$$

The goal is to implement $e^{-i\theta Z_1 \cdots Z_k}$ using only single-qubit gates and CNOT gates.

Recall that each operator Z_j has eigenvalues ± 1, so the product $Z_1 \cdots Z_k$ also has eigenvalues ± 1. The value of this product depends on whether an even or odd number of the qubits are in the $Z = -1$ eigenstate. In this sense, the eigenvalue of $Z_1 \cdots Z_k$ encodes the *parity* of the individual Z eigenvalues across the k qubits.

A standard method is to compute this parity onto a target qubit with a CNOT ladder, apply a single $R_z(2\theta)$ on that target, and then uncompute:

$$e^{-i\theta Z_1 \cdots Z_k} = \left(\text{CNOT ladder}\right) \left(I^{\otimes(k-1)} \otimes R_z(2\theta)\right) \left(\text{CNOT ladder}\right)^\dagger. \tag{10.11}$$

A four-qubit example of the ladder circuit is shown in Fig. 10.3.

Intuitively, the CNOT ladder coherently accumulates the product of the Z eigenvalues onto the target qubit. The $R_z(2\theta)$ gate then applies a phase conditioned on whether this product is $+1$ or -1, and the inverse ladder restores the original computational basis.

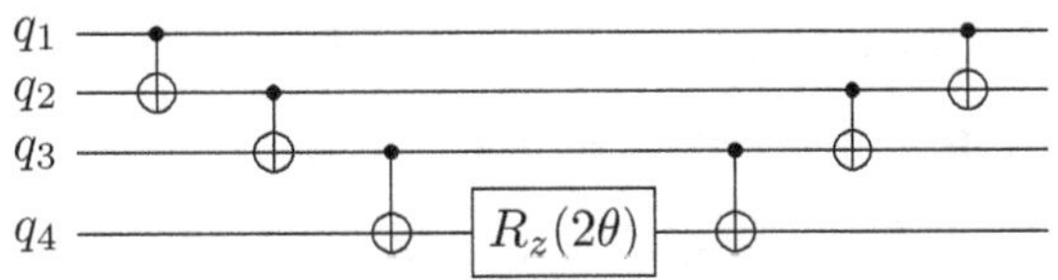

Figure 10.3: Ladder Circuit for $e^{-i\theta Z_1 Z_2 Z_3 Z_4}$

Exercise 10.5 Prove the parity gadget identity Eq. 10.11 by computing its action on a computational-basis state $|z_1 \cdots z_k\rangle$, where $z_j \in \{0,1\}$.

5 A Four-Qubit Pauli Exponential: $\boldsymbol{P} = X_1 Y_2 Z_3 X_4$

Let $\boldsymbol{P} = X_1 Y_2 Z_3 X_4$. Based on the single-qubit cases above, we first diagonalize each local Pauli into Z:

$$X \mapsto Z \text{ via } H, \qquad Y \mapsto Z \text{ via } HS^\dagger, \qquad Z \mapsto Z \text{ via } I.$$

Thus we use the local basis change

$$\boldsymbol{V}(\boldsymbol{P}) = \left(H_1\right)\left((H_2 S_2^\dagger)\right)\left(I_3\right)\left(H_4\right),$$

which maps $\boldsymbol{P}$ to the Z-string $Z_1 Z_2 Z_3 Z_4$ in the rotated basis. We then apply the CNOT ladder to implement $e^{-i\theta Z_1 Z_2 Z_3 Z_4}$ and finally undo the basis change. The full circuit is shown in Fig. 10.4.

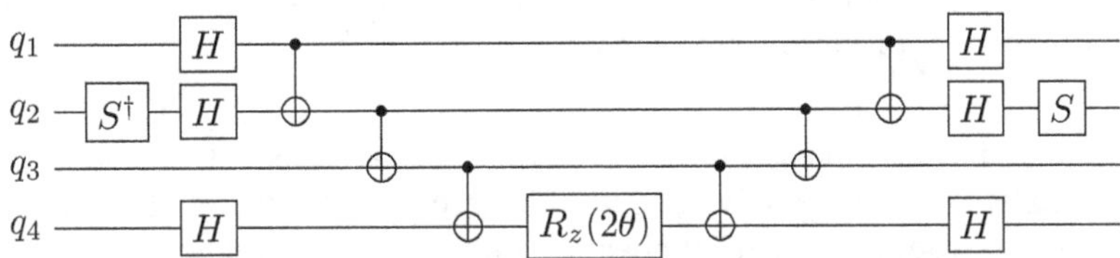

Figure 10.4: Ladder Circuit for $e^{-i\theta(X_1 Y_2 Z_3 X_4)}$

6 General Case $\boldsymbol{P} = P_1 P_2 \cdots P_n$

From the Z, X, and Y examples, we learn that each single-qubit Pauli can be converted into Z by a local basis change. From the multi-qubit Z-string example, we learn that once we have a Z-string on the support qubits, a single $R_z(2\theta)$ together with a CNOT ladder implements the desired exponential. Combining these ideas yields the general circuit.

Define $\boldsymbol{V}(\boldsymbol{P})$ as in Eq. 10.7. Then

$$e^{-i\theta \boldsymbol{P}} = \boldsymbol{V}(\boldsymbol{P}) \, e^{-i\theta \boldsymbol{Z}_{\mathrm{supp}(\boldsymbol{P})}} \, \boldsymbol{V}(\boldsymbol{P})^\dagger. \tag{10.12}$$

Here $\boldsymbol{Z}_{\mathrm{supp}(\boldsymbol{P})}$ is the Z-string obtained by replacing each non-identity factor in $\boldsymbol{P}$ by Z (and leaving identities as I). More precisely, define the *support set* of $\boldsymbol{P}$ as

$$\mathrm{supp}(\boldsymbol{P}) = \{\, j \in \{1,\ldots,n\} : P_j \neq I \,\},$$

then

$$\boldsymbol{Z}_{\mathrm{supp}(\boldsymbol{P})} = \prod_{j \in \mathrm{supp}(\boldsymbol{P})} Z_j,$$

with the understanding that $Z_{\text{supp}(P)}$ acts as identity on qubits outside $\text{supp}(P)$.

A standard implementation of $e^{-i\theta P}$ is:

1. Apply $V(P)^\dagger$ (local single-qubit gates), which maps P to the Z-string $Z_{\text{supp}(P)}$ in the computational basis.

2. Apply a CNOT ladder along the support qubits to compute the Z-parity onto a chosen target qubit.

3. Apply $R_z(2\theta)$ to the target.

4. Uncompute the CNOT ladder.

5. Apply $V(P)$ to return to the original basis.

■ **Example 10.2** Consider the two-qubit Pauli string $P = X_1 Z_2$. Then $\text{supp}(P) = \{1, 2\}$ and $Z_{\text{supp}(P)} = Z_1 Z_2$. The basis-change operator is

$$V(P) = H_1 \otimes I_2,$$

since X is mapped to Z by H and Z requires no change. A standard implementation applies $V(P)^\dagger$, computes the parity of (Z_1, Z_2) onto a chosen target qubit using one CNOT (e.g., control 1 target 2), applies $R_z(2\theta)$ on the target, and then uncomputes and applies $V(P)$. ■

Exercise 10.6 Let $P = X_1 Y_2 I_3 Z_4$. Write down $\text{supp}(P)$, the corresponding $Z_{\text{supp}(P)}$, and the local basis-change circuit $V(P)$. Then specify a CNOT ladder (choose a target qubit) that implements $e^{-i\theta P}$.

This implementation yields an $O(k)$-depth construction (up to connectivity constraints), using only single-qubit gates and CNOTs, where $k = \max_j |\text{supp}(P_j)|$ denotes the maximum number of non-identity Pauli factors appearing in any Pauli string P_j.

Importantly, in product-formula simulation of Pauli-sum Hamiltonians, each Trotter time slice reduces to a sequence of Pauli-string exponentials, and the constructions in this section provide the basic compilation primitive for implementing those steps.

Chapter Takeaways

1. Hamiltonian simulation asks for an efficient circuit approximating $U(t) = e^{-iHt}$; the required accuracy should be stated in a task-dependent way (unitary error vs. errors in measured observables).

2. Product-formula (Trotter–Suzuki) methods approximate $e^{-i(\sum_j H_j)t}$ by interleaving evolutions under the parts; first-order formulas are simplest, while higher-order formulas reduce error per step at the cost of more exponentials.

3. The cost of product formulas is controlled by commutators (non-commutativity) and by how the error accumulates with the number of segments; this motivates the depth–accuracy tradeoff and clarifies when

Trotterization is competitive.

4. Pauli Hamiltonians admit a direct gate-level implementation: each Pauli string exponential $e^{-i\theta P}$ can be compiled by (i) basis changes to map P to $Z \cdots Z$, (ii) an entangling parity computation, and (iii) a single-qubit $R_z(2\theta)$ rotation, followed by uncomputation.

5. Complexity has intrinsic limits: the no-fast-forwarding principle rules out generic sublinear-in-t simulation for broad classes of Hamiltonians, so any substantial improvement must exploit additional structure.

Problem Set 10

10.1 Let $H = X + Y$, where X and Y are Pauli operators.

(a) Since $H^2 = 2I$, the unitary e^{-iHt} admits a compact closed-form expression. Use the extended Euler formula to derive an explicit formula for e^{-iHt}.

(b) Approximate e^{-iHt} using first-order Trotterization. For $t = 20$, plot the error

$$\left\| U_{\text{Trotter}}(t;r) - e^{-iHt} \right\|$$

as a function of the number of time slices r.

10.2 Let $H = H_1 + H_2$ with $\|H_1\|, \|H_2\| \leq 1$.

(a) Show that the leading error term in one first-order step scales with $[H_1, H_2]$.

(b) Explain why first-order Trotterization is exact for any r when $[H_1, H_2] = 0$.

10.3 For each Hamiltonian below, write the corresponding r-step approximation to the propagator e^{-iHt} with $\Delta t = t/r$, and draw a quantum circuit for a single Trotter step (which is then repeated r times). Do this for both the first-order (Lie–Trotter) and second-order (Trotter–Suzuki) product-formula approximations.

(a) $H = \alpha X + \beta Z$

(b) $H = \alpha I \otimes X + \beta Z \otimes Z$

(c) $H = \alpha X \otimes X + \beta Z \otimes Z$

For each Pauli term, use the direct implementation circuits described in § 10.3.

10.4 Let $H = Z_1 Z_2 + X_2$ on two qubits.

(a) Write the first-order and second-order product-formula approximations to e^{-iHt} using $\Delta t = t/r$.

(b) Using the direct Pauli-exponential circuits from § 10.3, draw a circuit for one time slice of each approximation.

   (c) For a fixed t (e.g., $t = 5$), numerically compare the operator-norm error $\|U_{\text{Trotter}}(t; r) - e^{-iHt}\|$ for first- and second-order formulas as r increases.

10.5 Consider the n-qubit Pauli string $\boldsymbol{P} = X_1 X_2 \cdots X_k$ (with identities on the remaining qubits).

   (a) Using the standard compilation pattern in § 10.3.2.6, estimate the number of single-qubit gates and CNOT gates required to implement $e^{-i\theta \boldsymbol{P}}$.

   (b) How do these counts change if hardware connectivity only allows nearest-neighbor CNOTs along a line?

10.6 ✳ Consider a simple spin model such as a 1D transverse-field Ising Hamiltonian on n qubits,

$$H = -J \sum_{j=1}^{n-1} Z_j Z_{j+1} - h \sum_{j=1}^{n} X_j.$$

   (a) Write one second-order (Strang) time slice for e^{-iHt} using $\Delta t = t/r$.

   (b) Choose an initial state (e.g., $|0\rangle^{\otimes n}$) and an observable (e.g., total magnetization $\sum_j Z_j$). Simulate $\langle Z_j \rangle(t)$ for small n and compare first- vs. second-order formulas at the same circuit depth budget.

10.7 ✳ (Observable accuracy vs. unitary accuracy.) Let $|\psi_0\rangle$ be a fixed initial state and O an observable with $\|O\| \leq 1$. Define

$$\langle O \rangle_{\text{exact}}(t) = \langle \psi_0 | e^{iHt} O e^{-iHt} | \psi_0 \rangle, \quad \langle O \rangle_{\text{PF}}(t) = \langle \psi_0 | U_{\text{PF}}(t)^\dagger O U_{\text{PF}}(t) | \psi_0 \rangle,$$

where $U_{\text{PF}}(t)$ denotes a product-formula approximation to e^{-iHt}.

   (a) Show that $|\langle O \rangle_{\text{PF}}(t) - \langle O \rangle_{\text{exact}}(t)|$ can be bounded in terms of $\|U_{\text{PF}}(t) - e^{-iHt}\|$.

   (b) Give a concrete example (small n) where the observable error is much smaller than the unitary error for moderate r.

10.8 ✳ (Multi-product formulas.) Product formulas can be combined to reduce systematic Trotter error by taking weighted linear combinations of expectation values obtained from several shallow circuits.

   (a) Briefly explain the basic idea of a multi-product formula (MPF) at the level of expectation values.

   (b) For a small Hamiltonian of your choice, compare (numerically) a single deeper second-order Trotter circuit with an MPF constructed from multiple shallower second-order circuits that achieves similar accuracy.

11. Polynomial Approximation Algorithms

Contents

This chapter introduces a modern polynomial-approximation (PA) toolkit that underlies many of the best known quantum algorithms for Hamiltonian simulation and related linear-algebraic tasks. The key idea is to move beyond viewing a quantum computer merely as a machine that multiplies unitaries. Instead, we learn how to (i) represent general linear operators inside larger unitaries, and then (ii) apply controlled polynomial transformations to their spectra. The resulting framework—block encoding, quantum signal processing (QSP), qubitization, and quantum singular value transformation (QSVT)—provides a unified language for implementing operator functions such as e^{-iHt}, matrix inversion, eigenvalue filtering, and singular-value transformations with near-optimal asymptotic scaling.

Historically, linear combination of unitaries (LCU) was one of the earliest systematic routes to these ideas. Even though the full PA framework ultimately subsumes LCU, it remains a helpful conceptual bridge: it shows how ancillas and controlled operations can convert "linear algebra on paper" into unitary dynamics on a quantum circuit. We therefore begin with LCU, then formalize the representation step via block encoding, and finally develop quantum signal processing, qubitization, and quantum singular value transformation as progressively more powerful layers of spectral control.

11.1 Product vs. Polynomial Approximation

Hamiltonian simulation hinges on implementing e^{-iHt} efficiently. Over the past decades, several algorithmic approaches have been developed, each based on distinct mathematical principles and suited to different classes of Hamiltonians [31, 36, 37, 38, 39, 40]. At a high level, these approaches fall into two broad categories: product-formula methods and polynomial-approximation methods.

The first category consists of *product-formula methods*, exemplified by Trotterization (see Chapter 10). These methods approximate e^{-iHt} by decomposing H into a sum of simpler terms and then replacing the exponential of the sum by a product of exponentials of the individual terms. Historically, this approach was natural and intuitive: quantum circuits directly implement unitary evolution, and product formulas reflect the idea that a quantum computer is an efficient unitary matrix multiplier.

The second category is based on *polynomial approximation* (PA). Many quantum tasks require applying a function $f(A)$ to an operator A—often a non-unitary operator, or a unitary whose relevant information is most naturally accessed through its spectrum rather than through time evolution. Moreover, low-order product formulas are not asymptotically optimal in general, especially when one demands high precision. Over the past decade, a coherent PA framework has emerged that addresses these limitations by combining two ingredients: a representation of A inside a larger unitary (so that the computation remains unitary overall), and an efficient method for applying carefully designed polynomial transformations to the encoded spectrum.

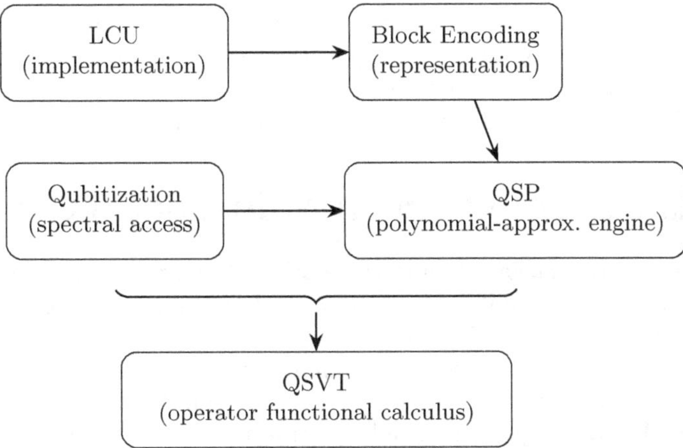

Figure 11.1: Overview of the Polynomial-Approximation Framework

In this chapter, it is useful to view the PA framework as a sequence of conceptual layers (see Fig. 11.1):

- *Linear combination of unitaries (LCU):* an early and intuitive method that realizes certain linear combinations of operators using ancillas, controlled unitaries, and postselection (or amplitude amplification), developed in early Hamiltonian-simulation and LCU constructions by Childs and Wiebe and by Berry, Childs, and Kothari [30, 31].

- *Block encoding:* a unified method for embedding a linear operator into a larger unitary acting on an extended Hilbert space, systematically presented by Gilyén *et al.* [41].

- *Quantum signal processing (QSP):* an optimal technique for implementing polynomial transformations of eigenphases using single-qubit phase rotations interleaved with controlled unitary operations, introduced by Low and Chuang [36].

- *Qubitization:* a construction that converts a block encoding into a structured walk operator whose eigenphases encode the singular values (and, in special cases, eigenvalues) of the embedded operator, developed by Low and Chuang [38].

- *Quantum singular value transformation (QSVT):* a general framework that combines these ideas to implement broad classes of polynomial functions of matrices, including those required for near-optimal Hamiltonian simulation and many other linear-algebraic primitives, developed by Gilyén *et al.* [41].

Conceptually, block encoding and qubitization provide the "handle" by which a general operator is lifted into a unitary setting, while quantum signal processing provides the "engine" that performs controlled polynomial transformations on the associated eigenphases. Quantum singular value transformation packages these steps into a single, versatile principle: once an operator is accessible through an appropriate block encoding, one can implement polynomial transformations of its singular values (and hence a large family of operator functions) using only unitary gates and a modest number of ancillas.

The remainder of this chapter develops these ideas systematically. We begin with LCU to build intuition, then formalize the representation step via block encoding, develop quantum signal processing as the core polynomial engine, and finally introduce qubitization and quantum singular value transformation as the unifying framework for modern Hamiltonian simulation and related algorithms.

11.2 ✳ Linear Combination of Unitaries (LCU)

Linear combination of unitaries (LCU) is a simple method for implementing a weighted sum of unitaries acting on a quantum state using an ancilla register and postselection [30]. Although the resulting transformation is generally non-unitary and probabilistic, LCU serves as a foundational primitive that motivates more powerful frameworks such as block encoding and quantum singular value transformation (QSVT). We begin with small examples to build intuition and then present the general construction.

11.2.1 Introduction with Examples

1 Implementing $(U + V)/2$

As a minimal example, let U and V be n-qubit unitary operators, and let $|\psi\rangle$ be an arbitrary n-qubit state. Our goal is to realize the (generally non-unitary) operator $U + V$ acting on $|\psi\rangle$.

Consider the circuit shown in Fig. 11.2, which uses one ancilla qubit initialized in $|0\rangle$, controlled applications of U and V, and a final Hadamard gate followed by a measurement of the ancilla.

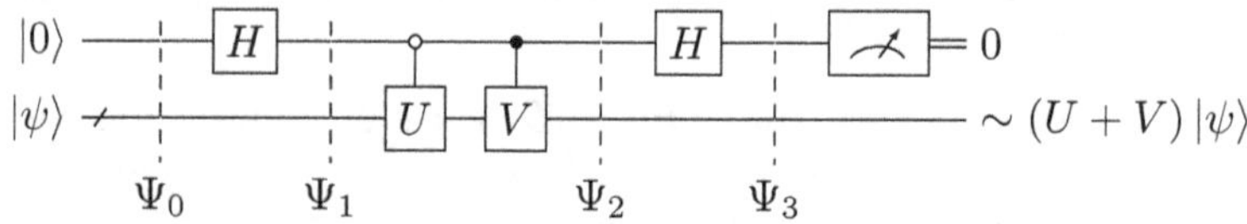

Figure 11.2: LCU Circuit for $(U + V)/2$

To see how the circuit works, start from the initial state $|\Psi_0\rangle = |0\rangle \, |\psi\rangle$. Applying a Hadamard gate to the ancilla produces

$$|\Psi_1\rangle = \frac{1}{\sqrt{2}}(|0\rangle + |1\rangle) \, |\psi\rangle \, .$$

Next, the circuit applies U when the ancilla is in $|0\rangle$ and V when the ancilla is in $|1\rangle$, yielding

$$|\Psi_2\rangle = \frac{1}{\sqrt{2}}(|0\rangle \, U \, |\psi\rangle + |1\rangle \, V \, |\psi\rangle).$$

Applying the final Hadamard gate on the ancilla gives

$$|\Psi_3\rangle = \frac{1}{2} \, |0\rangle \, (U + V) \, |\psi\rangle + \frac{1}{2} \, |1\rangle \, (U - V) \, |\psi\rangle \, .$$

If we now measure the ancilla and postselect on outcome $|0\rangle$, the unnormalized system state becomes $\frac{1}{2}(U+V)|\psi\rangle$, which is the desired linear combination up to a known scalar factor. The corresponding success probability is

$$p_0 = \left\| \frac{1}{2}(U+V)|\psi\rangle \right\|^2,$$

and the normalized postselected output state is

$$|\psi_0\rangle = \frac{(U+V)|\psi\rangle}{\|(U+V)|\psi\rangle\|}.$$

For completeness, if the ancilla measurement outcome is $|1\rangle$, the system is projected onto the normalized state

$$|\psi_1\rangle = \frac{(U-V)|\psi\rangle}{\|(U-V)|\psi\rangle\|},$$

which follows from the same interference mechanism.

> **Exercise 11.1** In the LCU circuit of Fig. 11.2, take $U = V = I$. Simplify the circuit and determine the resulting operation on $|\psi\rangle$.

> **Exercise 11.2** In the LCU circuit of Fig. 11.2, take $U = Z$ and $V = I$. Simplify the circuit and identify the effective operation on $|\psi\rangle$ when the ancilla is postselected in $|0\rangle$.

> **Exercise 11.3** The LCU circuit in Fig. 11.2 employs $\bar{c}U$ and cV, that is, a $|0\rangle$-controlled U and a $|1\rangle$-controlled V. What happens if we instead use cU and $\bar{c}V$, or cU and cV?

2   Implementing $\alpha U + \beta V$

We now generalize the construction to $\alpha U + \beta V$, where α and β are real and nonnegative coefficients. (The case of complex coefficients will be treated in § 11.2.2.) The key idea is to replace the Hadamard gates in Fig. 11.2 with a state-preparation unitary that encodes the desired weights.

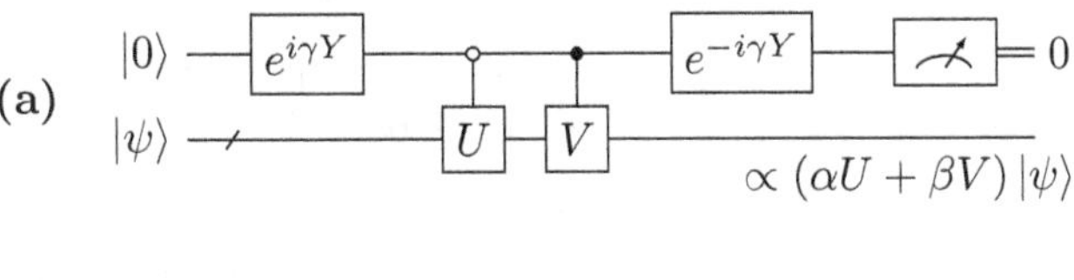

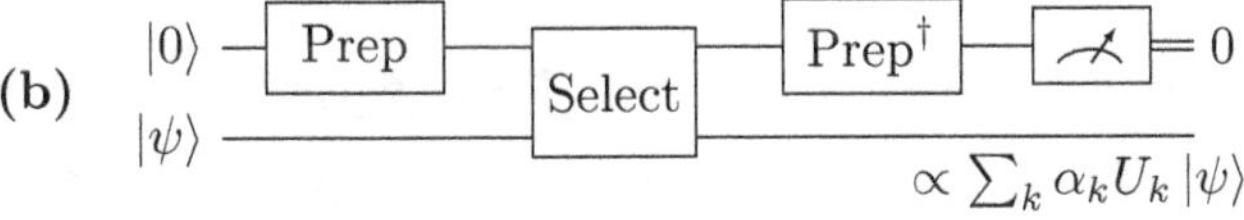

Figure 11.3: LCU Circuit for $\alpha U + \beta V$ and Its Abstract Form

Define

$$s = \alpha + \beta, \qquad a = \sqrt{\frac{\alpha}{s}}, \qquad b = \sqrt{\frac{\beta}{s}}, \qquad \text{so that} \qquad a^2 + b^2 = 1.$$

As shown in Fig. 11.3(a), we first prepare the ancilla in the superposition

$$|0\rangle \xrightarrow{\text{Prep}} a\,|0\rangle + b\,|1\rangle .$$

For a single qubit, Prep can be implemented by a Y-rotation $e^{i\gamma Y}$ (or $R_y(2\gamma)$) with $\cos\gamma = a$.

Next, we apply U when the ancilla is $|0\rangle$ and V when the ancilla is $|1\rangle$, and finally apply $\text{Prep}^\dagger$ to interfere the two branches. Starting from $|0\rangle\,|\psi\rangle$, the evolution is

$$|0\rangle\,|\psi\rangle \xrightarrow{\text{Prep}} \left(a\,|0\rangle + b\,|1\rangle\right)|\psi\rangle \tag{11.1a}$$

$$\xrightarrow{\text{Select}(U,V)} a\,|0\rangle\,U\,|\psi\rangle + b\,|1\rangle\,V\,|\psi\rangle \tag{11.1b}$$

$$\xrightarrow{\text{Prep}^\dagger} |0\rangle\,\frac{\alpha U + \beta V}{s}\,|\psi\rangle + |1\rangle\,|\Phi_\perp\rangle , \tag{11.1c}$$

where $|\Phi_\perp\rangle$ denotes some state orthogonal to $|0\rangle$ on the ancilla. Measuring the ancilla and postselecting on outcome $|0\rangle$ therefore implements the linear map

$$\frac{\alpha U + \beta V}{s}$$

on the system register, or equivalently, $\alpha U + \beta V$ up to the known normalization factor s.

The corresponding success probability is

$$p_0 = \left\| \frac{(\alpha U + \beta V)\,|\psi\rangle}{s} \right\|^2 ,$$

which depends on the input state $|\psi\rangle$ and on the interference between the two branches.

$\ast$ **Proof of Eq. 11.1c**

Suppose after applying $\text{Prep}^\dagger$ the joint state is

$$|\Psi_3\rangle = \left(\text{Prep}^\dagger \otimes I\right)\left(a\,|0\rangle\,U\,|\psi\rangle + b\,|1\rangle\,V\,|\psi\rangle\right) = |0\rangle\,|\Phi\rangle + |1\rangle\,|\Phi_\perp\rangle .$$

To justify Eq. 11.1c, we want to calculate $|\Phi\rangle$, which is obtained by projecting the ancilla qubit of $|\Psi_3\rangle$ onto $|0\rangle$, namely,

$$|\Phi\rangle = (\langle 0| \otimes I)\,|\Psi_3\rangle$$

$$= (\langle 0| \otimes I)(\text{Prep}^\dagger \otimes I)\left(a\,|0\rangle\,U\,|\psi\rangle + b\,|1\rangle\,V\,|\psi\rangle\right).$$

Since $(\langle 0| \otimes I)(\text{Prep}^\dagger \otimes I) = (\langle 0|\,\text{Prep}^\dagger) \otimes I$, we obtain

$$|\Phi\rangle = \left((\langle 0|\,\text{Prep}^\dagger) \otimes I\right)\left(a\,|0\rangle\,U\,|\psi\rangle + b\,|1\rangle\,V\,|\psi\rangle\right).$$

Because $\mathrm{Prep}\,|0\rangle = a\,|0\rangle + b\,|1\rangle$, it follows that $\langle 0|\,\mathrm{Prep}^\dagger = (\mathrm{Prep}\,|0\rangle)^\dagger = a\,\langle 0| + b\,\langle 1|$. Therefore,

$$
\begin{aligned}
|\Phi\rangle &= \big((a\,\langle 0| + b\,\langle 1|) \otimes I\big)\big(a\,|0\rangle\,U\,|\psi\rangle + b\,|1\rangle\,V\,|\psi\rangle\big) \\
&= a^2 U\,|\psi\rangle + b^2 V\,|\psi\rangle \\
&= \frac{\alpha U + \beta V}{s}\,|\psi\rangle .
\end{aligned}
$$

Thus the post-$\mathrm{Prep}^\dagger$ state can be decomposed as $|0\rangle\,\frac{\alpha U+\beta V}{s}\,|\psi\rangle + |1\rangle\,|\Phi_\perp\rangle$.

This example illustrates the core LCU pattern: coherently prepare a superposition that encodes the coefficients, apply a controlled selection of unitaries, and then interfere the branches so that one measurement outcome extracts the desired linear combination. This structure is summarized in the abstract circuit shown in Fig. 11.3(b).

Exercise 11.4 In the LCU circuit of Fig. 11.3, explain why the ancilla is prepared using amplitudes $\sqrt{\alpha}$ and $\sqrt{\beta}$ rather than α and β directly.

Exercise 11.5 In the LCU circuit of Fig. 11.3, the ancilla qubit is used only to *control* the application of U and V. One might therefore expect the ancilla to return deterministically to its initial state $|0\rangle$ after applying $\mathrm{Prep}^\dagger$. Explain why this intuition is incorrect.

Key Takeaways

1. LCU creates interference between controlled branches so that one measurement outcome extracts a desired linear combination (up to a known scale factor).

2. The extracted map is generally non-unitary, so the circuit succeeds only with probability $p_0 = \|A\,|\psi\rangle / \alpha\|^2$.

3. The "failure" outcome corresponds to a different linear combination (e.g., $U - V$) and is typically discarded or handled by amplification (OAA).

3 Implementing $\alpha_0 I + \alpha_1 U + \alpha_2 U^2 + \alpha_3 U^3$

Next, we extend the abstract LCU circuit in Fig. 11.3(b) to implement

$$
A = \sum_{j=0}^{3} \alpha_j U^j,
$$

assuming that the coefficients α_j are nonnegative real numbers.

Since there are four terms in the sum, two ancilla qubits are required to encode the indices in binary. This leads to the circuit shown in Fig. 11.4(a). The Prep stage

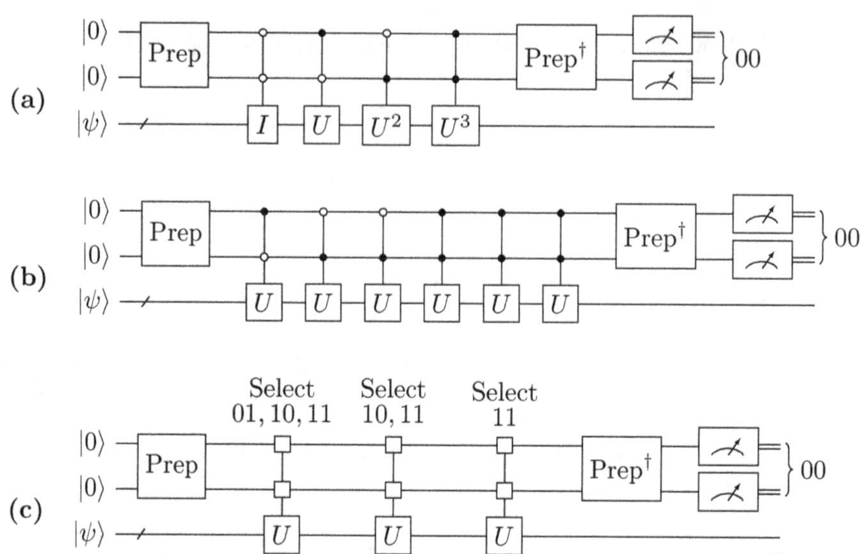

Figure 11.4: LCU Circuit for $\alpha_0 I + \alpha_1 U + \alpha_2 U^2 + \alpha_3 U^3$

prepares the ancilla superposition

$$|00\rangle \xrightarrow{\text{Prep}} \frac{\sqrt{\alpha_0}\,|00\rangle + \sqrt{\alpha_1}\,|01\rangle + \sqrt{\alpha_2}\,|10\rangle + \sqrt{\alpha_3}\,|11\rangle}{\sqrt{\alpha}},$$

or equivalently,

$$|0\rangle \xrightarrow{\text{Prep}} \frac{\sum_{k=0}^{3} \sqrt{\alpha_k}\,|k\rangle}{\sqrt{\alpha}},$$

where $\alpha = \sum_{k=0}^{3} \alpha_k$. The subsequent Select and Prep† stages follow the same pattern as in the previous examples. Although the $\alpha_0 I$ term does not apply any operation to the system register, it plays an essential role in the interference and normalization of the linear combination.

If the power U^k is implemented by applying U sequentially k times rather than as a single gate, the circuit can be rewritten as in Fig. 11.4(b). These controlled-U operations can be further combined, leading to the more compact construction shown in Fig. 11.4(c). In this representation, for example, the first U is applied whenever the ancilla is in one of the states $|01\rangle$, $|10\rangle$, or $|11\rangle$, ensuring that exactly k applications of U occur when the ancilla encodes the value k.

11.2.2 General Formulation of LCU

With the preceding examples, we are now ready to formalize LCU as a general algorithmic primitive. The LCU framework expresses a (generally non-unitary) operator as a weighted sum of efficiently implementable unitaries,

$$A = \sum_{k=0}^{N-1} \alpha_k U_k. \tag{11.2}$$

We assume that the coefficients α_k are nonnegative real numbers, without loss of generality. If a coefficient α_k is complex, its phase can be absorbed into the corresponding unitary by redefining $U_k \mapsto e^{i\theta_k} U_k$, where $\theta_k = \arg(\alpha_k)$, and using $|\alpha_k|$ in the state preparation step.

1   Circuit Implementation

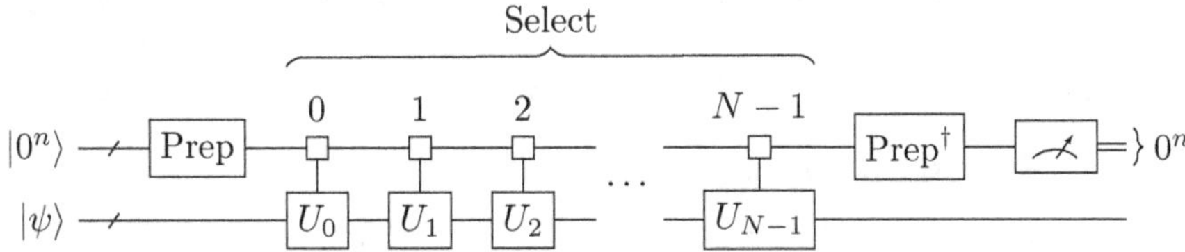

Figure 11.5: General LCU Circuit

The quantum circuit implementing this LCU construction is shown in Fig. 11.5. We assume an n-qubit ancilla register, allowing up to $N = 2^n$ distinct unitaries to be selected and coherently controlled.

Define the normalization constant

$$\alpha = \sum_{k=0}^{N-1} \alpha_k,$$

and let the state-preparation unitary Prep act on the ancilla register as

$$|0^n\rangle \xrightarrow{\text{Prep}} \sum_{k=0}^{N-1} \sqrt{\frac{\alpha_k}{\alpha}} \, |k\rangle.$$

Starting from the joint initial state $|0^n\rangle |\psi\rangle$, the combined evolution of the ancilla and system registers proceeds as

$$|0^n\rangle |\psi\rangle \xrightarrow{\text{Prep}} \sum_{k=0}^{N-1} \sqrt{\frac{\alpha_k}{\alpha}} \, |k\rangle |\psi\rangle \tag{11.3a}$$

$$\xrightarrow{\text{Select}} \sum_{k=0}^{N-1} \sqrt{\frac{\alpha_k}{\alpha}} \, |k\rangle U_k |\psi\rangle \tag{11.3b}$$

$$\xrightarrow{\text{Prep}^\dagger} |0^n\rangle \frac{A}{\alpha} |\psi\rangle + |\Phi_\perp\rangle, \tag{11.3c}$$

where $|\Phi_\perp\rangle$ denotes a state whose ancilla component is orthogonal to $|0^n\rangle$. This decomposition separates the component proportional to $A |\psi\rangle$ from all remaining terms that have no support on the ancilla state $|0^n\rangle$.

The proof of Eq. 11.3 closely parallels that of Eq. 11.1c in § 11.2.1.2, and we therefore omit it here. A detailed derivation is left as an exercise for the reader (Problem 11.3).

2   Measurement and Success Probability

From Eq. 11.3, we see that measuring the ancilla register and postselecting on the outcome $|0^n\rangle$ implements the linear map A/α on the system register. Thus, the LCU circuit realizes the operator A up to a known normalization factor, albeit in a probabilistic manner.

The corresponding success probability is

$$p_0 = \left\| \frac{A\,|\psi\rangle}{\alpha} \right\|^2 , \tag{11.4}$$

which depends on the input state $|\psi\rangle$ and on the spectral properties of A. In practical LCU-based algorithms, this probability is typically boosted to near unity using amplitude amplification techniques, which we will introduce in § 9.3.2.

3   Key Relations and Block-Encoding Viewpoint

Let U_{LCU} denote the overall unitary transformation implemented by the LCU circuit. By construction,

$$U_{\mathrm{LCU}} = (\mathrm{Prep}^\dagger \otimes I)\, \mathrm{Select}\, (\mathrm{Prep} \otimes I), \tag{11.5}$$

where Prep prepares the ancilla superposition encoding the coefficients α_k, and Select applies the unitary U_k conditioned on the ancilla state $|k\rangle$.

The action of U_{LCU} on the initial state can be summarized as

$$U_{\mathrm{LCU}} |0^n\rangle\, |\psi\rangle = |0^n\rangle\, \frac{A}{\alpha}\, |\psi\rangle + |\Phi_\perp\rangle , \tag{11.6}$$

from which it follows that

$$\langle 0^n|\, U_{\mathrm{LCU}}\, |0^n\rangle = \frac{A}{\alpha}. \tag{11.7}$$

Equation 11.7 shows that the operator A is embedded as the top-left block of the unitary U_{LCU}, up to normalization. In other words, U_{LCU} constitutes a block encoding of A, which we will develop in § 11.3.

Key Takeaways

---

1. Given $A = \sum_k \alpha_k U_k$ with $\alpha_k \geq 0$, LCU builds a unitary $U_{\mathrm{LCU}} = (\mathrm{Prep}^\dagger \otimes I)\, \mathrm{Select}\, (\mathrm{Prep} \otimes I)$.

2. The $|0^n\rangle$ ancilla branch contains $(A/\alpha)\,|\psi\rangle$, where $\alpha = \sum_k \alpha_k$.

3. LCU is best viewed as a constructor: it turns an LCU decomposition into a reusable unitary that "encodes" A.

11.2.3   Polynomial Functions of Unitaries via LCU

We can extend the construction shown in Fig. 11.4(c) to implement a general polynomial function of a unitary operator,

$$f(U) = \sum_{k=0}^{N-1} \alpha_k U^k ,$$

using the LCU framework. The resulting circuit implementation is illustrated in Fig. 11.6. Any polynomial with efficiently computable coefficients can be realized in this manner, provided the corresponding state preparation and controlled unitary operations are efficient.

In the circuit, the phase factors $e^{i\theta_k}$ explicitly account for the complex phases of the polynomial coefficients and are absorbed into the controlled unitary blocks. As

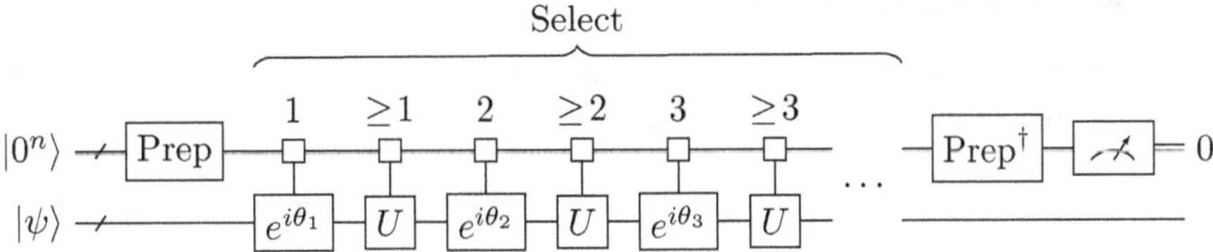

Figure 11.6: LCU Circuit for Polynomial Functions of a Unitary Operator

a result, the remaining weights α_k used in the state preparation procedure can be taken to be real and nonnegative.

An alternative implementation is presented in Problem 11.6 and Fig. 11.4. This construction is inspired by the quantum phase estimation (QPE) circuit and is particularly efficient when powers of the form U^{2^k} can be implemented efficiently as elementary operations.

Truncated Taylor series arise as a natural and important special case of this general construction. When f is an analytic function, it can be approximated by truncating its Taylor expansion, yielding a polynomial of the above form. This observation underlies the original LCU-based Hamiltonian simulation algorithms, where the time-evolution operator is approximated as

$$e^{iHt} \approx \sum_{k=0}^{N-1} \frac{(it)^k}{k!} H^k.$$

In this setting, the truncation order N controls the approximation error, while the LCU circuit implements the resulting polynomial in H.

11.2.4 ✳ Oblivious Amplitude Amplification

A major limitation of the LCU approach, as shown in Eq. 11.3, is that the success probability

$$p_0 = \left\| \frac{A \, |\psi\rangle}{\alpha} \right\|^2$$

of obtaining the desired ancilla outcome $|0^n\rangle$ typically decreases as the number of terms in the decomposition increases. (Here p_0 depends on the input state $|\psi\rangle$, but we suppress this dependence for notational simplicity.) For truncated Taylor-series expansions of the time-evolution operator, the normalization constant $\alpha = \sum_k \alpha_k$ grows with the truncation order, leading to very small p_0 and a prohibitive repetition overhead if one simply measures and postselects.

Oblivious amplitude amplification (OAA, see § 9.3.2) resolves this issue by *coherently* amplifying the amplitude on the good ancilla subspace (ancilla $|0^n\rangle$), without measuring the ancilla and without requiring prior knowledge of the system state $|\psi\rangle$ [31, 36]. The procedure is termed "oblivious" because the reflections used in the amplification depend only on the known LCU unitary U_{LCU} and the fixed ancilla-defined subspace, and are independent of $|\psi\rangle$.

An Illustrative Example

Consider the simplest nontrivial case $A = \alpha_0 U_0 + \alpha_1 U_1$, implemented using a single ancilla qubit. The LCU unitary prepares

$$U_{\mathrm{LCU}}\big(|0\rangle\,|\psi\rangle\big) = |0\rangle \otimes \frac{A\,|\psi\rangle}{\alpha} + |1\rangle \otimes |\Phi_\perp\rangle \equiv |G\rangle + |B\rangle,$$

where $\alpha = \alpha_0 + \alpha_1$ and $|\Phi_\perp\rangle$ is orthogonal to all states with ancilla $|0\rangle$. At this stage, the LCU construction is probabilistic: measuring the ancilla would yield the desired outcome with probability p_0.

OAA begins from this unmeasured state. Using reflections about $\mathcal{S}$ (the ancilla-$|0\rangle$ subspace) and their conjugates under U_{LCU}, one constructs an iterate that rotates amplitude toward $\mathcal{S}$. See § 9.3.2 for the general construction of OAA.

As a simple numerical case, assume $\|A\,|\psi\rangle\| = 1$ and $\alpha = 2$, so that $p_0 = 1/4$. Since $\sin\theta = \sqrt{p_0} = 1/2$ (i.e., $\theta = \pi/6$), a single application of the OAA iterate (Eq. 9.9) rotates the state by $2\theta = \pi/3$, yielding a final amplitude $\sin(3\theta) = \sin(\pi/2) = 1$ on the good component. Thus, after one iteration the state is (up to a global phase) $|G\rangle$, giving a deterministic implementation of A in this idealized setting.

In general, the optimal number of iterations depends on the initial value of p_0, but robust or fixed-point variants of OAA (see § 9.3.1) can achieve near-unit success probability without requiring precise knowledge of p_0.

Key Takeaways

1. Postselection makes LCU probabilistic, and p_0 can be very small when $\alpha = \sum_k \alpha_k$ is large (e.g., truncated Taylor series).

2. Oblivious amplitude amplification boosts the good ancilla amplitude coherently, without measuring and without depending on $|\psi\rangle$.

3. After OAA, the same LCU unitary becomes a near-deterministic way to apply A (up to normalization).

11.3　＊Block Encoding

In the previous section, we saw that LCU provides a systematic way to implement a weighted sum of unitaries using ancilla registers, controlled selection, and postselection. While powerful, the LCU construction is inherently probabilistic and framed in terms of postselected linear maps. In this section, we introduce *block encoding*, which recasts the same idea in a more structural and reusable form.

Block encoding provides a uniform way to represent general operators as subblocks of unitaries, enabling further processing using fully unitary circuits. It serves as the representation layer for modern quantum algorithms based on polynomial transformations.

11.3.1 From LCU to Block Encoding

Recall from § 11.2.2 that an operator

$$A = \sum_{k=0}^{N-1} \alpha_k U_k$$

can be implemented probabilistically via LCU by preparing an ancilla superposition, applying a controlled Select operation, and postselecting on the ancilla state $|0^n\rangle$. The key observation is that the entire LCU circuit defines a unitary operator acting jointly on the ancilla and system registers, even though the induced action on the system alone is non-unitary.

Block encoding makes this observation explicit. Instead of viewing the LCU circuit as a postselected procedure, we regard it as a single unitary whose top-left block is proportional to the desired operator.

1 Definition of Block Encoding

Let A be an operator acting on an m-qubit system. A unitary U acting on $m + n$ qubits is said to be an (α, n) *block encoding* of A if

$$\langle 0^n | U | 0^n \rangle = \frac{A}{\alpha}, \tag{11.8}$$

where $\alpha > 0$ is a known normalization factor. Equivalently, in a basis ordered by the ancilla register, the unitary U can be written schematically as

$$U = \begin{bmatrix} A/\alpha & * \\ * & * \end{bmatrix}. \tag{11.9}$$

The defining feature of a block encoding is that the desired operator A appears as a sub-block of a larger unitary, without measurement or postselection. The ancilla register provides the extra degrees of freedom required to embed a generally non-unitary operator into a unitary one.

Any unitary operator acting on a system register is trivially a $(1, 0)$ block encoding of itself.

In Eq. 11.8, $\langle 0^n | U | 0^n \rangle$ denotes a partial matrix element taken over the ancilla Hilbert space. It is shorthand for

$$(\langle 0^n | \otimes I)\, U\, (|0^n\rangle \otimes I),$$

where the identity operator acts on the system register. The resulting object is therefore an operator acting on the system Hilbert space.

As an example, the unitary

$$U = \begin{bmatrix} 1 & 0 & 0 & 0 \\ 0 & 0 & 1 & 0 \\ 0 & 0 & 0 & 1 \\ 0 & 1 & 0 & 0 \end{bmatrix} = \begin{bmatrix} \Pi & * \\ * & * \end{bmatrix}$$

is a $(1, 1)$ block encoding of the projector

$$\Pi = \begin{bmatrix} 1 & 0 \\ 0 & 0 \end{bmatrix}.$$

In this case, defining

$$V = |0\rangle \otimes I = \begin{bmatrix} I \\ 0 \end{bmatrix},$$

we can directly verify Eq. 11.8 via $V^\dagger U V = \Pi$.

Exercise 11.6 Express $|0^2\rangle \otimes I_2$ explicitly as a matrix. More generally, find the matrix dimensions of $|0^n\rangle \otimes I_{2^m}$.

Exercise 11.7 Let U_A be an (α, n) block encoding of A. Show that $U_A^\dagger$ is also an (α, n) block encoding of $A^\dagger$.

Key Takeaways

1. A block encoding stores A/α as a selected sub-block of a unitary U using an n-qubit ancilla and the projector $|0^n\rangle\langle 0^n|$.

2. The defining relation is $(\langle 0^n| \otimes I)\, U\, (|0^n\rangle \otimes I) = A/\alpha$.

3. Once A is block-encoded, later algorithms manipulate U unitarily, without postselection.

2   LCU as a Block Encoding

The LCU construction from the previous section naturally produces a block encoding. Consider the unitary U_{LCU} given by Eq. 11.5. As shown in § 11.2.2,

$$\langle 0^n| U_{\text{LCU}} |0^n\rangle = \frac{A}{\alpha}.$$

Thus, U_{LCU} is an (α, n) block encoding of A.

From this perspective, LCU is best understood not as a standalone algorithm, but as a method for constructing block encodings. This reinterpretation is crucial, because once an operator is available in block-encoded form, a wide range of transformations become possible without further postselection.

For example, a truncated Taylor series can be efficiently block-encoded via LCU (see § 11.2.3). In addition, sums of Pauli operators,

$$A = \sum_j c_j P_j,$$

can be block-encoded using LCU with normalization $\alpha = \sum_j |c_j|$. This case covers most Hamiltonians encountered in quantum simulation.

11.3.2   Block Encoding Beyond LCU

Block encoding is not limited to LCU constructions. Below are several important examples.

1 Halmos Dilation

Consider a (not necessarily unitary) contraction A satisfying $\|A\|_2 \leq 1$. The Halmos dilation provides a systematic way to embed A into a unitary operator acting on a larger Hilbert space. In its standard form, the dilation is written as

$$U_A = \begin{bmatrix} A & \sqrt{I - AA^\dagger} \\ \sqrt{I - A^\dagger A} & -A^\dagger \end{bmatrix}, \tag{11.10}$$

where I denotes the identity operator of appropriate dimension. When A is square, U_A is a unitary operator acting on the direct sum space $\mathcal{H} \oplus \mathcal{H}$.

In particular, if A is Hermitian, then $A = A^\dagger$ and the dilation simplifies to

$$U_H = \begin{bmatrix} H & \sqrt{I - H^2} \\ \sqrt{I - H^2} & -H \end{bmatrix}, \tag{11.11}$$

which is itself Hermitian and unitary. In this case, U_H is a reflection and provides a Hermitian block encoding of H.

The condition $\|A\|_2 \leq 1$ is essential for the Halmos dilation to be well defined, since it ensures that $I - AA^\dagger$ and $I - A^\dagger A$ are positive semidefinite and therefore admit Hermitian square roots.

Block encodings are not unique. For a contraction A, alternative unitary dilations can be obtained by permuting blocks or adjusting signs. For example,

$$\begin{bmatrix} A & \sqrt{I - AA^\dagger} \\ -\sqrt{I - A^\dagger A} & A^\dagger \end{bmatrix}, \qquad \begin{bmatrix} \sqrt{I - AA^\dagger} & A \\ A^\dagger & -\sqrt{I - A^\dagger A} \end{bmatrix}$$

are also valid unitary dilations of A.

When A is rectangular, it is often more natural to place A in an off-diagonal block, as in the second form above, so that the row and column block partitions remain consistent. This viewpoint underlies the singular-value block encodings.

Exercise 11.8 Provide a Hermitian block encoding of the matrix

$$A = \begin{bmatrix} 0.5 & 0 \\ 0 & 0.5 \end{bmatrix}.$$

Exercise 11.9 Assume A is Hermitian and $\|A\| \leq 1$. For the Hermitian Halmos dilation Eq. 11.11 verify that $U_H^2 = I$. Conclude that U_H is a reflection.

Exercise 11.10 Let A be a contraction ($\|A\| \leq 1$) and U_A be its Halmos dilation. Show directly that the columns of U_A are orthonormal, hence $U_A^\dagger U_A = I$.

2 Block Encoding in Qubitization

A key application of block encoding is *qubitization*, which we will develop in § 11.5.

3   Singular-Value Block Encoding

Singular-value block encodings provide the interface between general linear operators and quantum singular value transformation (QSVT). In particular, they explain why QSVT naturally acts on singular values (and, in the Hermitian case, on eigenvalues). We will discuss this construction in detail in § 11.6.

4   Sparse Matrix Block Encoding

Sparse-matrix block encodings play a central role in quantum algorithms for linear systems [40, 42, 43], including the Harrow–Hassidim–Lloyd (HHL) algorithm and its variants, which we will explore in § 12.3. They also underpin quantum algorithms for related linear-algebraic tasks, such as certain differential-equation solvers, matrix-function estimation (including matrix exponentiation in structured settings), and algorithms for continuous-time Markov processes.

5   ✳ Density Matrix Block Encoding

Let ρ be a density matrix on an n-qubit system,

$$\rho = \sum_k p_k \, |\psi_k\rangle\langle\psi_k|,$$

where $p_k \geq 0$ and $\sum_k p_k = 1$. Define a purification

$$|\Psi\rangle = \sum_k \sqrt{p_k}\, |k\rangle\, |\psi_k\rangle.$$

Assume there exists a unitary U_ρ such that

$$U_\rho |0\rangle |0^n\rangle = |\Psi\rangle.$$

Then

$$(\langle 0| \otimes I)\, U_\rho^\dagger (|0\rangle \otimes I) = \rho,$$

so $U_\rho^\dagger$ provides a $(1,1)$ block encoding of ρ.

This construction underlies quantum principal component analysis and density matrix exponentiation, where repeated applications of ρ enable spectral estimation and simulation of mixed-state dynamics.

> **Exercise 11.11** What are the dimensions of ρ, $|\Psi\rangle$, and U_ρ here?

> **Exercise 11.12** Why is $U_\rho^\dagger$, not U_ρ, a block encoding of ρ?

11.3.3   Core Relations for Block Encodings

The following relations summarize the block-encoding identities used repeatedly throughout this chapter.

A unitary U_A is an (α, n) block encoding of an operator A if

$$\langle 0^n| U_A |0^n\rangle = \frac{A}{\alpha};$$

schematically, in a basis ordered by the ancilla register,

$$U_A = \begin{bmatrix} A/\alpha & * \\ * & * \end{bmatrix}.$$

An input state $|\psi\rangle$ is embedded by appending the ancilla register in $|0^n\rangle$, i.e.,

$$|0^n\rangle\,|\psi\rangle \equiv \begin{bmatrix} |\psi\rangle \\ 0 \end{bmatrix}. \tag{11.12}$$

Multiplying the block matrix by the extended vector gives

$$U_A\,|0^n\rangle\,|\psi\rangle = \begin{bmatrix} A/\alpha & * \\ * & * \end{bmatrix}\begin{bmatrix} |\psi\rangle \\ 0 \end{bmatrix} = \begin{bmatrix} (A/\alpha)\,|\psi\rangle \\ * \end{bmatrix} = \begin{bmatrix} (A/\alpha)\,|\psi\rangle \\ 0 \end{bmatrix} + \begin{bmatrix} 0 \\ * \end{bmatrix}$$

$$= |0^n\rangle\,\frac{A\,|\psi\rangle}{\alpha} + |\perp\rangle, \tag{11.13}$$

where $|\perp\rangle$ is orthogonal to the $|0^n\rangle$ ancilla subspace.

Applying $(|0^n\rangle\langle 0^n| \otimes I)$ yields

$$(|0^n\rangle\langle 0^n| \otimes I)\,U_A\,|0^n\rangle\,|\psi\rangle = |0^n\rangle\,\frac{A\,|\psi\rangle}{\alpha}. \tag{11.14}$$

Exercise 11.13 Let U_A be an (α, n) block encoding of A.

(a) Show that for any scalar $c \in \mathbb{R}$ with $|c| \le 1$, U_A is also an $(\alpha/|c|, n)$ block encoding of cA.

(b) Suppose you have an (α, n) block encoding of A and want an $(1, n)$ block encoding of A/α. Explain why this is "free" at the level of normalization but *not* free when interpreting success amplitudes in postselection-based viewpoints.

Exercise 11.14 Let U_A be an (α, n) block encoding of A. Prove that $\|A\| \le \alpha$. (Hint: use $\|(\langle 0^n| \otimes I)U_A(|0^n\rangle \otimes I)\| \le \|U_A\| = 1$.)

Key Takeaways

1. A block encoding is a unitary U_A whose $|0^n\rangle$-to-$|0^n\rangle$ block equals A/α.

2. Starting from $|0^n\rangle\,|\psi\rangle$, one application of U_A coherently "applies" A/α on the $|0^n\rangle$ branch.

3. By adding reflections about $|0^n\rangle$ and QSP phases, we can turn this access into polynomial transformations of A (QSVT).

11.4　✳ Quantum Signal Processing (QSP)

Quantum signal processing (QSP) is named by analogy with classical signal processing: one applies a sequence of phase shifts and filters, resulting in an effective transfer function of the form $y = P(x)$.

QSP performs an analogous task on quantum "signals." The signal is the eigenphase θ (or equivalently the eigenvalue $e^{i\theta}$) of a unitary operator, and a carefully designed single-qubit gate sequence produces a unitary whose $(0,0)$ matrix element encodes a desired polynomial transformation.

We explain QSP in two steps. This section introduces the core mechanism: how interleaving fixed rotations about one axis (typically Y) and variable rotations about another axis (typically Z) yields polynomial transformations of a scalar parameter $x \in [-1, 1]$. The next section explains how block encoding and qubitization lift this mechanism from scalars to operators.

11.4.1 The Basic QSP Gate Sequence

We begin with the basic QSP circuit shown in Fig. 11.7, consisting of alternating Z-rotations and Y-rotations.

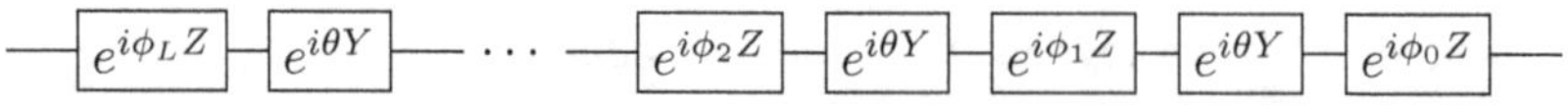

$$-\boxed{e^{i\phi_L Z}}-\boxed{e^{i\theta Y}}- \cdots -\boxed{e^{i\phi_2 Z}}-\boxed{e^{i\theta Y}}-\boxed{e^{i\phi_1 Z}}-\boxed{e^{i\theta Y}}-\boxed{e^{i\phi_0 Z}}-$$

Figure 11.7: Basic Quantum Signal Processing (QSP) Circuit

The Y-rotations are

$$e^{i\theta Y} = \cos\theta\, I + i\sin\theta\, Y = \begin{bmatrix} \cos\theta & \sin\theta \\ -\sin\theta & \cos\theta \end{bmatrix},$$

where θ is real, and the Z-rotations are

$$e^{i\phi_\ell Z} = \begin{bmatrix} e^{i\phi_\ell} & 0 \\ 0 & e^{-i\phi_\ell} \end{bmatrix},$$

where $\{\phi_\ell\}_{\ell=0}^{L}$ are real parameters and L is the depth of the sequence.

The full QSP unitary is

$$U_{\boldsymbol{\phi}}(\theta) = e^{i\phi_0 Z} \prod_{\ell=1}^{L} \left(e^{i\theta Y} e^{i\phi_\ell Z} \right), \tag{11.15}$$

and the central object of interest is the $(0,0)$ matrix element

$$P(\cos\theta) = \langle 0 | \, U_{\boldsymbol{\phi}}(\theta) \, | 0 \rangle,$$

which will be a polynomial in $x = \cos\theta$.

 The apparent reversal of the gate order between Fig. 11.7 and Eq. 11.15 reflects the convention that operators act on states from right to left, while circuit diagrams are read from left to right.

1 Depth-2 Sequence

For a depth-2 sequence,

$$U_{\boldsymbol{\phi}}(\theta) = e^{i\phi_0 Z} e^{i\theta Y} e^{i\phi_1 Z} e^{i\theta Y} e^{i\phi_2 Z},$$

a direct calculation gives

$$\left(U_{\boldsymbol{\phi}}(\theta) \right)_{00} = e^{i(\phi_0 + \phi_2)} \left(\cos\phi_1 \cos 2\theta + i \sin\phi_1 \right).$$

Using $\cos 2\theta = 2\cos^2\theta - 1$, we obtain

$$P(x) = \cos\phi_1(2x^2 - 1) + i\sin\phi_1, \qquad x = \cos\theta,$$

which is a degree-2 polynomial in x.

2 Chebyshev Polynomials

If all phases $\phi_\ell = 0$, the sequence yields

$$P(\cos\theta) = \cos(L\theta),$$

or equivalently

$$P(x) = \cos(L\arccos x) = T_L(x),$$

where $T_L(x)$ is the Chebyshev polynomial of the first kind.

3 Chebyshev Polynomial Expansion

In general, the QSP matrix element takes the form of a finite Chebyshev series

$$P(x) = \sum_{j=0}^{L} c_j T_j(x), \tag{11.16}$$

where the coefficients c_j depend on the chosen phases $\{\phi_\ell\}$.

Since $\{T_j(x)\}$ form an orthogonal basis on $[-1, 1]$, QSP enables approximation of a wide class of target functions by solving, classically, for a suitable phase sequence ϕ and depth L achieving the desired precision.

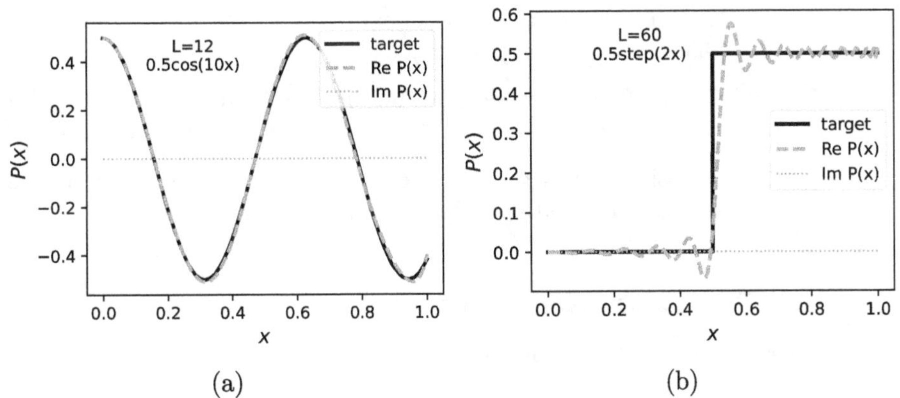

Figure 11.8: QSP Synthesis of Cosine and Step Functions

Figure 11.8 shows two examples simulated with QSPPACK [44]: approximations to $0.5\cos(10x)$ and $0.5\,\mathrm{step}(2x)$, where the scale factor ensures the target remains well within the admissible range of QSP polynomials.

Exercise 11.15 For $L = 1$, compute $P(x) = \langle 0|\, U_\phi(\theta)\, |0\rangle$ explicitly and show that it is an affine function of $x = \cos\theta$.

Exercise 11.16 Assume $\phi_\ell = -\phi_{L-\ell}$ (antisymmetric phases). Show that $P(x)$ in Eq. 11.17 can be chosen real for all $x \in [-1,1]$.

Exercise 11.17 Reformulate the above construction using X-rotations, $U = e^{i\theta X}$, instead of Y-rotations, and verify that the polynomial structure is preserved.

Common Polynomial Approximation Schemes

Polynomial approximations are a central tool in both classical and quantum algorithms, and different expansion schemes optimize different notions of accuracy.

Taylor Series. A Taylor expansion approximates a function locally around a point by matching its derivatives. It provides high accuracy near the expansion point but can perform poorly far from it, making it unsuitable for global approximations on large intervals.

Chebyshev Series. Chebyshev polynomials form an orthogonal basis on the interval $[-1,1]$ and minimize the maximum (uniform) approximation error over the domain. As a result, Chebyshev expansions yield near-minimax approximations and are particularly well suited for QSP-based spectral transformations.

Lagrange Interpolation. Lagrange polynomials interpolate a function exactly at a prescribed set of nodes. While accurate at those points, interpolation can suffer from large oscillations between nodes unless the nodes are carefully chosen (e.g., Chebyshev nodes).

Key Takeaways

1. QSP interleaves a signal-dependent rotation (e.g., $e^{i\theta Y}$) with tunable phase rotations (e.g., $e^{i\phi_\ell Z}$).

2. The $(0,0)$ matrix element becomes a bounded polynomial $P(x)$ in $x = \cos\theta$ (or a Laurent polynomial in $z = e^{i\theta}$ in generalized QSP).

3. Choosing phases ϕ is a classical synthesis problem; running the circuit applies the resulting polynomial transformation coherently.

11.4.2 The Chebyshev QSP Theorem

1 Rotation-Based Formulation

The original Chebyshev QSP theorem [36, 38] characterizes the matrix elements of all unitaries generated by sequences of alternating Z- and Y-rotations (Eq. 11.15). Specifically, it states that any such unitary can be written as

$$U_\phi(\theta) = \begin{bmatrix} P(x) & iQ(x)\sqrt{1-x^2} \\ iQ^*(x)\sqrt{1-x^2} & P^*(x) \end{bmatrix}, \qquad x = \cos\theta, \qquad (11.17)$$

where $P(x)$ is a polynomial of degree at most L, $Q(x)$ is a polynomial of degree at most $L-1$, the parity of $P(x)$ is $L \bmod 2$, and the unitarity constraint

$$|P(x)|^2 + (1-x^2)|Q(x)|^2 = 1$$

holds for all $x \in [-1,1]$.

Every QSP circuit produces an admissible polynomial pair (P, Q) satisfying these conditions, and conversely, every such pair can be synthesized by an appropriate choice of phases ϕ. Thus, QSP provides a complete classification and synthesis theory for SU(2) unitaries whose matrix elements are polynomial functions of a scalar signal $x \in [-1, 1]$.

If $P(x)$ is required to be real-valued, the corresponding phase sequence may be chosen antisymmetric, satisfying $\phi_\ell = -\phi_{L-\ell}$.

2 Reflector-Based Formulation

An alternative but equivalent formulation is often used to simplify circuit implementations. In this formulation, the elementary rotation $e^{i\theta Y}$ is replaced by the Hermitian reflector

$$U = \begin{bmatrix} \cos\theta & \sin\theta \\ \sin\theta & -\cos\theta \end{bmatrix} = \begin{bmatrix} x & \sqrt{1-x^2} \\ \sqrt{1-x^2} & -x \end{bmatrix}. \tag{11.18}$$

This leads to the QSP sequence

$$U_\phi(\theta) = e^{i\phi_0 Z} \prod_{\ell=1}^{L} \left(U\, e^{i\phi_\ell Z} \right). \tag{11.19}$$

This simplification is justified because Z commutes with all operators of the form $e^{i\phi_\ell Z}$. As a result, any extra factors of Z can be absorbed into adjacent phase rotations by redefining $\phi_\ell \mapsto \phi_\ell + \frac{\pi}{2}$. Consequently, the formulation in Eq. 11.19 is equivalent to the original construction Eq. 11.15 up to an overall global phase of $(-i)^L$.

3 Synthesizing Non-Parity or Complex Functions

While the standard Chebyshev QSP enforces a parity constraint on $P(x)$, a general function $f(x)$ may be decomposed into symmetric and antisymmetric parts

$$f_{\text{sym}}(x) = \frac{f(x) + f(-x)}{2}, \qquad f_{\text{antisym}}(x) = \frac{f(x) - f(-x)}{2},$$

which can be synthesized separately and combined using LCU techniques.

A canonical example is Hamiltonian simulation, where we approximate e^{-ixt} by separately synthesizing $\cos(xt)$ and $\sin(xt)$, typically with a scale factor $\alpha < 1$ for numerical stability:

$$\alpha e^{-ixt} = \alpha \cos(xt) - i\alpha \sin(xt).$$

Exercise 11.18 Using Python, optionally with a QSP library, synthesize an approximation to $f(x) = \alpha/x$, which underlies quantum linear systems algorithms.

4 ✳ Proof Sketch of the Chebyshev QSP Theorem

A concise proof proceeds by induction on the depth L.

Base Case ($L = 0$).

The unitary reduces to $U = e^{i\phi_0 Z}$, which has the required form with $P(x) = e^{i\phi_0}$ and $Q(x) = 0$.

Inductive Step.

Assume that after ℓ layers the unitary has the form

$$U_\ell(x) = \begin{bmatrix} P_\ell(x) & iQ_\ell(x)\sqrt{1-x^2} \\ iQ_\ell^*(x)\sqrt{1-x^2} & P_\ell^*(x) \end{bmatrix},$$

with the stated degree and parity properties. Multiplication by $e^{i\theta Y}$ mixes P_ℓ and Q_ℓ through combinations of $x = \cos\theta$ and $\sqrt{1-x^2}$, increasing the polynomial degree by at most one, while multiplication by $e^{i\phi Z}$ applies phase factors without changing the degree.

Unitarity Constraint.

Since each factor in the sequence is unitary, the resulting matrix form enforces

$$|P(x)|^2 + (1-x^2)|Q(x)|^2 = 1 \qquad \text{for all } x \in [-1,1].$$

Completeness.

Conversely, given any admissible polynomial pair (P, Q) satisfying the degree, parity, and unitarity conditions, one can recursively factor the corresponding $SU(2)$-valued polynomial into elementary Y- and Z-rotations, thereby recovering a phase sequence ϕ that realizes the desired QSP unitary.

> ### Key Takeaways
>
> 1. A length-L QSP sequence produces polynomials $P(x)$ ($\deg \leq L$) and $Q(x)$ ($\deg \leq L-1$) with a fixed parity: even $L \Rightarrow P$ even, odd $L \Rightarrow P$ odd.
>
> 2. Unitarity forces the constraint $|P(x)|^2 + (1-x^2)|Q(x)|^2 = 1$ for all $x \in [-1,1]$, hence $|P(x)| \leq 1$ on $[-1,1]$.
>
> 3. Conversely, any admissible P (with the degree/parity/boundedness conditions) can be synthesized by choosing phases ϕ.

11.4.3 Generalized QSP Theorems

Generalized QSP theorems remove the parity restriction by working directly with the eigenvalue $z = e^{i\theta}$ [45].

1 Analytic QSP

Analytic QSP defines

$$W = \begin{bmatrix} z & 0 \\ 0 & 1 \end{bmatrix}.$$

The generalized QSP circuit takes the form

$$U_A(z) = A_0 \prod_{\ell=1}^{L} (W A_\ell),$$

where $A_\ell \in SU(2)$ are arbitrary single-qubit unitaries.

A convenient parameterization is

$$A_\ell = e^{i\alpha_\ell Z} e^{i\phi_\ell Y} e^{i\beta_\ell Z}.$$

Since Z-rotations commute with W, the parameters β_ℓ may be absorbed into $\alpha_{\ell+1}$, leaving $2(L+1)+1$ independent parameters, including a global phase.

The analytic QSP theorem states that

$$U_A(z) = \begin{bmatrix} P(z) & * \\ Q(z) & * \end{bmatrix},$$

where $P(z)$ and $Q(z)$ are complex polynomials of degree at most L satisfying

$$|P(z)|^2 + |Q(z)|^2 = 1 \qquad \text{for all } |z| = 1.$$

Conversely, any such polynomial pair (P, Q) can be realized by an appropriate generalized QSP circuit.

2   Laurent QSP

A similar formulation, Laurent QSP, uses

$$W = \begin{bmatrix} z & 0 \\ 0 & z^{-1} \end{bmatrix}.$$

In this formulation, $P(z)$ and $Q(z)$ are Laurent polynomials of degree at most L:

$$P(z) = \sum_{j=-L}^{L} p_j z^j, \qquad Q(z) = \sum_{j=-L}^{L} q_j z^j.$$

The Chebyshev QSP theorem is recovered by restricting to $x = (z + z^{-1})/2$.

11.4.4   Synthesis Procedure

In Chebyshev QSP, synthesis consists of determining a phase sequence ϕ such that the resulting polynomial $P(x)$ approximates a target function. In practice, one typically (i) chooses a polynomial approximation (often a Chebyshev series), (ii) enforces the unitarity constraint by constructing a compatible partner polynomial, and (iii) factors the resulting SU(2)-valued polynomial into elementary rotations. Stable numerical algorithms and software packages exist for this purpose [44].

For generalized QSP, the synthesis procedure typically involves designing a Laurent polynomial $P(z)$ approximating the target, constructing a compatible $Q(z)$ satisfying unitarity, and computing the corresponding sequence of single-qubit unitaries $\{A_\ell\}$.

Open challenges include numerical stability at high degree, optimal depth for nonsmooth functions, and robustness under finite-precision constraints.

11.5   ✳ QSP of Generic Operators via Qubitization

Quantum signal processing (QSP) is, at its core, a single-qubit synthesis framework: by interleaving a fixed "signal-dependent" operation with tunable phase rotations, one generates polynomial (or Laurent-polynomial) transformations of an underlying scalar parameter (the *signal*). The QSP theorems characterize which transformations are achievable and show how to synthesize them using only a short sequence of single-qubit rotations.

To extend this framework from a scalar signal to a generic operator A, we must expose the spectral data of A (eigenvalues in the Hermitian case, singular values in the general case) as a signal parameter that controls an effective SU(2) rotation. This is accomplished by two ingredients. Block encoding provides a structural embedding of A into a larger unitary. Qubitization then turns this embedding into a direct sum of invariant two-dimensional subspaces, in each of which the dynamics reduces to a single-qubit rotation whose angle encodes the relevant spectral quantity. QSP can then be applied uniformly across all subspaces, thereby implementing a matrix function of A.

Together, block encoding and qubitization enable systematic spectral transformations via QSP. This framework, introduced by Low and Chuang [38] and further developed in [41, 46], underlies modern algorithms for Hamiltonian simulation, eigenvalue transformation, and quantum singular value transformation.

Overview

---

1. Block encoding provides a unitary access model for A (with scale α and n ancillas).

2. Qubitization turns that access into a direct sum of SU(2) rotations whose angles encode spectral data.

3. QSP then applies a designed polynomial to those angles, yielding a controlled transformation of A's spectrum.

Notations

Throughout this section, n denotes the number of ancilla qubits used by the block encoding, and $\alpha > 0$ is the known normalization factor in the block encoding.

U_A is an (α, n) block encoding of an operator A if

$$(\langle 0^n | \otimes I)\, U_A\, (|0^n\rangle \otimes I) = A/\alpha, \qquad U_A = \begin{bmatrix} A/\alpha & * \\ * & * \end{bmatrix}$$

with respect to the decomposition $|0^n\rangle \otimes \mathcal{H}_{\mathrm{sys}} \,\oplus\, (|0^n\rangle)^{\perp} \otimes \mathcal{H}_{\mathrm{sys}}$.

In qubitization for a Hermitian H, we typically apply QSP to the *scaled* operator H/α, whose spectrum lies in $[-1, 1]$.

$Z_{\mathrm{sig}} = 2\,|0^n\rangle\langle 0^n| - I_{\mathrm{anc}}$ is the rank-one reflection about $|0^n\rangle$, and the walk operator is $W = (Z_{\mathrm{sig}} \otimes I)\, U_H$.

11.5.1 Single-Qubit Qubitization

We begin with the simplest setting in which a single ancilla qubit suffices, i.e., $n = 1$. Consider a two-term Hermitian operator (e.g., a Hamiltonian) $H = \alpha_1 U_1 + \alpha_2 U_2$, whose block encoding can be implemented using the LCU construction in § 11.2.1.2.

In its conventional normalized form, we take a unitary U_H such that

$$(\langle 0| \otimes I)\, U_H \,(|0\rangle \otimes I) = H/\alpha, \qquad U_H = \begin{bmatrix} H/\alpha & * \\ * & * \end{bmatrix}, \tag{11.20}$$

for some known $\alpha > 0$ (often $\alpha = \sum_j |c_j|$ in a Pauli-LCU decomposition, or a known upper bound on $\|H\|$). The key point is that the *signal* accessible to QSP will be the scaled eigenvalue $x_k = \lambda_k/\alpha \in [-1, 1]$, not λ_k itself.

1 Spectral Encoding on the Signal Qubit

Let H have eigenpairs

$$H\,|\phi_k\rangle = \lambda_k\,|\phi_k\rangle.$$

Because U_H is a Hermitian block encoding of H/α, its restriction to the span of $\{|0\rangle\,|\phi_k\rangle, |1\rangle\,|\phi_k\rangle\}$ is fixed by (i) the block-encoding condition in Eq. 11.20 and (ii) unitarity and Hermiticity of U_H. Concretely, writing $x_k := \lambda_k/\alpha \in [-1, 1]$, we may choose phases so that

$$U_H\,|0\rangle\,|\phi_k\rangle = |0\rangle\, x_k\,|\phi_k\rangle + |1\rangle\,\sqrt{1 - x_k^2}\,|\phi_k\rangle, \tag{11.21a}$$

$$U_H\,|1\rangle\,|\phi_k\rangle = |0\rangle\,\sqrt{1 - x_k^2}\,|\phi_k\rangle - |1\rangle\, x_k\,|\phi_k\rangle. \tag{11.21b}$$

The coefficient x_k in the $|0\rangle$ branch is exactly the encoded block H/α, while the remaining amplitudes are determined (up to irrelevant phase conventions) by the requirement that U_H be a Hermitian unitary.

Equation 11.21 shows that, when restricted to a fixed eigenstate $|\phi_k\rangle$, the action of U_H is entirely confined to the two-dimensional subspace spanned by $\{|0\rangle\,|\phi_k\rangle, |1\rangle\,|\phi_k\rangle\}$. In this invariant subspace, U_H acts as

$$U_k = \begin{bmatrix} x_k & \sqrt{1 - x_k^2} \\ \sqrt{1 - x_k^2} & -x_k \end{bmatrix}, \tag{11.22}$$

with the basis ordered by the signal qubit. The system register remains untouched. This is analogous in spirit to phase kickback: spectral information about H is coherently imprinted onto an ancilla, but here the imprint is an SU(2) geometry (a reflection/rotation structure) rather than a pure phase.

> **Exercise 11.19** In the single-signal-qubit setting, derive Eq. 11.21 from the statement that U_H is a Hermitian $(\alpha, 1)$ block encoding of H/α. (You may assume $\|H\| \le \alpha$ and work inside the eigenbasis of H.)

2 Conversion to a Rotation

The matrix U_k in Eq. 11.22 is Hermitian (a reflection), not a rotation. To match the standard QSP setting, we convert it into an SU(2) rotation by left-multiplying with Pauli-Z on the signal qubit:

$$W_k = Z\,U_k = \begin{bmatrix} x_k & \sqrt{1 - x_k^2} \\ -\sqrt{1 - x_k^2} & x_k \end{bmatrix}. \tag{11.23}$$

This satisfies

$$W_k = e^{i\theta_k Y}, \qquad \cos\theta_k = x_k = \lambda_k/\alpha, \qquad \theta_k = \arccos(\lambda_k/\alpha).$$

Thus, each eigenvalue λ_k of H (after normalization by α) becomes a rotation angle on the signal qubit.

At the operator level, define the *walk operator*

$$W = (Z_{\text{sig}} \otimes I_{\text{sys}})\, U_H, \tag{11.24}$$

where in this single-ancilla setting $Z_{\text{sig}} = Z$. Since $|0\rangle$ is a $+1$ eigenstate of Z, we still have $(\langle 0| \otimes I)\, W\, (|0\rangle \otimes I) = H/\alpha$, so W preserves the same encoded block (up to the same normalization).

The terminology *walk operator* reflects its product-of-reflections structure: U_H acts as a reflection on each invariant subspace, and multiplying by Z_{sig} converts this into a rotation, analogous to the step operator of a discrete-time quantum walk.

 In qubitized QSP, the scalar signal must lie in $[-1, 1]$ because it is interpreted as $x = \cos\theta$ for an SU(2) rotation. Accordingly, if U_H is an (α, n) block encoding of a Hermitian operator H, then QSP acts on the normalized spectrum $x = \lambda/\alpha \in [-1, 1]$, where λ is an eigenvalue of H. Increasing α shrinks the signal range, which may be accounted for by rescaling the target function (e.g., e^{-iHt} becomes $x \mapsto e^{-ix(\alpha t)}$).

Key Takeaway

Given an $(\alpha, 1)$ Hermitian block encoding of H, qubitization maps each eigenvalue λ_k to a single-qubit rotation $W_k = e^{i\theta_k Y}$ with $\cos\theta_k = \lambda_k/\alpha$. This reduction to SU(2) rotations is what makes QSP applicable to generic operators.

Exercise 11.20 Let $W = (Z_{\text{sig}} \otimes I)U_H$. Show that in each invariant subspace, W_k has eigenvalues $e^{\pm i\theta_k}$ with $\cos\theta_k = \lambda_k/\alpha$.

3   Constructing QSP

Once qubitization yields the signal rotations $W_k = e^{i\theta_k Y}$, QSP proceeds exactly as in § 11.4.1. Interleaving the signal rotation with Z-axis phase rotations produces

$$U_\phi(x_k) = e^{i\phi_0 Z} \prod_{\ell=1}^{L} \left(e^{i\theta_k Y} e^{i\phi_\ell Z} \right), \qquad x_k = \lambda_k/\alpha. \tag{11.25}$$

By choosing the phases $\{\phi_\ell\}$ appropriately, the $(0, 0)$ matrix element implements a polynomial $P(x_k)$ subject to the QSP admissibility constraints. Because x_k is the eigenvalue of H/α, this corresponds to applying the matrix function $P(H/\alpha)$ to the system register (and, with appropriate postprocessing or embedding, to implementing functions of H itself). In Hamiltonian simulation, for example, one typically targets $f(\lambda) = e^{-i\lambda t}$ by instead approximating the scaled function $x \mapsto e^{-i(\alpha x)t} = e^{-ix(\alpha t)}$ on $[-1, 1]$.

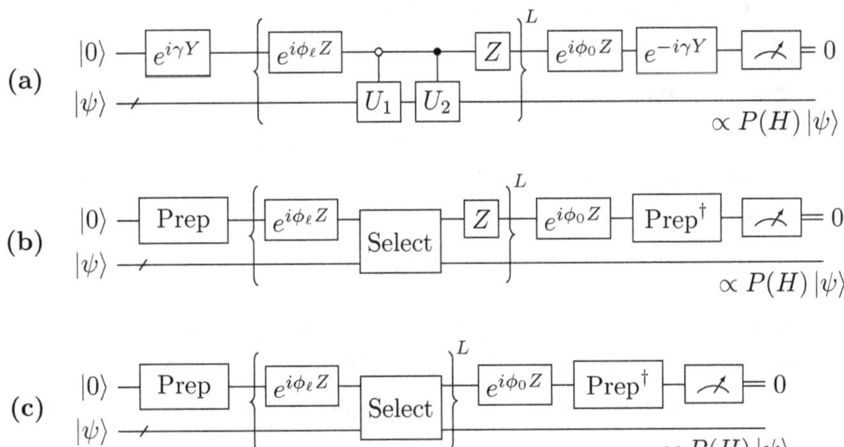

Figure 11.9: Single-Qubit QSP via Qubitization

4   Circuit Implementation

A QSP sequence alternates (i) phase rotations about Z on the signal qubit and (ii) applications of the walk operator W on the joint signal–system register, as illustrated in Fig. 11.9. Part (a) shows the explicit circuit for $H = \alpha_1 U_1 + \alpha_2 U_2$, obtained by inserting QSP phase rotations into the LCU-based block-encoding circuit of Fig. 11.3.

Part (b) shows the abstract structure: the block encoding U_H satisfying Eq. 11.20, followed by Z_{sig} to form W as in Eq. 11.24, and then repeated interleaving with $e^{i\phi_\ell Z}$.

Part (c) depicts a further simplification in which explicit Z_{sig} factors are absorbed into neighboring phase rotations, yielding a reflector-based form consistent with § 11.4.2.2:

$$U_\phi(x_k) = e^{i\phi_0 Z} \prod_{\ell=1}^{L} \left(U_k e^{i\phi_\ell Z} \right), \tag{11.26}$$

which is equivalent to Eq. 11.25 up to a known global phase.

 Even when the walk operator is absorbed at the circuit level, it remains the natural conceptual object: it exposes the SU(2) rotation geometry and is the standard organizing principle in analyses of eigenvalue and singular-value transformations.

5   Subspace Interpolation

A central structural feature underlying qubitization is that the signal–system Hilbert space decomposes into a direct sum of invariant two-dimensional subspaces, each associated with an eigenvalue of H. For each eigenpair $H |\phi_k\rangle = \lambda_k |\phi_k\rangle$, the states $|0\rangle |\phi_k\rangle$ and $|1\rangle |\phi_k\rangle$ span a subspace invariant under both U_H and W. Equivalently,

$$U_H = \bigoplus_k U_k, \qquad W = \bigoplus_k W_k,$$

where U_k and W_k take the explicit forms in Eq. 11.22 and Eq. 11.23. Because the same phase sequence ϕ is applied to all subspaces, the resulting circuit applies the same polynomial transformation $P(x)$ to every scaled eigenvalue $x_k = \lambda_k/\alpha$, thereby implementing $P(H/\alpha)$.

6   QSP for a Generic Unitary

The discussion above emphasizes Hermitian inputs, which is the natural setting for Hamiltonian simulation and eigenvalue transformations. Another common setting arises when the input operator is already a unitary U with spectral decomposition

$$U \left| \psi_k \right\rangle = e^{i\theta_k} \left| \psi_k \right\rangle.$$

Here the eigenphase θ_k itself can serve as the signal, so QSP can be applied without first constructing a Hermitian block encoding.

One convenient viewpoint is to associate each eigenvalue $e^{i\theta_k}$ with the single-qubit signal operator

$$W_k = e^{i\theta_k Z} = \begin{bmatrix} e^{i\theta_k} & 0 \\ 0 & e^{-i\theta_k} \end{bmatrix}. \tag{11.27}$$

In this formulation, controlled-U and controlled-$U^\dagger$ supply the signal-dependent Z rotation, while the tunable QSP phases are implemented as rotations about a noncommuting axis (e.g., $e^{i\phi_\ell Y}$). The resulting circuit (Fig. 11.10) implements a Laurent-polynomial transformation of the eigenvalues $e^{i\theta_k}$, consistent with the generalized QSP framework in § 11.4.3.

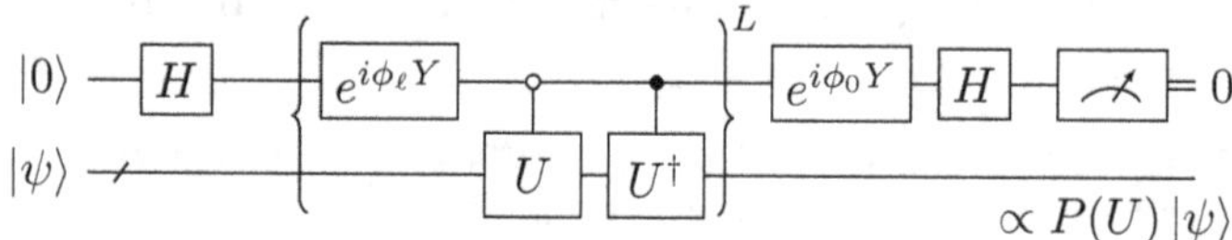

Figure 11.10: QSP for a Generic Unitary

If controlled-$U^\dagger$ is unavailable, one can still perform QSP using the signal primitive

$$W = \begin{bmatrix} U & 0 \\ 0 & I \end{bmatrix},$$

which leads to *analytic QSP* and yields ordinary polynomials $P(z)$ in $z = e^{i\theta}$ (no z^{-1} terms). Access to both U and $U^\dagger$ enables the more general *Laurent* setting, where $P(z)$ may include both positive and negative powers and thus can represent general Fourier-type transformations on the unit circle.

11.5.2   Multi-Qubit Qubitization

In many applications, implementing a block encoding of H requires an n-qubit ancilla register, for example to index terms in an LCU decomposition, to store sparsity information, or to provide workspace for oracle queries. Qubitization extends directly to this multi-ancilla setting.

1   Effective Two-Dimensional Signal Space

Let U_H be an (α, n) block encoding of a Hermitian operator H:

$$\left(\langle 0^n | \otimes I \right) U_H \left(|0^n\rangle \otimes I \right) = H/\alpha, \qquad \alpha \geq \|H\|. \tag{11.28}$$

Fix an eigenpair $H|\phi_k\rangle = \lambda_k|\phi_k\rangle$ and define $x_k = \lambda_k/\alpha \in [-1, 1]$. By the block-encoding condition, U_H maps the joint state $|0^n\rangle|\phi_k\rangle$ into a superposition of $|0^n\rangle|\phi_k\rangle$ and an orthogonal ancilla component tensored with the same system eigenstate:

$$U_H|0^n\rangle|\phi_k\rangle = x_k|0^n\rangle|\phi_k\rangle + \sqrt{1 - x_k^2}\,|0^n_{\perp,k}\rangle|\phi_k\rangle, \qquad (11.29)$$

where $|0^n_{\perp,k}\rangle$ is a normalized ancilla state orthogonal to $|0^n\rangle$. Thus, for each k, the two joint states $|0^n\rangle|\phi_k\rangle$ and $|0^n_{\perp,k}\rangle|\phi_k\rangle$ span a two-dimensional invariant subspace that carries the same effective geometry as in the single-ancilla case.

To define the walk operator, introduce the reflection about the distinguished ancilla basis state,

$$Z_{\text{sig}} = 2|0^n\rangle\langle 0^n| - I_{\text{anc}}, \qquad (11.30)$$

and set

$$W = (Z_{\text{sig}} \otimes I_{\text{sys}})\,U_H. \qquad (11.31)$$

Since $Z_{\text{sig}}|0^n\rangle = +|0^n\rangle$, we still have $(\langle 0^n| \otimes I)\,W\,(|0^n\rangle \otimes I) = H/\alpha$, so the same scaled operator remains encoded. Moreover, restricting W to the kth invariant subspace yields an SU(2) rotation $W_k = e^{i\theta_k Y}$ with $\cos\theta_k = x_k = \lambda_k/\alpha$.

For $n = 1$, Z_{sig} reduces to Pauli-Z. For $n > 1$, Z_{sig} is a rank-one reflection about $|0^n\rangle$ and is not equal to the parity operator $Z_1 Z_2 \cdots Z_n$ in general.

From the perspective of QSP, the conceptual core does not change: although the block encoding may use many ancillas, the relevant signal space per eigenvalue is still two-dimensional. The additional ancillas are implementation resources for U_H; they do not create a higher-dimensional "signal qudit."

2 Circuit Implementation

At the operator level, the multi-ancilla QSP sequence may be written as

$$U_\phi = e^{i\phi_0 Z_{\text{sig}}} \prod_{\ell=1}^{L} \left(\underbrace{(Z_{\text{sig}} U_H)}_{W} \, e^{i\phi_\ell Z_{\text{sig}}} \right). \qquad (11.32)$$

As in the single-qubit case, the Z_{sig} factors can be absorbed into neighboring phase rotations, yielding the reflector-based form

$$U_\phi = e^{i\phi_0 Z_{\text{sig}}} \prod_{\ell=1}^{L} \left(U_H e^{i\phi_\ell Z_{\text{sig}}} \right). \qquad (11.33)$$

The resulting circuit is shown in Fig. 11.11, mirroring the single-qubit construction in Fig. 11.9(c).

The logical phase operations $e^{i\phi_\ell Z_{\text{sig}}}$ act as a relative phase between $|0^n\rangle$ and its orthogonal complement. A standard implementation uses an auxiliary qubit to compute whether the ancilla register is in $|0^n\rangle$, apply a single-qubit phase rotation to the auxiliary qubit, and then uncompute (a compute–rotate–uncompute "phase gadget"). This realizes the desired relative phase on the ancilla register via phase kickback while returning the auxiliary qubit to $|0\rangle$ after each step.

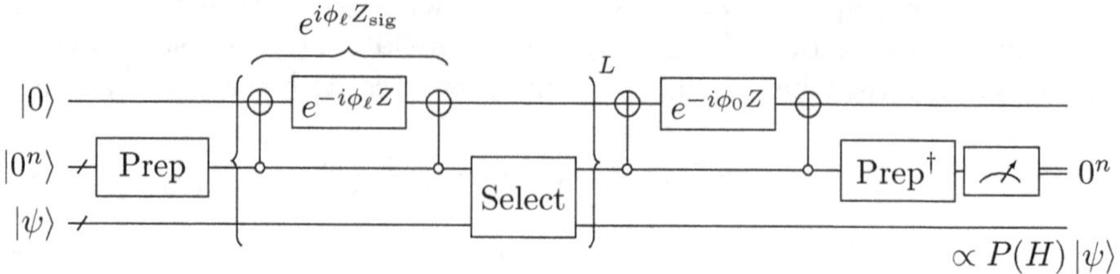

Figure 11.11: Multi-Qubit QSP via Qubitization

Exercise 11.21 Explain why applying an $e^{-i\phi_\ell Z}$ rotation to the auxiliary qubit (rather than $e^{i\phi_\ell Z}$) can realize the effective operation $e^{i\phi_\ell Z_{\mathrm{sig}}}$ on the signal ancilla register via phase kickback.

Exercise 11.22 Modify the auxiliary phase gadget so that it uses X-axis rotations instead of Z-axis rotations on the auxiliary qubit, and explain why the resulting circuit is equivalent up to a basis change on that qubit.

Key Takeaways

1. Even with n ancillas, qubitization is still "qubit": for each eigenvalue λ_k of H, the dynamics lives in a two-dimensional invariant subspace.

2. The reflection $Z_{\mathrm{sig}} = 2\,|0^n\rangle\langle 0^n| - I$ defines the logical signal qubit; it is not the parity operator $Z_1 \cdots Z_n$ unless $n = 1$.

3. Extra ancillas are an implementation cost (to realize U_H), not an increase in the signal dimension seen by QSP.

11.5.3   Near-Optimal Complexity for Qubitized QSP

A principal advantage of QSP is its near-optimal query complexity for a broad class of spectral transformation tasks. Once an operator is available through block encoding and qubitization, the problem reduces to approximating a scalar target function on a bounded interval (typically $[-1, 1]$) by a QSP-admissible polynomial (or Laurent polynomial, in the unitary-eigenvalue setting). Approximating a target function to error ε typically requires a degree

$$L = O\big(\text{intrinsic scale} + \log(1/\varepsilon)\big),$$

and each QSP step uses a constant number of calls to the qubitized signal unitary, so the total query complexity is linear in L.

1   Hamiltonian Simulation as the Canonical Example

Suppose U_H is an (α, n) block encoding of a Hermitian H, so the signal variable is $x = \lambda/\alpha \in [-1, 1]$. Simulating e^{-iHt} can be reduced to approximating the scalar function $x \mapsto e^{-i(\alpha x)t} = e^{-ix(\alpha t)}$ on $[-1, 1]$ using QSP. The Chebyshev expansion

$$e^{-ix(\alpha t)} = J_0(\alpha t) + 2\sum_{k=1}^{\infty}(-i)^k J_k(\alpha t)\, T_k(x)$$

has coefficients $J_k(\alpha t)$ that decay rapidly once $k \gtrsim \alpha t$, so truncation at degree $L = O(\alpha t)$ yields constant precision, and achieving error ε requires an additional $O(\log(1/\varepsilon))$ terms from the exponentially small tail. This gives the familiar dependence

$$L = O\big(\alpha t + \log(1/\varepsilon)\big)$$

for simulating e^{-iHt} when the block encoding is normalized by α.

2　Beyond Time Evolution: General Spectral Transformations

The same mechanism extends far beyond time evolution. Whenever a task can be phrased as applying a scalar function to the spectrum after block encoding and qubitization, QSP yields a near-optimal implementation. Examples include spectral filtering, thresholding, matrix inversion, fractional matrix powers, and other analytic matrix functions. For instance, approximating $f(x) = 1/x$ on $[1/\kappa, 1]$ (as in quantum linear systems algorithms) leads to

$$L = O\big(\kappa \log(1/\varepsilon)\big),$$

matching known lower bounds up to constant factors in standard black-box models. In this sense, QSP provides an essentially optimal framework for implementing operator-valued spectral transformations once spectral access is available through block encoding and qubitization.

11.6　∗Quantum Singular Value Transformation (QSVT)

Quantum signal processing (QSP), together with block encoding and qubitization, provides a general framework for implementing polynomial transformations of an encoded linear operator. Operationally, one alternates applications of a qubitized walk operator with phase operations acting on a designated signal subspace, thereby synthesizing a polynomial "transfer function" applied uniformly across invariant two-dimensional subspaces.

Quantum singular value transformation (QSVT) extends this framework from eigenvalues (unitary or Hermitian settings) to singular values of an *arbitrary* linear operator. Once an operator A is embedded into a larger unitary via block encoding, its singular values induce invariant two-dimensional subspaces of associated walk operators, within which the dynamics reduces to effective SU(2) rotations. The same phase-synthesis machinery developed for QSP can then be applied uniformly across these subspaces to implement prescribed polynomial transformations of the singular values, yielding a transformation of the form $P^{(\mathrm{SV})}(A)$ on the original input–output mapping.

QSVT, introduced by Gilyén *et al.* [41], unifies and streamlines a wide range of quantum-algorithmic primitives, including Grover-type amplitude amplification, Hamiltonian simulation, and linear-systems solvers. It is therefore helpful to view QSVT as a general-purpose "spectral transformation" framework that subsumes many seemingly different constructions under a single polynomial viewpoint.

(R) | This section builds on QC Math [1], Section 13.2: *Singular Value Decomposition.*

Singular Value Decomposition (SVD): a Refresher

For a normal (in particular, Hermitian) operator H, the spectral decomposition takes the form

$$H = \sum_k \lambda_k \, |\psi_k\rangle\langle\psi_k|, \quad \text{or} \quad H\,|\psi_k\rangle = \lambda_k\,|\psi_k\rangle, \quad \text{or} \quad H = U\Lambda U^\dagger,$$

where $\{|\psi_k\rangle\}$ is an orthonormal basis and λ_k are the eigenvalues. By contrast, the singular value decomposition (SVD) applies to an arbitrary linear operator $A : \mathcal{H}_{\text{in}} \to \mathcal{H}_{\text{out}}$ and is given by

$$A = \sum_k \sigma_k \, |w_k\rangle\langle v_k|, \quad \text{or} \quad A\,|v_k\rangle = \sigma_k\,|w_k\rangle, \quad \text{or} \quad A = W\Sigma V^\dagger,$$

where $\{|w_k\rangle\} \subset \mathcal{H}_{\text{out}}$ and $\{|v_k\rangle\} \subset \mathcal{H}_{\text{in}}$ are orthonormal sets.

Singular values are always real and nonnegative, and the SVD exists for all linear operators, including rectangular matrices. When A is Hermitian and positive semidefinite, the left and right singular vectors coincide and the singular values coincide with the eigenvalues.

If σ_k is a singular value of A, then σ_k^2 is an eigenvalue of both $AA^\dagger$ and $A^\dagger A$, with $|w_k\rangle$ and $|v_k\rangle$ as the corresponding eigenvectors, respectively.

For a Hermitian operator H, a matrix function is defined spectrally by

$$f(H) = \sum_k f(\lambda_k)\,|\psi_k\rangle\langle\psi_k|.$$

For a general operator A, the analogous singular-value functional calculus is defined via the SVD by

$$f_{(\text{SV})}(A) = \sum_k f(\sigma_k)\,|w_k\rangle\langle v_k| = W\,f(\Sigma)\,V^\dagger,$$

where $f(\Sigma)$ applies f entrywise to the nonnegative diagonal entries of Σ. Quantum singular value transformation provides an efficient quantum procedure for implementing such singular-value transformations when A is accessed through a block encoding.

11.6.1   Walkthrough with an Example

Since the mathematics of QSVT can be technically involved, we begin with a concrete example illustrating how QSVT extends the qubitization-based QSP construction in § 11.5, and how singular value transformation emerges from essentially the same circuit structure. The key difference is not the polynomial machinery itself, but the nature of the encoded signal and the invariant subspaces on which the QSP sequence acts.

In standard (qubitized) QSP, the signal is an eigenvalue of a unitary or Hermitian operator, and qubitization yields a two-dimensional invariant subspace associated with each eigenstate. By contrast, QSVT applies to a general linear operator, and the corresponding invariant subspaces are associated with pairs of left and right singular vectors. In this case, the same geometric mechanism produces two-dimensional invariant subspaces whose rotation angles encode singular values rather than eigenvalues.

1　Block Encoding

Consider the $\mathbb{C}^{3\times 2}$ matrix

$$A = \begin{bmatrix} 1 & 0 \\ 0 & 1 \\ 1 & 1 \end{bmatrix},$$

with singular value decomposition $A = W\Sigma V^\dagger$, where

$$\Sigma = \begin{bmatrix} \sqrt{3} & 0 \\ 0 & 1 \\ 0 & 0 \end{bmatrix}, \quad W = \frac{1}{\sqrt{6}}\begin{bmatrix} 1 & -\sqrt{3} & \sqrt{2} \\ 1 & \sqrt{3} & \sqrt{2} \\ 2 & 0 & -\sqrt{2} \end{bmatrix}, \quad V = \frac{1}{\sqrt{2}}\begin{bmatrix} 1 & -1 \\ 1 & 1 \end{bmatrix}.$$

The two nonzero singular values are $\sigma_1 = \sqrt{3}$ and $\sigma_2 = 1$, with corresponding singular vectors given by the appropriate columns of W and V, for example

$$|w_1\rangle = \frac{1}{\sqrt{6}}\begin{bmatrix} 1 \\ 1 \\ 2 \end{bmatrix}, \quad |v_1\rangle = \frac{1}{\sqrt{2}}\begin{bmatrix} 1 \\ 1 \end{bmatrix}, \quad |v_2\rangle = \frac{1}{\sqrt{2}}\begin{bmatrix} -1 \\ 1 \end{bmatrix}.$$

Since $\|A\| = \sigma_{\max}(A) = \sqrt{3} > 1$, we rescale the operator as $\tilde{A} = A/\alpha$ with $\alpha = \sqrt{3}$ so that $\|\tilde{A}\| \leq 1$. We then construct a unitary completion via a Halmos dilation (see § 11.3.2),

$$U_A = \begin{bmatrix} C & \tilde{A} \\ \tilde{A}^\dagger & -B \end{bmatrix}, \quad C = \sqrt{I_3 - \tilde{A}\tilde{A}^\dagger}, \quad B = \sqrt{I_2 - \tilde{A}^\dagger\tilde{A}}. \tag{11.34}$$

Numerically, this yields the explicit 5×5 unitary

$$U_A = \begin{array}{c} \\ |0\rangle_W \\ |1\rangle_W \\ |2\rangle_W \\ \\ |0\rangle_V \\ |1\rangle_V \end{array}\begin{array}{cc} \begin{matrix} |0\rangle_W \quad |1\rangle_W \quad |2\rangle_W \ : \ |0\rangle_V \quad |1\rangle_V \end{matrix} \\ \left[\begin{array}{ccc:cc} \frac{2+\sqrt{6}}{6} & \frac{2-\sqrt{6}}{6} & -\frac{1}{3} & \frac{\sqrt{3}}{3} & 0 \\ \frac{2-\sqrt{6}}{6} & \frac{2+\sqrt{6}}{6} & -\frac{1}{3} & 0 & \frac{\sqrt{3}}{3} \\ -\frac{1}{3} & -\frac{1}{3} & \frac{1}{3} & \frac{\sqrt{3}}{3} & \frac{\sqrt{3}}{3} \\ \hdashline \frac{\sqrt{3}}{3} & 0 & \frac{\sqrt{3}}{3} & -\frac{\sqrt{6}}{6} & \frac{\sqrt{6}}{6} \\ 0 & \frac{\sqrt{3}}{3} & \frac{\sqrt{3}}{3} & \frac{\sqrt{6}}{6} & -\frac{\sqrt{6}}{6} \end{array}\right] \end{array},$$

where the basis ordering corresponds to the output space $\mathbb{C}^3$ (labeled by $|j\rangle_W$) followed by the input space $\mathbb{C}^2$ (labeled by $|j\rangle_V$).

For a rectangular operator A, it is often most natural to place A in an off-diagonal block, so that the input and output spaces are embedded as orthogonal subspaces and treated symmetrically.

Define the inclusion maps $\mathcal{W} : \mathbb{C}^3 \hookrightarrow \mathbb{C}^5$ and $\mathcal{V} : \mathbb{C}^2 \hookrightarrow \mathbb{C}^5$ by

$$\mathcal{W} = \begin{bmatrix} I_3 \\ 0 \end{bmatrix}, \quad \mathcal{V} = \begin{bmatrix} 0 \\ I_2 \end{bmatrix}.$$

These identify the output and input spaces of A with orthogonal subspaces of the dilation space, yielding the block-encoding selection relation

$$\mathcal{W}^\dagger U_A \mathcal{V} = \frac{A}{\alpha}. \tag{11.35}$$

Exercise 11.23 Verify Eq. 11.35 by carrying out the block matrix multiplication $\mathcal{W}^\dagger U_A \mathcal{V}$.

Exercise 11.24 Let $A : \mathcal{H}_{\text{in}} \to \mathcal{H}_{\text{out}}$ be rectangular.

(a) Explain why placing A in an off-diagonal block of a unitary dilation can be more natural than the top-left placement when $\dim \mathcal{H}_{\text{in}} \neq \dim \mathcal{H}_{\text{out}}$.

(b) Using inclusion maps $\mathcal{W}, \mathcal{V}$, show how Eq. 11.35 generalizes the "$\langle 0^n | U_A | 0^n \rangle$" viewpoint.

Implementation of U_A

In practice, the block-encoding unitary U_A is not implemented by explicitly constructing the full Halmos dilation matrix. Instead, U_A is realized implicitly through structured circuit primitives that coherently encode the action of A/α using ancilla registers.

From the circuit perspective, U_A is simply the unitary that makes the block-encoding relation Eq. 11.35 true.

A common route is via linear combination of unitaries (LCU). If A can be expressed as $A = \sum_j \alpha_j U_j$ with efficiently implementable unitaries U_j, then U_A is implemented using ancilla state preparation, a controlled-Select operation applying the U_j, and uncomputation. The resulting circuit is unitary and satisfies the required selection relation, while the remaining blocks needed for unitarity arise automatically.

In other settings (e.g., sparse-matrix algorithms), U_A is implemented using oracle access to matrix entries together with controlled arithmetic and swap operations. For rectangular operators, different system registers represent the input and output spaces, and ancillas mediate transitions between them, but the circuit never explicitly constructs the full dilation matrix.

From an algorithmic perspective, Halmos dilation is best viewed as a conceptual tool that guarantees the existence of a unitary completion, rather than as a gate-level implementation recipe. QSVT requires coherent access to the block encoding, not to the complementary blocks.

---

2 Qubitization and Two-Dimensional Invariant Subspaces

Since U_A acts on a larger Hilbert space ($\mathbb{C}^5$ in this example), we define two orthogonal projectors that select the embedded output and input subspaces, respectively:

$$\Pi_W = \mathcal{W}\mathcal{W}^\dagger = \begin{bmatrix} I_3 & 0 \\ 0 & 0 \end{bmatrix}, \qquad \Pi_V = \mathcal{V}\mathcal{V}^\dagger = \begin{bmatrix} 0 & 0 \\ 0 & I_2 \end{bmatrix}. \tag{11.36}$$

We also define the augmented singular vectors by embedding the original singular vectors into the dilation space,

$$|\widetilde{w}_k\rangle = \mathcal{W}|w_k\rangle = \begin{bmatrix} |w_k\rangle \\ 0 \end{bmatrix}, \qquad |\widetilde{v}_k\rangle = \mathcal{V}|v_k\rangle = \begin{bmatrix} 0 \\ |v_k\rangle \end{bmatrix}, \qquad k = 1, 2, \ldots, r,$$

where r is the number of nonzero singular values of A ($r = 2$ in this example).

Using $\mathcal{W}^\dagger U_A \mathcal{V} = A/\alpha$, $A\,|v_k\rangle = \sigma_k\,|w_k\rangle$, $\Pi_W\,|\widetilde{w}_k\rangle = |\widetilde{w}_k\rangle$, and $\Pi_V\,|\widetilde{v}_k\rangle = |\widetilde{v}_k\rangle$, we obtain the key coupling relations

$$\Pi_W U_A |\widetilde{v}_k\rangle = \frac{\sigma_k}{\alpha}\,|\widetilde{w}_k\rangle\,, \qquad \Pi_V U_A^\dagger |\widetilde{w}_k\rangle = \frac{\sigma_k}{\alpha}\,|\widetilde{v}_k\rangle\,. \tag{11.37}$$

The Companion Vectors and Invariant Subspaces

Assume first that $\sigma_k < \alpha$, so that $1 - (\sigma_k/\alpha)^2 > 0$. Define the companion vectors

$$|\widetilde{w}_k^\perp\rangle = \frac{(I - \Pi_W)U_A |\widetilde{v}_k\rangle}{\sqrt{1 - (\sigma_k/\alpha)^2}}\,, \qquad |\widetilde{v}_k^\perp\rangle = \frac{(I - \Pi_V)U_A^\dagger |\widetilde{w}_k\rangle}{\sqrt{1 - (\sigma_k/\alpha)^2}}\,,$$

which are normalized and orthogonal to $|\widetilde{w}_k\rangle$ and $|\widetilde{v}_k\rangle$, respectively. By construction, we can decompose

$$U_A |\widetilde{v}_k\rangle = \Pi_W U_A |\widetilde{v}_k\rangle + (I - \Pi_W)U_A |\widetilde{v}_k\rangle = \frac{\sigma_k}{\alpha}\,|\widetilde{w}_k\rangle + \sqrt{1 - \left(\frac{\sigma_k}{\alpha}\right)^2}\,|\widetilde{w}_k^\perp\rangle\,,$$

and similarly

$$U_A^\dagger |\widetilde{w}_k\rangle = \Pi_V U_A^\dagger |\widetilde{w}_k\rangle + (I - \Pi_V)U_A^\dagger |\widetilde{w}_k\rangle = \frac{\sigma_k}{\alpha}\,|\widetilde{v}_k\rangle + \sqrt{1 - \left(\frac{\sigma_k}{\alpha}\right)^2}\,|\widetilde{v}_k^\perp\rangle\,.$$

Because U_A is unitary, its restriction to the span of $\{|\widetilde{v}_k\rangle, |\widetilde{v}_k^\perp\rangle\}$ must be an isometry onto the span of $\{|\widetilde{w}_k\rangle, |\widetilde{w}_k^\perp\rangle\}$. In particular, the images of the orthonormal basis $\{|\widetilde{v}_k\rangle, |\widetilde{v}_k^\perp\rangle\}$ must form an orthonormal basis of the output span, which forces

$$U_A |\widetilde{v}_k^\perp\rangle = \sqrt{1 - \left(\frac{\sigma_k}{\alpha}\right)^2}\,|\widetilde{w}_k\rangle - \frac{\sigma_k}{\alpha}\,|\widetilde{w}_k^\perp\rangle\,, \quad U_A^\dagger |\widetilde{w}_k^\perp\rangle = \sqrt{1 - \left(\frac{\sigma_k}{\alpha}\right)^2}\,|\widetilde{v}_k\rangle - \frac{\sigma_k}{\alpha}\,|\widetilde{v}_k^\perp\rangle\,.$$

These equations show that U_A maps $\mathrm{span}\{|\widetilde{v}_k\rangle, |\widetilde{v}_k^\perp\rangle\}$ into $\mathrm{span}\{|\widetilde{w}_k\rangle, |\widetilde{w}_k^\perp\rangle\}$, and $U_A^\dagger$ maps $\mathrm{span}\{|\widetilde{w}_k\rangle, |\widetilde{w}_k^\perp\rangle\}$ back into $\mathrm{span}\{|\widetilde{v}_k\rangle, |\widetilde{v}_k^\perp\rangle\}$. Since U_A is invertible with inverse $U_A^\dagger$, the two spans must coincide as a single two-dimensional invariant subspace for both U_A and $U_A^\dagger$. We denote this invariant subspace by

$$\mathcal{S}_k := \mathrm{span}\{|\widetilde{v}_k\rangle, |\widetilde{v}_k^\perp\rangle\} = \mathrm{span}\{|\widetilde{w}_k\rangle, |\widetilde{w}_k^\perp\rangle\}\,. \tag{11.38}$$

 In general, the qubitization subspace $\mathcal{S}_k$ is not equal to either the embedded output space or the embedded input space.

When $\sigma_k = \alpha$, the normalization factor $\sqrt{1 - (\sigma_k/\alpha)^2}$ vanishes and the above companion-vector definition does not apply. In that case, U_A maps $|\widetilde{v}_k\rangle$ exactly to $|\widetilde{w}_k\rangle$ (up to phase), so the induced two-dimensional dynamics becomes degenerate in the sense that there is no nontrivial rotation angle to transform.

The Induced 2×2 Action

Representing U_A on $\mathcal{S}_k$ using the ordered input basis $\{|\widetilde{v}_k\rangle, |\widetilde{v}_k^\perp\rangle\}$ and output basis $\{|\widetilde{w}_k\rangle, |\widetilde{w}_k^\perp\rangle\}$, we obtain

$$[U_A]_{\mathcal{S}_k} = \begin{bmatrix} \sigma_k/\alpha & \sqrt{1 - (\sigma_k/\alpha)^2} \\ \sqrt{1 - (\sigma_k/\alpha)^2} & -\sigma_k/\alpha \end{bmatrix} = \begin{bmatrix} \cos\theta_k & \sin\theta_k \\ \sin\theta_k & -\cos\theta_k \end{bmatrix}\,, \tag{11.39}$$

where $\cos\theta_k = \sigma_k/\alpha$ and $\sin\theta_k = \sqrt{1 - (\sigma_k/\alpha)^2}$. Equivalently,

$$[U_A]_{\mathcal{S}_k} = Z\, e^{i\theta_k Y},$$

which is the same two-dimensional "signal operator" form encountered in qubitized QSP, except that the signal parameter is now the singular value ratio σ_k/α.

In this example, $\alpha = \sqrt{3}$, so $\sigma_1/\alpha = 1$ and $\sigma_2/\alpha = 1/\sqrt{3}$. Therefore,

$$[U_A]_{\mathcal{S}_1} = \begin{bmatrix} 1 & 0 \\ 0 & -1 \end{bmatrix} = Z, \quad [U_A]_{\mathcal{S}_2} = \begin{bmatrix} 1/\sqrt{3} & \sqrt{2/3} \\ \sqrt{2/3} & -1/\sqrt{3} \end{bmatrix}, \qquad \cos\theta_2 = \frac{1}{\sqrt{3}}.$$

Exercise 11.25 In the rectangular setting, verify the coupling relations Eq. 11.37 starting from the block-encoding relation Eq. 11.35.

3   Walk Operators and QSVT Sequences

By now we should recognize that Eq. 11.39 has exactly the same two-dimensional structure as the qubitized-QSP signal operator discussed in § 11.5 (cf. Eq. 11.22). However, there is an important structural distinction in the rectangular setting: the signal subspace alternates between the embedded input and output subspaces.

Using the projectors Π_W and Π_V from Eq. 11.36, define the associated reflections

$$Z_W = 2\Pi_W - I = \begin{bmatrix} I_3 & 0 \\ 0 & -I_2 \end{bmatrix}, \qquad Z_V = 2\Pi_V - I = \begin{bmatrix} -I_3 & 0 \\ 0 & I_2 \end{bmatrix}. \tag{11.40}$$

We then define two walk operators

$$\Omega_W = Z_W U_A, \qquad \Omega_V = Z_V U_A^\dagger,$$

which convert the reflector form Eq. 11.39 into a proper SU(2) rotation on each invariant subspace $\mathcal{S}_k$, suitable for QSP-style phase modulation. (In the literature, closely related objects are often denoted using projector symbols such as Π and $\widetilde{\Pi}$.)

To incorporate phase parameters, define the phase rotations generated by the reflections in Eq. 11.40,

$$R_W(\phi_\ell) = e^{i\phi_\ell Z_W}, \qquad R_V(\phi_\ell) = e^{i\phi_\ell Z_V}.$$

A QSVT sequence alternates phase rotations and walk operators while switching between the two signal subspaces,

$$\cdots R_V(\phi_\ell)\, \Omega_V\, R_W(\phi_{\ell+1})\, \Omega_W \cdots. \tag{11.41}$$

This is the direct rectangular analogue of the qubitized-QSP pattern "phase–walk–phase–walk–$\cdots$", except that we now have two reflections (Z_W and Z_V) and hence two kinds of phase rotations.

Finally, within each invariant subspace $\mathcal{S}_k$, the restrictions of Ω_W and Ω_V act as signal-dependent rotations with angle determined by $\cos\theta_k = \sigma_k/\alpha$, while $R_W(\phi)$ and $R_V(\phi)$ act as Z-phase modulations on the appropriate side. As a result, the effective evolution on $\mathcal{S}_k$ mirrors the QSP product structure (Eq. 11.19), enabling the same phase-synthesis machinery to implement polynomial transformations of the singular values.

Exercise 11.26 For the reduced 2×2 form in Eq. 11.39, compute the corresponding walk-operator restrictions

$$[\Omega_W]_{\mathcal{S}_k}, \qquad [\Omega_V]_{\mathcal{S}_k}$$

and show that they are proper rotations with the same angle θ_k.

11.6.2 The QSVT Theorem

With the intuition from the preceding example, we now state the central QSVT guarantee in a form adapted to rectangular operators.

Let U_A be an (α, n) block encoding of $A \in \mathbb{C}^{N_{\text{out}} \times N_{\text{in}}}$, and let $A = W\Sigma V^\dagger$ be its singular value decomposition. Assume we are given inclusion maps $\mathcal{W}$ and $\mathcal{V}$ into the dilation space of U_A such that

$$\mathcal{W}^\dagger U_A \mathcal{V} = \frac{A}{\alpha} = W\frac{\Sigma}{\alpha}V^\dagger. \tag{11.42}$$

Let $\Pi_W = \mathcal{W}\mathcal{W}^\dagger$ and $\Pi_V = \mathcal{V}\mathcal{V}^\dagger$ be the associated orthogonal projectors, and define the corresponding reflections

$$Z_W = 2\Pi_W - I, \qquad Z_V = 2\Pi_V - I. \tag{11.43}$$

We also define the logical phase rotations

$$R_W(\phi) = e^{i\phi Z_W}, \qquad R_V(\phi) = e^{i\phi Z_V}, \tag{11.44}$$

and the two walk operators

$$\Omega_W = Z_W U_A, \qquad \Omega_V = Z_V U_A^\dagger. \tag{11.45}$$

Admissible Polynomials

Let $P(x)$ be a real polynomial of degree at most L whose parity matches L (odd degree $\Rightarrow$ odd polynomial, even degree $\Rightarrow$ even polynomial), and assume $|P(x)| \leq 1$ for all $x \in [-1, 1]$. Then there exists a phase sequence $\phi = (\phi_1, \phi_2, \ldots, \phi_L)$ such that a circuit composed of $\{R_W(\phi_\ell), R_V(\phi_\ell), \Omega_W, \Omega_V\}$ implements a unitary U_ϕ whose appropriate block realizes the singular-value transformation determined by P.

Odd Case (Output–Input Map)

If L is odd (so P is odd), one can choose U_ϕ such that

$$\mathcal{W}^\dagger U_\phi \mathcal{V} = W P\!\left(\frac{\Sigma}{\alpha}\right) V^\dagger = \sum_k P\!\left(\frac{\sigma_k}{\alpha}\right) |w_k\rangle\langle v_k|. \tag{11.46}$$

A convenient implementation in walk-operator form is

$$U_\phi = R_W(\phi_1)\,\Omega_W \prod_{\ell=1}^{(L-1)/2} \Big(R_V(\phi_{2\ell})\,\Omega_V\,R_W(\phi_{2\ell+1})\,\Omega_W\Big). \tag{11.47}$$

In this case, the transformed operator is an output–input map (it maps right singular vectors to left singular vectors), matching the rectangular structure of A.

Even Case (Input-Side Transform)

If L is even (so P is even), the natural transformed operator is supported on the input singular subspace, and one can choose U_ϕ such that

$$V^\dagger U_\phi V = V\, P\!\left(\frac{\Sigma}{\alpha}\right) V^\dagger = \sum_k P\!\left(\frac{\sigma_k}{\alpha}\right) |v_k\rangle\langle v_k|. \qquad (11.48)$$

A convenient implementation in walk-operator form is

$$U_\phi = \prod_{\ell=1}^{L/2} \Big(R_V(\phi_{2\ell-1})\, \Omega_V\, R_W(\phi_{2\ell})\, \Omega_W \Big). \qquad (11.49)$$

Cost and Ancillas

In both cases, U_ϕ uses $O(L)$ applications of U_A and $U_A^\dagger$ (equivalently, of Ω_W and Ω_V), and it preserves the ancilla size n required by the original block encoding. The additional work consists only of implementing the logical reflections (or equivalently, their phase rotations) on the same signal subspaces selected by Π_W and Π_V.

> **Exercise 11.27** Sanity check: consider the trivial case where all $\phi_\ell = 0$. Verify that the odd case Eq. 11.47 results in $P(x) = x$ and the even case Eq. 11.49 results in $P(x) = 1$.

While readers are referred to the original paper by Gilyén *et al.* [41] for a rigorous proof, it is worth emphasizing a key conceptual point. QSVT does not require explicit knowledge of the singular vectors of A. Once a block encoding is given, the associated reflections and walk operators automatically decompose the relevant Hilbert space into invariant two-dimensional subspaces indexed by the singular values.

Within each such subspace, the action of the QSVT sequence depends only on the scalar parameter σ_k/α. The singular vectors merely determine how an input state decomposes across these subspaces. Consequently, the polynomial transformation is applied uniformly and coherently across all singular values without ever isolating the singular vectors individually.

Key Takeaways

1. QSVT applies QSP-style phase modulation to singular values σ_k/α of a block-encoded operator A.

2. The walk/reflection structure creates invariant 2D subspaces indexed by $(|w_k\rangle, |v_k\rangle)$, so the circuit depends only on σ_k/α.

3. Odd P yields an output–input map $W\, P(\Sigma/\alpha)\, V^\dagger$; even P yields an input-side transform $V\, P(\Sigma/\alpha)\, V^\dagger$.

11.6.3 Circuit Implementation

The QSVT operator sequences for odd and even polynomials (Eqs. 11.47 and 11.49) can be implemented by the circuits shown in Fig. 11.12. A key point is that the

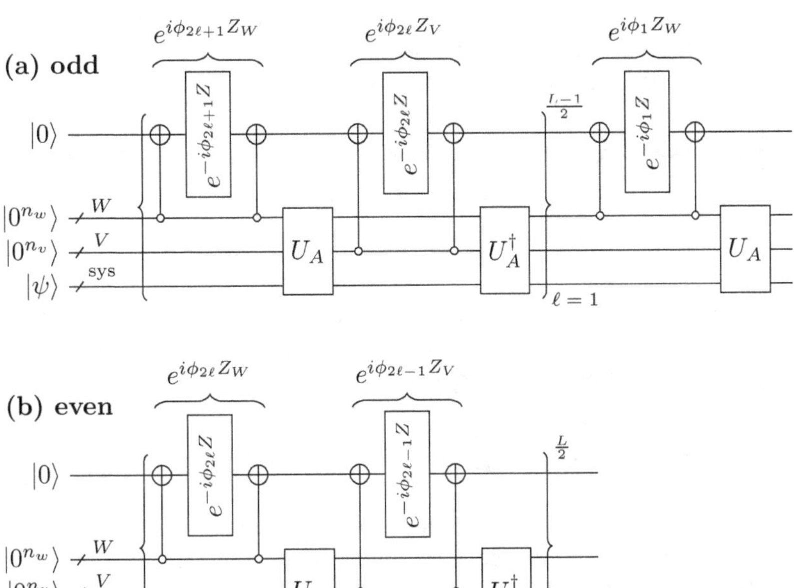

Figure 11.12: Quantum Singular Value Transformation (QSVT) Circuit

abstract walk operators $\Omega_W = Z_W U_A$ and $\Omega_V = Z_V U_A^\dagger$ can be replaced at the circuit level by U_A and $U_A^\dagger$, together with suitable phase rotations.

Indeed, using $Z_W = -i\, e^{i\frac{\pi}{2} Z_W}$, the reflection Z_W appearing in Ω_W can be absorbed into a neighboring logical phase rotation $R_W(\phi)$ by shifting the phase parameter $\phi \mapsto \phi + \frac{\pi}{2}$. An entirely analogous argument applies to Z_V. As a result, the QSVT sequence is often written in the simplified (circuit-level) form

$$\cdots R_V(\phi_\ell)\, U_A^\dagger\, R_W(\phi_{\ell+1})\, U_A \cdots, \tag{11.50}$$

which is the form most commonly used in implementations.

In Fig. 11.12, we assume that the signal register coincides with the block-encoding ancilla register. This register is partitioned into two subregisters of sizes n_w and n_v, labeled W and V, corresponding to the inclusion maps $\mathcal{W}$ and $\mathcal{V}$ in Eq. 11.42, which embed the output and input spaces of A into the dilation space of U_A, respectively.

To implement the logical phase rotations, an additional ancilla qubit (the top wire in the circuit) is introduced. This qubit serves as an auxiliary phase gadget and is not part of the signal register itself, in direct analogy with the multi-qubit QSP circuit in Fig. 11.11.

For example, $R_W(\phi) = e^{i\phi Z_W}$ applies a phase factor $e^{+i\phi}$ to states supported on the distinguished signal subspace selected by Π_W (equivalently, the ancilla basis state $|0^{n_w}\rangle$ on the W subregister), and a phase factor $e^{-i\phi}$ to the orthogonal complement. To realize this operation, a multi-controlled NOT coherently detects $|0^{n_w}\rangle$ on the W subregister and writes the result onto the auxiliary qubit. A single-qubit Z-rotation is then applied to the auxiliary qubit, after which the detection is uncomputed. This compute–rotate–uncompute pattern induces the desired relative phase on the signal register via phase kickback, while returning the auxiliary qubit to $|0\rangle$ after each step of the QSVT sequence. The phase rotations $R_V(\phi) = e^{i\phi Z_V}$ are implemented in the same manner, with control conditioned on the V subregister.

> **Exercise 11.28** In many QSVT/QSP-for-unitaries circuits, both U and $U^\dagger$ appear.
>
> (a) Explain, at the level of invariant subspaces, why alternating between "forward" and "backward" steps is natural in the rectangular (input/output) setting.
>
> (b) Give one scenario where access to $U^\dagger$ is essentially free and another where it is not. Hint: consider cases where U is given as a gate sequence and black-box oracle models.

11.7 ∗ Grand Unification of Major Quantum Algorithms

One of the most striking consequences of QSVT is that it provides a single, coherent framework for many quantum algorithms that are often presented as unrelated. Problems as diverse as unstructured search (Grover), Hamiltonian simulation, factoring (Shor), linear systems solving, ground-state preparation, and Gibbs state preparation can be recast as instances of applying a suitable polynomial (or analytic) transformation to the singular values or eigenvalues of an embedded operator.

11.7.1 Overview

At a high level, the unifying structure is as follows. One first encodes a problem-specific operator—such as a Hamiltonian, a unitary evolution operator, or a data matrix—into a larger unitary using block encoding and often qubitization. The algorithmic task is then reduced to applying a scalar function $f(x)$ to the spectrum of the encoded operator. QSVT supplies a uniform quantum mechanism for implementing such spectral transformations using coherent access to the block encoding and a sequence of phase-modulated reflections.

Below we sketch how several prominent algorithms fit into this template, emphasizing the target spectral functions.

Algorithm Class	Encoded Spectrum	Target Function $f(x)$
Hamiltonian Simulation	Eigenvalues of H	$\cos(tx)$, $\sin(tx)$
Grover Search	Overlap singular value	$\sin((2k+1)\arcsin x)$
Ground-State Preparation	Energy eigenvalues	Step-like filter $\Theta(x_0 - x)$
Phase Estimation / Factoring	Eigenphases of U	Indicator / window functions
Linear Systems (HHL)	Singular values of A	$1/x$ (restricted domain)
Gibbs State Preparation	Energy eigenvalues	$e^{-\beta x}$

Table 11.1: Major Quantum Algorithms Viewed Through the Lens of QSVT

Key Takeaways

1. Many quantum algorithms follow the same pattern: block-encode an operator, then apply a designed spectral function $f(x)$.

2. QSP/QSVT supply the hardware-friendly part: implementing the polynomial (or Laurent-polynomial) approximation to f using repeated calls to the encoding.

3. The remaining design work is classical: choose f and approximate it with the right degree, parity, and boundedness.

11.7.2 Representative Examples

1 Hamiltonian Simulation

Hamiltonian simulation, the task of approximating e^{-iHt}, is a canonical example of spectral transformation [47]. After rescaling and obtaining an (α, n) block encoding of H, the relevant signal variable is $x = \lambda/\alpha \in [-1, 1]$, where λ ranges over the eigenvalues of H. Time evolution corresponds to applying trigonometric functions to the spectrum:

$$e^{-iHt} = \cos(Ht) - i\sin(Ht).$$

QSVT (or QSP in Hermitian settings) implements these transformations by approximating $\cos(\alpha t x)$ and $\sin(\alpha t x)$ with bounded polynomials on $[-1, 1]$. This yields simulation algorithms whose query complexity is governed by the approximation degree (typically $O(\alpha t + \log(1/\varepsilon))$), rather than by Trotterization error bounds.

2 Grover Search as Singular-Value Amplification

Grover's algorithm evolves within a two-dimensional invariant subspace spanned by the target state $|t\rangle$ and its orthogonal complement. Writing the initial state as

$$|s\rangle = c\,|t\rangle + \sqrt{1 - c^2}\,|t^\perp\rangle, \qquad c = \langle t|s\rangle,$$

and defining $c = \sin\theta$, a single Grover iterate acts as a rotation by angle 2θ in this subspace. After k iterations, the target amplitude is

$$\langle t|G^k|s\rangle = \sin\big((2k + 1)\theta\big) = \sin\big((2k + 1)\arcsin c\big).$$

From the QSVT viewpoint, the parameter $x = c$ can be interpreted as a singular value of an embedded map capturing the overlap with the marked subspace. Grover's procedure therefore implements the odd function

$$f_{\mathrm{Grover}}(x) = \sin\big((2k + 1)\arcsin x\big),$$

which is an odd polynomial in x of degree $2k + 1$ (equivalently, a simple Chebyshev-related transformation). In this sense, Grover's search is a specific instance of singular-value amplification realized through a particular spectral polynomial.

In the standard "fixed-step" Grover iterate, the amplitude $\sin((2k+1)\theta)$ increases and then decreases as k grows, so choosing k incorrectly can overshoot the optimum. The QSVT framework is not restricted to this specific polynomial: by selecting a different bounded odd polynomial $P(x)$ (implemented via an appropriate phase schedule), one can obtain fixed-point or monotone singular-value amplification that increases the success amplitude without oscillations, thereby avoiding overshoot while retaining near-optimal query complexity.

3   Spectral Filtering: Ground-State Preparation

Ground-state preparation is an instance of spectral filtering: one aims to suppress excited-state components of a Hamiltonian H while retaining the ground state. After rescaling and block encoding, this corresponds to approximating a step-like filter

$$\Theta(x_0 - x) = \begin{cases} 1, & x \le x_0, \\ 0, & x > x_0, \end{cases}$$

where x_0 lies just above the ground-state energy (in the rescaled units). Since Θ is discontinuous, QSVT instead implements smooth bounded polynomials that transition rapidly from values near 1 to values near 0, thereby attenuating higher-energy eigencomponents.

4   Phase-Estimation-Based Algorithms and Factoring

Phase-estimation-based algorithms, including Shor's factoring algorithm, can be viewed as spectral filtering problems. In the standard setting, the central object is a unitary U with eigenstates $|\psi_k\rangle$ satisfying $U |\psi_k\rangle = e^{i\theta_k} |\psi_k\rangle$, where the eigenphases θ_k encode problem-specific structure (e.g., an order in modular arithmetic). The algorithmic goal is to separate components whose phases lie in different regions of $[0, 2\pi)$.

More generally, given spectral access through a block encoding (Hermitian or not), one often wants to distinguish eigenvalues or singular values lying in a prescribed range. This is naturally described by window functions,

$$f_{\text{win}}(\lambda) = \begin{cases} 1, & \lambda \in \mathcal{I}, \\ 0, & \lambda \notin \mathcal{I}, \end{cases}$$

where λ denotes an eigenphase, eigenvalue, or singular value. Since ideal windows are discontinuous, one instead uses bounded polynomial (or Laurent-polynomial) approximations that sharply separate $\mathcal{I}$ from its complement.

This same viewpoint connects to singular value estimation. While one can estimate singular values by applying quantum phase estimation (QPE) to an associated walk operator whose eigenphases encode σ_k/α, many modern approaches implement the needed spectral primitives directly via QSVT-style constructions, thereby avoiding explicit eigenphase readout when only a downstream spectral transformation is required.

5   Linear Systems Solvers as Approximate Inversion

Quantum linear systems solvers [42, 43] aim to prepare a state proportional to $A^{-1} |b\rangle$ for a suitably well-conditioned matrix A. After block encoding A, this becomes the task of applying an inverse-like transformation

$$f_{\text{inv}}(x) = \frac{1}{x}$$

to the singular values of A on a restricted interval bounded away from 0 (typically $x \in [1/\kappa, 1]$ after normalization). Since $1/x$ is unbounded near 0, QSVT implements a bounded polynomial approximation to $1/x$ over the relevant spectral range.

This clarifies the relationship to the original HHL approach. HHL uses QPE to resolve eigenvalues and then applies a controlled rotation proportional to $1/\lambda$. In contrast, QSVT-based solvers implement the same input–output goal through a direct polynomial singular-value transformation, often reducing ancilla overhead and making the dependence on precision essentially logarithmic.

6 Gibbs State Preparation as Exponential Filtering

Gibbs state preparation concerns preparing a system in thermal equilibrium with respect to a Hamiltonian H, described by

$$\rho_{\text{Gibbs}} = \frac{e^{-\beta H}}{\text{r}\!\left(e^{-\beta H}\right)}.$$

It may be viewed as a real-valued analogue of Hamiltonian simulation: instead of applying oscillatory functions e^{-iHt}, one seeks to apply the decaying exponential

$$f_{\text{Gibbs}}(x) = e^{-\beta x}$$

to the spectrum of H (after rescaling to a bounded interval). QSVT enables polynomial or Fourier-based approximations to $e^{-\beta x}$, allowing coherent suppression of high-energy components, typically combined with standard techniques such as purification, amplitude amplification, or postselection.

11.7.3 A Unified Design Paradigm

Taken together, these examples suggest that much of the diversity of quantum algorithms reflects different *encoded operators* and different *target spectral functions*, rather than fundamentally different quantum primitives.

Once a problem-specific operator is block encoded (or qubitized), the remaining design work is largely classical: construct a polynomial or analytic approximation to the desired scalar function with the required degree, parity, and boundedness properties. QSVT then supplies a uniform quantum procedure to apply this transformation coherently across all invariant two-dimensional subspaces, without resolving individual eigenstates or singular vectors.

Finally, QSVT does not claim that earlier algorithms such as Shor's or Grover's are literally implemented by QSVT circuits. Rather, it shows that they can be understood as special cases of a broader paradigm: coherent spectral transformation. In many settings, this viewpoint enables alternative constructions with greater modularity and near-optimal complexity.

Chapter Takeaways

1. Two complementary approximation routes dominate quantum linear-algebra design: *product formulas* (compose many simple steps, as in Trotterization) vs. *polynomial methods* (approximate a target function by a low-degree polynomial and implement it coherently).

2. Linear combination of unitaries (LCU) turns a weighted sum $\sum_j \alpha_j U_j$ into a *flagged* circuit branch using an ancilla state that encodes the coefficients; postselection (or amplification) then yields the desired linear combination.

3. LCU naturally extends from sums to *polynomials* of unitaries by composing LCU gadgets, making it an early workhorse for Hamiltonian simulation and other matrix-function tasks.

4. Oblivious amplitude amplification (OAA) boosts the success probability of LCU-style postselection *without* depending on the unknown input state, which is essential when the LCU gadget is used as a reusable subroutine.

5. Block encoding packages operator access in a standard interface: a unitary U is a block encoding of A/α if a designated block of U equals A/α; this abstraction supports clean reductions between problems.

6. Block encodings arise both from LCU constructions and from other access models (e.g., sparse-matrix oracles), and they satisfy core algebraic closure rules (scaling, addition, product, and controlled use) that enable modular algorithm design.

7. Quantum signal processing (QSP) implements a target *polynomial transformation* by an alternating single-ancilla gate sequence; the Chebyshev-based theorems characterize which polynomials are feasible and how the approximation degree controls cost.

8. QSP can be lifted from single-qubit primitives to generic operators via *qubitization*, which converts a block encoding into a walk-like unitary whose eigenphases encode the singular/eigenvalues of interest, enabling near-optimal complexity for many tasks.

9. Quantum singular value transformation (QSVT) is the culminating generalization: given a block encoding of A, QSVT applies a chosen polynomial to the singular values of A (and hence implements a wide range of matrix functions) with cost essentially set by polynomial degree.

10. Viewed together, LCU $\rightarrow$ block encoding $\rightarrow$ (qubitization $\rightarrow$) QSP/QSVT provides a unified toolkit that explains, streamlines, and often improves the complexity of major quantum algorithms built on Hamiltonian simulation and linear algebra.

Problem Set 11

11.1 Show that the LCU circuit for $U + V$ (Fig. 11.2) reduces to the Hadamard test circuit when $V = I$. In particular, demonstrate the following:

(a) Suppose U is a reflector, i.e., $U^2 = I$. Show that the spectrum of U is contained in $\{+1, -1\}$ and that the operators $\Pi_+ = \frac{1}{2}(I + U)$ and

$\Pi_- = \frac{1}{2}(I - U)$ are the orthogonal projectors onto the $+1$ and -1 eigenspaces of U, respectively.

(b) Show that postselecting the ancilla in $|0\rangle$ and $|1\rangle$ implements the projectors $\Pi_+ = \frac{1}{2}(I + U)$ and $\Pi_- = \frac{1}{2}(I - U)$ on the system register, respectively.

(c) Interpret the LCU circuit for $U = Z$ in terms of the reflector and projector structure identified above.

(d) Interpret the LCU circuit for $U = \mathrm{SWAP}$ in terms of the same reflector and projector mechanism, and explain the corresponding decomposition of the system Hilbert space.

(e) Explain how the case $U = \mathrm{SWAP}$ relates to the standard SWAP test for estimating the overlap between two states. In particular, show that the ancilla outcome corresponding to $\Pi_{\mathrm{sym}} = \frac{1}{2}(I + \mathrm{SWAP})$ projects onto the symmetric subspace and occurs with probability $p(0) = \frac{1}{2}\big(1 + |\langle\psi|\phi\rangle|^2\big)$ when the input is $|\psi\rangle \otimes |\phi\rangle$.

11.2 Analyze the LCU circuit of Fig. 11.3 when the ancilla is initialized in the general qubit state $\cos\theta\,|0\rangle + \sin\theta\,|1\rangle$. Determine the effective operation applied to the system register after postselecting the ancilla in $|0\rangle$, and express the resulting linear combination of U and V in terms of θ.

11.3 Prove Eq. 11.3. Hint: Follow the proof of Eq. 11.1c in § 11.2.1.2.

11.4 (OAA as a subroutine in LCU.)

(a) In a simple two-term LCU with coefficients $c_1, c_2 \geq 0$ and $\alpha = c_1 + c_2$, write down the ancilla state that encodes these weights.

(b) Explain why the postselection success probability is typically $\Theta(\|A\,|\psi\rangle\|^2/\alpha^2)$.

(c) Describe (at a high level) how OAA changes this success probability and what resources it consumes (calls to U and reflections about $|0^n\rangle$).

11.5 (OAA success probability bounds from operator norms.) Assume the block-encoding form

$$U(|0^n\rangle\,|\psi\rangle) = |0^n\rangle\,\frac{A\,|\psi\rangle}{\alpha} + |\Phi_\perp(\psi)\rangle.$$

(a) Show that $p_0(\psi) = \|A\,|\psi\rangle\|^2/\alpha^2$.

(b) If $\|A\| \leq 1$, prove that $p_0(\psi) \leq 1/\alpha^2$ for all $|\psi\rangle$.

(c) Give an example of a state $|\psi\rangle$ that attains (or nearly attains) this upper bound when A has a singular value close to 1.

11.6 The LCU circuit shown in Fig. 11.13, where Prep prepares the ancilla state $\sum_k \frac{\sqrt{\alpha_k}}{\alpha}\,|k\rangle$, implements the polynomial function $\sum_k \alpha_k U^k$ on the system register.

(a) Prove that the circuit correctly implements the stated polynomial function. Hint: review how the phase-estimation subcircuit shown in Fig. 7.2 operates.

(b) In Fig. 11.13, the coefficients α_k are assumed to be nonnegative. In general, however, the coefficients may be complex. Show how the corresponding phase factors can be absorbed into the controlled applications of U^k by modifying the circuit accordingly.

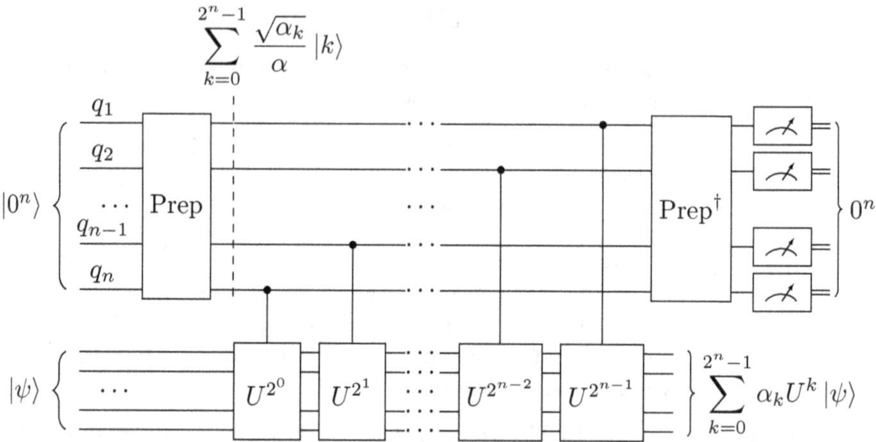

Figure 11.13: LCU Circuit for Polynomial Functions of a Unitary Operator

11.7 Suppose U_A is an (α, n) block encoding of A and U_B is a (β, n) block encoding of B on the same system size (and same ancilla size n). Show that

$$\left(\langle 0^n | \otimes I\right) U_A U_B \left(|0^n\rangle \otimes I\right) = \frac{AB}{\alpha\beta}$$

does not generally hold. Give a short condition under which it *does* hold (e.g., an "obliviousness" condition / block-structure condition).

11.8 Design a gate decomposition for the reflection

$$Z_{\text{sig}} = 2\,|0^n\rangle\langle 0^n| - I$$

using multi-controlled Z (or multi-controlled X plus conjugations). Estimate the Toffoli (or T-) count as a function of n under a standard decomposition model.

11.9 Let $x = \sin\theta$ be the initial overlap with a target subspace.

(a) Recall that the fixed-step Grover iterate yields an amplitude $\sin((2k+1)\theta)$ for iteration k. Explain why overshoot occurs if k is chosen too large.

(b) Propose a condition on a degree-L odd polynomial $P(x)$ ensuring monotone amplification on $x \in [x_{\min}, 1]$.

(c) Give a short argument (no full construction required) for why QSVT/QSP can realize such a P while keeping $|P(x)| \leq 1$ on $[-1, 1]$.

11.10 In the rectangular QSVT theorem, the odd-degree case produces an output–input map, while the even-degree case produces an input-side transform.

(a) Using the SVD, explain why an odd function of singular values naturally maps $|v_k\rangle$ to $|w_k\rangle$, while an even function naturally stays on the $|v_k\rangle$ side.

(b) For a toy singular value $\sigma/\alpha = \cos\theta$, compute the induced 2×2 action for one QSVT layer and identify which matrix entry corresponds to $P(\cos\theta)$.

11.11 Assume U_H is an (α, n) block encoding of a Hermitian H with $\|H\| \leq \alpha$.

(a) Starting from

$$e^{-ixt} = J_0(t) + 2\sum_{k \geq 1}(-i)^k J_k(t)\, T_k(x),$$

explain why truncating at $k \leq L$ gives constant precision once $L = \Theta(t)$.

(b) State (you may cite without proof) a standard tail bound for Bessel functions and conclude that $L = O(\alpha t + \log(1/\varepsilon))$ suffices for error ε.

11.12 Choose one target function below on $x \in [-1, 1]$ and synthesize a QSP phase sequence:

$$f_1(x) = 0.5\sin(8x), \quad f_2(x) = 0.5\tanh(3x), \quad f_3(x) = e^{-i\tau x} \text{ for a chosen } \tau.$$

(a) Pick a degree L and justify it qualitatively.

(b) Report the maximum approximation error on a dense grid.

(c) Briefly discuss any numerical instability you observe as L increases.

11.13 You are given an (α, n) block encoding of a Hermitian H with spectrum in $[-\alpha, \alpha]$. Design (at the level of steps, not low-level gates) a QSVT/QSP-based procedure for one task below:

(a) Energy thresholding: project approximately onto the span of eigenstates with eigenvalues $\leq E_0$.

(b) Approximate sign function: implement $\mathrm{sgn}(H)$ on the support where $|H| \geq \Delta$.

(c) Inversion on a promise gap: implement H^{-1} on eigenvalues in $[1/\kappa, 1]$ and suppress the rest.

For your chosen task, state the target scalar function $f(x)$, the domain promise (gap), and the expected degree scaling in $\log(1/\varepsilon)$ and in the relevant condition parameter (e.g., κ or $1/\Delta$).

12. Quantum Linear System Algorithms (QLSA)

Contents

Solving linear systems is a central primitive in scientific computing and data analysis, appearing in tasks ranging from discretized differential equations to least-squares fitting. In quantum algorithms, the umbrella term *quantum linear system algorithms (QLSA)* refers to methods that, given access to a matrix A and a right-hand side b, prepare a *solution state* $|x\rangle$ whose amplitudes encode a normalized version of $x = A^{-1}b$. The Harrow–Hassidim–Lloyd (HHL) algorithm is the original and best-known entry point to this family [48].

The QLSA output model is both the power and the limitation of these methods. Preparing $|x\rangle$ can enable efficient estimation of selected global properties of the solution—such as expectation values, overlaps, or sampling access—but it does not, in general, provide an efficient route to printing all entries of x as classical data.

This chapter adopts a modern, modular view of QLSA. We access A through a block encoding and implement inversion as a *coherent spectral (or singular-value) transformation*. Concretely, we use Quantum Singular Value Transformation (QSVT, Chapter 11) to approximate the reciprocal function on a promised spectral interval, and we use Oblivious Amplitude Amplification (OAA) to boost the flagged success probability. We then (i) formulate the task and its assumptions, (ii) develop the QSVT+OAA inversion procedure and its complexity, (iii) construct sparse-matrix block encodings via standard access oracles, and (iv) summarize representative uses and practical limitations of QLSA in practice.

12.1　Linear Systems as a Quantum Task

12.1.1　Purpose and Input–Output Models

1　Classical Linear Systems

The classical linear systems problem asks us to solve

$$Ax = b,$$

where $A \in \mathbb{C}^{N \times N}$ is a known matrix and $b \in \mathbb{C}^N$ is a known vector. In most numerical settings, the goal is to output a classical description of the solution vector x, either exactly (in symbolic settings) or approximately (in floating-point arithmetic). From this perspective, the act of *reading out x* is part of the task: the final answer is a list of N numbers.

Closely related classical tasks include least-squares fitting and the numerical solution of linear differential equations. For least-squares problems with rectangular or overdetermined systems ($A \in \mathbb{C}^{m \times N}$, $m \geq N$), one seeks

$$x_{\mathrm{LS}} = \arg\min_x \|Ax - b\|_2,$$

which reduces to applying the Moore–Penrose pseudoinverse A^+ (for example, via the normal equations $A^\dagger A x = A^\dagger b$), so that $x_{\mathrm{LS}} = A^+ b$.

Similarly, discretizing linear ordinary or partial differential equations (e.g., Poisson's equation, the heat equation, or Schrödinger-type equations) typically yields a sparse linear system $Ax = b$, where A encodes the discretized differential operator and b incorporates sources, initial conditions, or boundary terms. In both cases, the core operation is matrix inversion or pseudoinversion applied to a transformed right-hand side.

In later applications of HHL-type routines, these problems often appear in a unified form: one implements (or approximates) the (pseudo)inverse action by applying a reciprocal-type transformation to the singular values of A (or of a related operator), thereby producing a solution state corresponding to $A^+ b$ or an analogous expression.

2 Quantum Linear Systems

In quantum algorithms, the target is fundamentally different. The HHL framework
(and its modern variants) aims to prepare a *quantum state* that encodes the solution
in its amplitudes. More precisely, we assume that the right-hand side can be prepared
as a normalized state

$$|b\rangle = \sum_{j=0}^{N-1} b_j \, |j\rangle, \qquad \sum_{j=0}^{N-1} |b_j|^2 = 1,$$

and the algorithm outputs a state proportional to

$$|x\rangle \; \propto \; A^{-1} |b\rangle.$$

Equivalently, if we define the (classical) solution vector $x = A^{-1}b$ and the normal-
ization factor $\|A^{-1} |b\rangle \|$, then the output state is

$$|x\rangle = \frac{A^{-1} |b\rangle}{\|A^{-1} |b\rangle \|}.$$

This "solution state" viewpoint is the first conceptual shift one must make in order
to understand what HHL can and cannot do.

3 Input (Access) Models

The quantum formulation is only meaningful once we specify how A and b are
accessed. In the modern viewpoint, we separate the linear-systems *primitive* (coherent
inversion) from the *input model* (how the data are provided).

- Preparing $|b\rangle$. We assume access to a state-preparation unitary U_b such that
  $U_b |0\rangle = |b\rangle$. The cost of this step, denoted T_b, is highly application-dependent.
  Favorable regimes include cases where $|b\rangle$ is generated by a short circuit, arises
  naturally within a quantum computation, or has special structure (e.g., sparsity
  or succinct description).

- Accessing A. We assume access to A through a block encoding: for some
  $\alpha \geq \|A\|$ we can implement a unitary U_A whose selected block equals A/α (up
  to controlled error). The cost of implementing this block encoding is denoted
  T_A. In structured settings—most notably the sparse-access model—U_A can be
  constructed explicitly from simple oracles, which we develop in § 12.3.

In short, HHL-type solvers provide a coherent inversion routine *assuming* efficient
access to $|b\rangle$ and to a block encoding of A/α; whether this yields an end-to-end
advantage depends critically on the chosen access model.

4 Output Models

Because the output is a quantum state, HHL is not designed to efficiently print the
full classical vector x. In general, extracting all N entries of x would require many
measurements and is typically incompatible with the goal of achieving a speedup.
Instead, HHL is useful when the downstream task only requires limited information
that can be accessed from repeated preparations of $|x\rangle$. Common examples include:

- Expectation values. For an observable (or more generally, an efficiently implementable operator) M, estimate

$$\langle x| M |x\rangle.$$

This captures many "global" properties of the solution without reconstructing it entry-by-entry.

- Overlaps and inner products. For a state $|c\rangle$ that can be prepared efficiently, estimate the overlap

$$\langle c|x\rangle,$$

  or related quantities derived from it.

- Sampling. Measure $|x\rangle$ in the computational basis to sample an index j according to the distribution $|x_j|^2$. This is meaningful when the application only needs randomized access to coordinates or when the sampling distribution itself is the object of interest.

In short, the quantum linear-systems task is best viewed as a state-preparation primitive: HHL prepares $|x\rangle$ so that one can compute other quantities of interest from it, rather than outputting the entire solution vector as classical data.

Key Takeaways

---

1. The output of HHL is a quantum state $|x\rangle = \frac{A^{-1}|b\rangle}{\|A^{-1}|b\rangle\|}$, not a classical list of entries.

2. HHL is useful when the goal is a *small-output* quantity, such as $\langle x| M |x\rangle$, an overlap $\langle c|x\rangle$, or sampling from $|x_j|^2$.

3. HHL also assumes efficient access to inputs: a preparation procedure for $|b\rangle$ and a block encoding of A/α.

4. Reading out all entries of x typically costs $\Omega(N)$ measurements and can eliminate any speedup.

12.1.2   Assumptions and Parameters

Any discussion of HHL must make explicit the access model for A and the promises under which matrix inversion is well behaved. In this chapter we adopt the modern viewpoint that the matrix is provided through a *block encoding* (discussed in § 11.3). Concretely, we assume that for some known scaling factor $\alpha \geq \|A\|$ we can implement a unitary U_A such that a designated block of U_A equals A/α (up to controlled error). Equivalently, we have efficient quantum access to the linear operator A in a form compatible with QSVT (see § 11.6). This assumption is not automatic: the cost of constructing such a block encoding depends on the problem setting (e.g., sparsity or other structure), and it is part of the overall resource accounting.

A second essential assumption is a *spectral promise* that excludes eigenvalues (or singular values) near 0. After scaling, a typical promise is that the singular values of A/α lie in

$$\sigma(A/\alpha) \subseteq [1/\kappa, 1],$$

or, in the Hermitian case, that the eigenvalues of A/α lie in

$$\mathrm{spec}(A/\alpha) \subseteq [-1, -1/\kappa] \cup [1/\kappa, 1],$$

where $\kappa \geq 1$ plays the role of a (promised) condition number. Intuitively, the promise says that the spectrum is bounded away from 0, so that inversion does not require unbounded amplification.

Finally, we must specify an accuracy target. We write ϵ for the desired precision, with the understanding that "precision" is application-dependent. In a state-output problem, it is often most natural to define ϵ in terms of how accurately the prepared state approximates the ideal solution state, or in terms of the induced error in the quantities one actually measures from $|x\rangle$ (such as expectation values or overlaps). For our purposes, it suffices to keep ϵ as a single parameter capturing the target approximation error of the linear-systems primitive.

Summarizing, the main parameters that govern the cost of quantum linear-system solvers are:

- the cost of implementing the block encoding of A/α,

- the condition number parameter κ (equivalently, the promised spectral gap away from 0),

- the target precision ϵ,

- the cost of preparing $|b\rangle$.

12.1.3 Conditioning, Stability, and Cost

The prominence of the condition number is not an artifact of a particular implementation; it is a reflection of the intrinsic instability of matrix inversion. To see this, suppose that A is diagonalizable with eigenpairs $\{(\lambda_j, |u_j\rangle)\}$ and expand

$$|b\rangle = \sum_j \beta_j |u_j\rangle .$$

Formally,

$$A^{-1} |b\rangle = \sum_j \frac{\beta_j}{\lambda_j} |u_j\rangle .$$

If some $|\lambda_j|$ is small, then the corresponding coefficient β_j/λ_j is large. Thus, components of $|b\rangle$ aligned with small-eigenvalue directions are greatly magnified by inversion. This has two immediate consequences.

First, inversion becomes numerically (and physically) unstable near $\lambda = 0$. Small perturbations in b or small approximation errors in implementing the inverse can be amplified, leading to a large relative change in the solution. This is the same phenomenon that underlies conditioning in classical numerical linear algebra.

Second, conditioning controls the algorithmic cost. In quantum implementations, one typically applies a bounded approximation to the reciprocal function on an interval that excludes 0, such as $[1/\kappa, 1]$ (or the corresponding signed interval in the Hermitian case). As κ increases, this interval approaches 0, and approximating $1/x$ uniformly becomes harder. Moreover, most constructions produce the desired transformation in a flagged subspace with a success probability that deteriorates as the smallest relevant (singular) value decreases. For these reasons, κ governs both the complexity of implementing the inverse map and the overhead required to amplify the success probability.

For the remainder of this chapter, we therefore treat κ not merely as a technical detail but as a central descriptor of problem difficulty: well-conditioned instances (κ modest) are the natural regime where quantum linear-system solvers can be effective, whereas ill-conditioned instances can negate any potential advantage.

Key Takeaways

---

1. Access model. The algorithm assumes a block encoding of A/α can be implemented at some cost T_A.

2. Spectral promise. Inversion requires a gap away from 0 (e.g., $x \in [1/\kappa, 1]$ after scaling), where κ acts as a condition-number parameter.

3. Precision. The target error ϵ controls how accurately the reciprocal is approximated, and it propagates to errors in $|x\rangle$ and in measured quantities.

12.2 ∗ From Block-Encoding to Inversion via QSVT

This section is the conceptual core: HHL is a coherent spectral (or singular-value) transformation implementing a bounded approximation to $1/x$. In the original HHL formulation, this idea is realized through phase estimation and a controlled rotation. In the modern viewpoint adopted throughout this chapter, once A is available as a block encoding, Quantum Singular Value Transformation (QSVT) can apply an explicitly designed transformation to the singular values (and, in the Hermitian case, to the eigenvalues). The role of "inversion" is therefore reduced to an approximation-theoretic task: construct a bounded function that behaves like $1/x$ on the promised spectral region, and then implement that function coherently using QSVT.

Overview

---

1. HHL is best viewed as *coherent inversion*: it prepares a *solution state* $|x\rangle \propto A^{-1} |b\rangle$ rather than outputting the full classical vector x.

2. Given a block encoding of A/α and a spectral promise $x \in [1/\kappa, 1]$, QSVT implements a bounded approximation to $1/x$ on the promised region.

3. The inverse appears in a *flagged* branch; OAA boosts the success probability to a constant so $|x\rangle$ can be prepared reliably.

12.2.1 Inversion as Coherent Spectral Transformation

Suppose we are given an (α, n) block encoding of a matrix A, meaning that we can implement a unitary U_A such that the operator A/α appears as a designated block of U_A (up to controlled error). After rescaling by α, we may assume $\|A/\alpha\| \leq 1$. The linear-systems task can then be phrased as follows: given $|b\rangle$, produce a state proportional to $A^{-1} |b\rangle$. At the operator level, this is equivalent to implementing a function of A that approximates A^{-1} on the relevant subspace.

To see why this is fundamentally a spectral transformation, consider first the Hermitian case, where A admits a spectral decomposition

$$A = \sum_j \lambda_j |u_j\rangle\langle u_j|.$$

If $|b\rangle = \sum_j \beta_j |u_j\rangle$, then

$$A^{-1} |b\rangle = \sum_j \frac{\beta_j}{\lambda_j} |u_j\rangle \,,$$

so inversion corresponds to applying the scalar map $\lambda \mapsto 1/\lambda$ to each eigencomponent. In the non-Hermitian setting, the same idea is expressed in terms of singular values. If $A = \sum_j \sigma_j |u_j\rangle\langle v_j|$ is a singular value decomposition, then the pseudoinverse A^+ (on the support of A) rescales the singular directions by $1/\sigma_j$. Either way, the computational problem is to implement a reciprocal transformation *coherently* across all relevant spectral components.

Historically, the original HHL algorithm for Hermitian matrices realizes the reciprocal transformation using phase estimation and a controlled rotation, rather than QSVT. In this chapter we adopt the modern viewpoint: once A is provided as a block encoding, QSVT offers a direct and modular way to implement the same coherent spectral transformation.

> **Exercise 12.1** In the Hermitian case, inversion acts as $\lambda \mapsto 1/\lambda$ on eigencomponents. In the general case, the pseudoinverse acts as $\sigma \mapsto 1/\sigma$ on singular components (with a cutoff). Write the two expressions $A^{-1}b$ and A^+b in the eigenvalue and singular-value decompositions, respectively.

Before discussing how QSVT implements a bounded approximation to the reciprocal, we illustrate the basic spectral rescaling behind inversion with a 2×2 example.

12.2.2 A Small Worked Example

Consider the 2×2 Hermitian matrix

$$A = \begin{pmatrix} 1 & 0 \\ 0 & 1/\kappa \end{pmatrix}, \qquad \kappa \geq 1,$$

whose eigenpairs are

$$\lambda_0 = 1, \ |u_0\rangle = |0\rangle\,, \qquad \lambda_1 = 1/\kappa, \ |u_1\rangle = |1\rangle\,.$$

Let the right-hand side be

$$|b\rangle = \cos\theta \, |0\rangle + \sin\theta \, |1\rangle\,, \qquad \theta \in [0, \pi/2].$$

The (unnormalized) inverse action is

$$A^{-1} |b\rangle = \cos\theta \, |0\rangle + \kappa \sin\theta \, |1\rangle\,.$$

This explicitly shows the defining feature of matrix inversion: components associated with smaller eigenvalues are magnified.

1 Normalization and the Solution State

The quantum output is the normalized solution state

$$|x\rangle = \frac{\cos\theta \, |0\rangle + \kappa \sin\theta \, |1\rangle}{\sqrt{\cos^2\theta + \kappa^2 \sin^2\theta}}\,.$$

If θ is small so that $|b\rangle$ is mostly aligned with the well-conditioned direction $|0\rangle$, then $|x\rangle \approx |0\rangle$. If θ has substantial weight on $|1\rangle$ and κ is large, then $|x\rangle$ becomes strongly biased toward $|1\rangle$, reflecting the strong amplification of the ill-conditioned direction.

The example highlights why small eigenvalues are problematic: the map $x \mapsto 1/x$ is unbounded near 0, which is why we must work with a promised interval excluding 0 and a bounded approximation suitable for QSVT.

Exercise 12.2 In the 2×2 example, compute $\|A^{-1}b\|$ and verify that the normalized state $|x\rangle$ depends on κ and θ only through the ratio $\kappa\tan\theta$. Interpret this dependence geometrically on the Bloch sphere.

12.2.3　Reciprocal Approximation on a Promised Interval

Because QSVT implements *bounded* transformations, we do not attempt to realize $f(x) = 1/x$ directly. Instead, we use the spectral promise that the relevant eigenvalues (Hermitian case) or singular values (general case) avoid 0. After scaling, this promise typically takes the form

$$x \in [1/\kappa, 1],$$

where κ is a condition-number parameter. On this interval, one can approximate the reciprocal by a polynomial or other admissible function (see illustration in Fig. 12.1). A common strategy is to approximate a *scaled* reciprocal

$$g(x) = \frac{\gamma}{x},$$

where $\gamma > 0$ is chosen so that $|g(x)| \leq 1$ on the target interval. For example, choosing $\gamma = 1/\kappa$ ensures

$$\left|\frac{\gamma}{x}\right| \leq 1 \quad \text{for all} \quad x \in [1/\kappa, 1].$$

This boundedness is not a mere technicality: it is what allows the reciprocal transformation to be embedded into a unitary process, with the inverse action emerging in a flagged subspace.

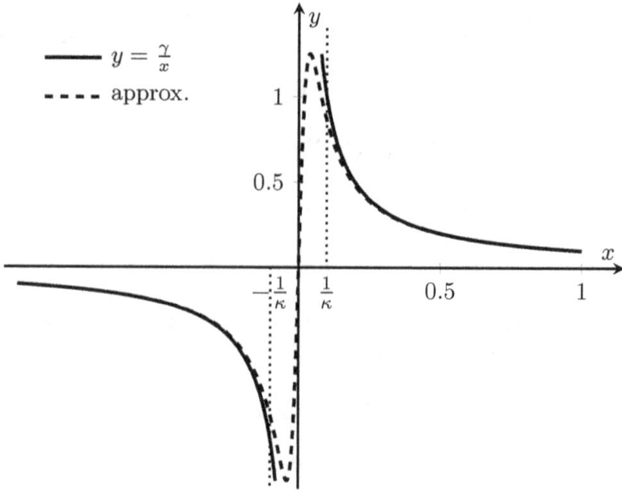

Figure 12.1: Approximation for the $1/x$ Function

In practice, the function implemented by QSVT is a polynomial (or, depending on the construction, an effective rational approximation) that satisfies the boundedness constraints required by QSVT on $[-1, 1]$. It is therefore useful to distinguish:

- the *ideal* target $x \mapsto 1/x$ on the promised region, and

- the *implementable* bounded approximation $x \mapsto g(x)$ realized by QSVT to within an error tolerance ϵ.

The error parameter ϵ quantifies how closely the implemented transformation approximates the reciprocal on $[1/\kappa, 1]$, and therefore controls the quality of the prepared solution state.

> **Exercise 12.3** Let $g(x) = \gamma/x$ on $x \in [1/\kappa, 1]$.
>
> - Find the largest γ such that $|g(x)| \leq 1$ on this interval.
>
> - Explain how this scaling is reflected in the success amplitude of a flagged implementation of $\widetilde{A^{-1}}$.

Key Takeaways

1. QSVT can only implement bounded functions on $[-1, 1]$, so we approximate a *scaled* reciprocal $g(x) = \gamma/x$ on $x \in [1/\kappa, 1]$.

2. Choosing $\gamma = 1/\kappa$ guarantees $|g(x)| \leq 1$ on the promised interval, making the transformation unitary-implementable.

3. This scaling typically makes the flagged success probability decrease with κ, which is why amplification (OAA) is needed.

12.2.4 Flagged Inversion and Success Probability

A subtle but essential point is that QSVT typically implements the desired transformation in a *flagged* manner. Informally, the QSVT unitary acts on an extended Hilbert space that includes one or more ancilla registers. When these ancillas are projected onto a particular "good" outcome, the induced transformation on the data register corresponds to the intended matrix function.

Specialized to linear systems, the effect is that after applying the QSVT-based circuit to an input $|b\rangle$, the overall state has the schematic form

$$\sqrt{1 - p}\, |0\rangle_{\text{flag}} |\text{junk}\rangle \; + \; \sqrt{p}\, |1\rangle_{\text{flag}} \left(|\widetilde{x}\rangle \right),$$

where $|\widetilde{x}\rangle \propto \widetilde{A^{-1}} |b\rangle$ approximates $A^{-1} |b\rangle$ on the promised region, and $|\text{junk}\rangle$ denotes an orthogonal failure component. Conditioning on the good flag outcome therefore prepares a normalized state close to the ideal solution state

$$|x\rangle = \frac{A^{-1} |b\rangle}{\|A^{-1} |b\rangle\|}.$$

The normalization factor $\|A^{-1} |b\rangle\|$ is not output explicitly; it is implicitly absorbed into the success amplitude $\sqrt{p}$.

If one relied on postselection alone, the expected number of repetitions to obtain the good outcome would scale as $1/p$. When p is small, this overhead becomes prohibitive. This is precisely where Oblivious Amplitude Amplification (OAA, § 9.3.2) enters. Since the success branch is coherently flagged and the transformation is implemented within a unitary framework, OAA boosts the amplitude of the good subspace without measuring and restarting the computation. Operationally, OAA increases the success probability to a constant using $O(1/\sqrt{p})$ applications of the flagged procedure (and its inverse), together with controlled reflections.

12.2.5   A Canonical QLSA Procedure (HHL via QSVT + OAA)

We now assemble the ingredients into a practical linear-systems procedure. Throughout, assume that the following inputs are available:

- an efficient procedure to prepare $|b\rangle$,

- an (α, n) block encoding U_A of A (equivalently, a block encoding of A/α),

- a spectral promise excluding 0, parameterized by a condition number κ after scaling,

- a target approximation error ϵ.

1   Step A: Prepare $|b\rangle$

Prepare the right-hand side state

$$|b\rangle = \sum_{j=0}^{N-1} b_j |j\rangle, \qquad \sum_{j=0}^{N-1} |b_j|^2 = 1.$$

We denote by T_b the cost of this preparation, since it is highly application-dependent and can dominate end-to-end performance.

2   Step B: Apply an Approximate Inverse

Design an admissible bounded approximation to $1/x$ on the promised interval (to error ϵ) and implement it using QSVT together with the block encoding of A/α. The resulting unitary produces a flagged branch proportional to $\widetilde{A^{-1}} |b\rangle$, as described in § 12.2.4.

3   Step C: Amplify the "Good" Outcome

Apply OAA (see § 9.3.2) to amplify the flagged success probability from p to a constant using $O(1/\sqrt{p})$ applications of the flagged inverse procedure (and its inverse), together with the required reflections. Measuring the flag then yields the good outcome with high probability, and the data register is approximately the desired solution state $|x\rangle$.

> **Exercise 12.4** Write a five-line pseudocode summary of the QSVT+OAA HHL procedure, including the required inputs (U_A, κ, ϵ, and $|b\rangle$) and the final measurement performed to obtain $|x\rangle$.

12.2.6   Complexity Discussion

Let T_A denote the cost of implementing the block encoding of A/α (including any ancillas needed by the access model), let L be the QSVT sequence length needed to approximate $1/x$ on $[1/\kappa, 1]$ to error ϵ, let p be the success probability of the flagged inverse step before amplification, and let T_b be the cost of preparing $|b\rangle$. A representative high-level scaling is

$$\text{Cost} \sim (T_b + L T_A) \cdot O\!\left(\frac{1}{\sqrt{p}}\right),$$

where the dependence on κ typically enters through both L (approximation difficulty) and p (the scaling needed to keep the implemented reciprocal bounded). The main message is that the end-to-end cost depends critically on access to A, conditioning (through κ), precision goals (through ϵ), and the cost of preparing $|b\rangle$.

> **Exercise 12.5** Explain briefly how each of the following affects the total cost: (a) T_b, (b) T_A, (c) L, and (d) p. For each, state whether it typically enters additively or multiplicatively in the overall scaling.

12.3 ✳ Block-Encoding Sparse Matrices

A central prerequisite for HHL-type solvers is an efficient block encoding of the system matrix A. In many applications, A is *sparse* in a known basis, and we are given oracle access that allows us to locate and query its nonzero entries efficiently [40, 42, 43]. This section summarizes a standard construction: given sparse access to an s-sparse matrix A, we can build a block encoding of A/α with $\alpha = O(s)$ when $\|A\|_{\max} \leq 1$ (and more generally $\alpha = O(s\|A\|_{\max})$). The main idea is to prepare coherent "row" and "column" states whose inner products reproduce matrix entries.

12.3.1 Sparse-Access Model

Let $A \in \mathbb{C}^{N \times N}$. We say that A is *s-sparse* if each row (and hence each column) contains at most s nonzero entries. We assume access to A through standard sparse-access oracles. Conceptually, the access consists of two capabilities:

- *Indexing nonzero locations.* Given a column index $j \in \{0, \ldots, N - 1\}$ and an integer $\ell \in \{1, \ldots, s\}$, an oracle returns the row index of the ℓ-th nonzero entry in column j (with padding if fewer than s entries are present). One may similarly index nonzero entries in rows.

- *Querying values.* Given indices (i, j), an oracle returns (an encoding of) the complex entry A_{ij} (or returns 0 if the entry is absent).

Precise oracle formats vary across references; for our purposes, it suffices that these queries can be implemented coherently in superposition and with $\mathrm{polylog}(N)$ overhead.

In our discussion below, we use the entrywise bound $\|A\|_{\max} = \max_{i,j} |A_{ij}|$. To ensure that the state-preparation amplitudes remain bounded, we normalize by $s\|A\|_{\max}$ when constructing a block encoding in complete generality. In many treatments one assumes $\|A\|_{\max} \leq 1$ for simplicity, in which case this normalization reduces to a block encoding with scaling factor $\alpha = O(s)$.

12.3.2   Sparse-Access Oracles and State Preparation

There are several equivalent ways to construct block encodings for sparse matrices. Here we adopt the circuit-oriented formulation of Camps and Lin *et al.* [49, 50], which is especially convenient when we want explicit, reusable circuit diagrams. This approach cleanly separates where the nonzero entries are located from what their values are, and it makes the normalization mechanism in the extracted block transparent. The construction is built from three components.

1   Diffusion Over a Sparsity-Index Register

Let the ℓ-register consist of m qubits, with computational basis $\{|\ell\rangle : \ell \in \{0, \ldots, s-1\}\}$. Define the diffusion operator

$$D_s := H^{\otimes m}, \qquad D_s |0^m\rangle = \frac{1}{\sqrt{s}} \sum_{\ell=0}^{s-1} |\ell\rangle .$$

This operator has two roles:

- *Uniformly choose a candidate location in superposition.* Applying D_s maps $|0^m\rangle$ to $\frac{1}{\sqrt{s}} \sum_\ell |\ell\rangle$, so the subsequent oracles O_C and O_A act coherently on *all* candidate nonzeros indexed by ℓ.

- *Create the $1/s$ scaling by interference.* In the block-encoding matrix element, we project onto $\langle 0^m|$ at the end. This projection is equivalent to applying $D_s^\dagger$ and taking the $|0^m\rangle$ component. Since $\langle 0^m|D_s^\dagger|\ell\rangle = 1/\sqrt{s}$, this produces the normalization factor needed for a block encoding of A/s (or, more generally, $A/(s\|A\|_{\max})$ after rescaling).

2   Index Oracle O_C (Locations of Nonzeros)

Let $c(j, \ell)$ denote the row index of the ℓth entry in a fixed length-s list of (potential) nonzeros in column j (padded if needed). The index oracle is the unitary

$$O_C : \; |\ell\rangle |j\rangle \longmapsto |\ell\rangle |c(j,\ell)\rangle .$$

In structured settings (banded, circulant, tridiagonal, etc.), $c(j, \ell)$ is typically computable by a short reversible arithmetic subroutine.

For example, if column j has three nonzero entries located at row indices $6, 7, 8$ (and we set $s = 3$), then we may define $c(j, 0) = 6$, $c(j, 1) = 7$, and $c(j, 2) = 8$.

In this column-indexed formulation, we do not need a separate "row oracle" $r(i, \ell)$: the row index is recovered by computing $c(j, \ell)$ from (j, ℓ), and the projection onto $\langle i|$ in the block-encoding matrix element selects exactly those ℓ for which $c(j, \ell) = i$.

3   Value Oracle O_A (Values of Nonzeros)

The value oracle is implemented as a controlled single-qubit rotation on a flag ancilla, arranged so that the $|0\rangle$-amplitude encodes the (normalized) matrix entry. In the simplest presentation we assume $\|A\|_{\max} \leq 1$ so that $|A_{ij}| \leq 1$ for all entries. Then

$$O_A : \; |0\rangle |\ell\rangle |j\rangle \longmapsto \left(A_{c(j,\ell),j} |0\rangle + \sqrt{1 - |A_{c(j,\ell),j}|^2} |1\rangle \right) |\ell\rangle |j\rangle , \tag{12.1}$$

where the first register is the one-qubit flag ancilla, the second register stores the sparsity index $\ell \in \{0, \ldots, s-1\}$, and the third register stores the column index $j \in \{0, \ldots, N-1\}$.

When $A_{c(j,\ell),j} \in [-1, 1]$ is real, Eq. 12.1 can be implemented by a controlled Y rotation on the flag qubit that maps $|0\rangle$ to $A_{c(j,\ell),j} |0\rangle + \sqrt{1 - A_{c(j,\ell),j}^2} |1\rangle$. When $A_{c(j,\ell),j}$ is complex, we implement a phase-aware variant: writing $A_{c(j,\ell),j} = re^{i\varphi}$ with $r \in [0, 1]$, we first apply a controlled Y rotation that maps $|0\rangle$ to $r |0\rangle + \sqrt{1 - r^2} |1\rangle$, and then apply a controlled phase rotation (e.g., a controlled $R_z(\varphi)$) so that the $|0\rangle$ component acquires the factor $e^{i\varphi}$. In either case, the net effect is that the $|0\rangle$-amplitude equals $A_{c(j,\ell),j}$.

12.3.3 Recovering Matrix Elements by an Inner Product

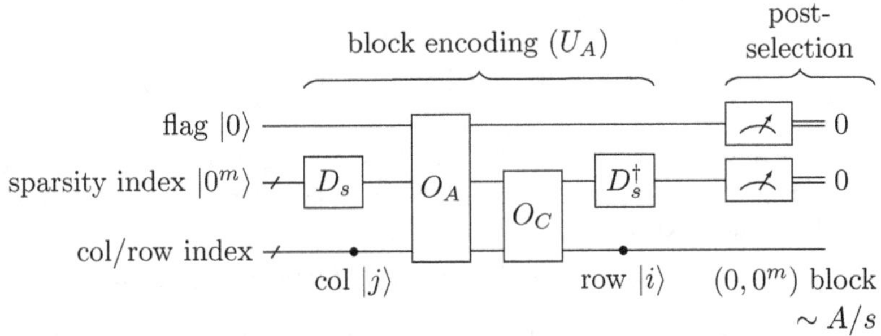

Figure 12.2: Generic Sparse-Matrix Block-Encoding Circuit

Fix indices (i, j). As shown in Fig. 12.2, we start from $|0\rangle |0^m\rangle |j\rangle$ and apply D_s on the ℓ-register, followed by O_A (value loading) and O_C (location computation):

$$|0\rangle |0^m\rangle |j\rangle \xrightarrow{I \otimes D_s \otimes I} \frac{1}{\sqrt{s}} \sum_{\ell=0}^{s-1} |0\rangle |\ell\rangle |j\rangle$$

$$\xrightarrow{O_A} \frac{1}{\sqrt{s}} \sum_{\ell=0}^{s-1} \Big(A_{c(j,\ell),j} |0\rangle + \cdots \Big) |\ell\rangle |j\rangle$$

$$\xrightarrow{O_C} \frac{1}{\sqrt{s}} \sum_{\ell=0}^{s-1} \Big(A_{c(j,\ell),j} |0\rangle + \cdots \Big) |\ell\rangle |c(j,\ell)\rangle .$$

Here the ellipsis denotes the $|1\rangle$-flag branch of Eq. 12.1, which will be removed by the projection onto $\langle 0|$.

To extract a matrix element, we take the overlap with the test bra $\langle 0| \langle 0^m| \langle i|$. Equivalently, we apply $D_s^\dagger$ on the ℓ register before projecting onto $\langle 0^m|$:

$$\langle 0| \langle 0^m| \langle i| \left(I \otimes D_s^\dagger \otimes I \right) O_C O_A \left(I \otimes D_s \otimes I \right) |0\rangle |0^m\rangle |j\rangle$$

$$= \frac{1}{s} \sum_{\ell=0}^{s-1} A_{c(j,\ell),j} \langle i|c(j,\ell)\rangle = \frac{A_{ij}}{s}, \tag{12.2}$$

provided that the padded list $\{c(j, \ell)\}_{\ell=0}^{s-1}$ contains each row index corresponding to a nonzero entry in column j exactly once (and otherwise contributes 0). This is the basic "overlap equals matrix entry" identity.

Exercise 12.6 State precisely what information O_A and O_C return in the sparse-access model. Then explain, in one paragraph, why these two oracles are sufficient to reconstruct any requested matrix element A_{ij} (assuming A is s-sparse).

Exercise 12.7 Write down the two states whose overlap encodes a matrix element A_{ij} (up to scaling). Verify the overlap algebraically for a 4×4 example where each row has at most $s = 2$ nonzeros.

12.3.4　A Block Encoding of A/s

Define the unitary

$$U_A := \left(I \otimes D_s^\dagger \otimes I\right)\left(I \otimes O_C\right)O_A\left(I \otimes D_s \otimes I\right), \tag{12.3}$$

where the identity operators act on the column register $|j\rangle$ (and any additional workspace registers used to implement the oracles). By Eq. 12.2, the $|0\rangle\,|0^m\rangle$-block of U_A equals A/s:

$$(\langle 0|\,\langle 0^m|\otimes I)\,U_A\,(|0\rangle\,|0^m\rangle \otimes I) = \frac{A}{s}.$$

Hence U_A is a block encoding of A/s. Without the assumption $\|A\|_{\max} \leq 1$, we may first rescale the loaded entries so that Eq. 12.1 holds with $A_{ij}/\|A\|_{\max}$ in place of A_{ij}, yielding a block encoding of $A/(s\|A\|_{\max})$.

■ **Example — Block-Encoding a Circulant Tridiagonal Matrix.** Consider the $N \times N$ circulant tridiagonal matrix

$$A = \begin{bmatrix} \alpha & \gamma & 0 & \cdots & 0 & \beta \\ \beta & \alpha & \gamma & \ddots & & 0 \\ 0 & \beta & \alpha & \ddots & \ddots & \vdots \\ \vdots & \ddots & \ddots & \ddots & \gamma & 0 \\ 0 & & \ddots & \beta & \alpha & \gamma \\ \gamma & 0 & \cdots & 0 & \beta & \alpha \end{bmatrix}. \tag{12.4}$$

Each column (and each row) has exactly three nonzero entries: the diagonal entry and its two nearest neighbors, with wrap-around at the boundaries. Hence A is s-sparse with $s = 3$, and $\|A\|_{\max} = \max\{|\alpha|, |\beta|, |\gamma|\}$.

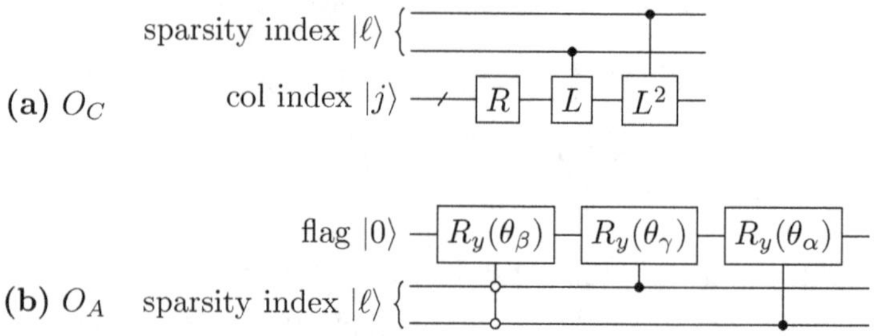

Figure 12.3: Circulant Tridiagonal Matrix Block-Encoding Circuits

Index oracle O_C (wrap-around via cyclic shifts).

We use an $m = 2$ qubit sparsity-index register and encode the three valid labels as

$$\ell \in \{00, 01, 10\} \quad \text{(with 11 unused/padded)}.$$

Define the row index of the selected entry in column j by

$$c(j, 00) = (j + 1) \bmod N, \qquad c(j, 01) = j, \qquad c(j, 10) = (j - 1) \bmod N.$$

Then the index oracle is $O_C : |\ell\rangle\,|j\rangle \mapsto |\ell\rangle\,|c(j, \ell)\rangle$.

A concrete realization is shown in Fig. 12.3(a). Here R and L are *cyclic shift* unitaries on the $n = \lceil \log_2 N \rceil$-qubit index register:

$$R\,|j\rangle = |(j + 1) \bmod N\rangle, \qquad L\,|j\rangle = |(j - 1) \bmod N\rangle,$$

and R^2, L^2 denote two-step cyclic shifts. The circuit applies an unconditional shift R (moving $j \mapsto j + 1$), then applies additional controlled shifts so that the net effect depends on ℓ:

$$\ell = 00 : R \Rightarrow (j + 1) \bmod N,$$
$$\ell = 01 : RL \Rightarrow j,$$
$$\ell = 10 : RL^2 \Rightarrow (j - 1) \bmod N.$$

Because the shifts are *cyclic*, the wrap-around cases ($j = 0$ or $j = N - 1$) are handled automatically by the modular arithmetic in R and L.

Value oracle O_A (three controlled rotations).

Assume $\|A\|_{\max} \leq 1$ for simplicity. The value oracle performs controlled single-qubit rotations on the flag ancilla so that the $|0\rangle$-amplitude equals the selected entry:

$$O_A : \ |0\rangle\,|\ell\rangle\,|j\rangle \mapsto \left(A_{c(j,\ell),j}\,|0\rangle + \sqrt{1 - |A_{c(j,\ell),j}|^2}\,|1\rangle \right) |\ell\rangle\,|j\rangle.$$

For the circulant tridiagonal matrix, the selected value depends only on which neighbor is chosen:

$$A_{c(j,00),j} = \beta, \qquad A_{c(j,01),j} = \alpha, \qquad A_{c(j,10),j} = \gamma.$$

Thus we can implement O_A using three ℓ-controlled R_y rotations on the flag qubit, as shown in Fig. 12.3(b), one rotation for each of the three labels $\ell \in \{00, 01, 10\}$. Concretely, for real entries in $[-1, 1]$, we choose angles so that $\cos(\theta/2) = A_{c(j,\ell),j}$, giving

$$\ell = 00 : R_y(\theta_\beta)\,|0\rangle, \qquad \theta_\beta = 2\arccos(\beta),$$
$$\ell = 01 : R_y(\theta_\alpha)\,|0\rangle, \qquad \theta_\alpha = 2\arccos(\alpha),$$
$$\ell = 10 : R_y(\theta_\gamma)\,|0\rangle, \qquad \theta_\gamma = 2\arccos(\gamma).$$

(For complex entries, we add a controlled phase rotation so that the $|0\rangle$-amplitude acquires the correct complex phase.)

Diffusion D_s.

Since $s = 3$ but $m = 2$, we choose a diffusion operator D_s satisfying

$$D_s\,|00\rangle = \frac{1}{\sqrt{3}}\left(|00\rangle + |01\rangle + |10\rangle\right),$$

so that the unused label $|11\rangle$ has zero amplitude and does not affect correctness.

Block encoding.

Let U_A be the composite

$$U_A := \left(I \otimes D_s^\dagger\right)\left(I \otimes O_C\right)O_A\left(I \otimes D_s\right),$$

with registers ordered as (flag ancilla)$\otimes$(sparsity index)$\otimes$(column/row index). By the overlap identity in Eq. 12.2, for all i, j,

$$\langle 0 | \langle 0^m | \langle i | \, U_A \, | 0 \rangle | 0^m \rangle | j \rangle = \frac{A_{ij}}{3}.$$

Therefore U_A is a block encoding of A/α with $\alpha = 3$ (or $\alpha = 3\|A\|_{\max}$ in general). ∎

Exercise 12.8 The circuit-level realizations of O_C and O_A are not unique. For the circulant tridiagonal matrix in § 12.3.4:

(a) Propose an alternative circuit for O_C that uses as few gates as possible (you may treat R and L as available primitives).

(b) Propose an alternative circuit for O_A that avoids any doubly controlled rotation.

12.3.5   Connections to HHL and QLSA Variants

Sparse matrices are one of the most important settings in which block encodings are both natural and efficient. Under a standard sparse-access model, the cost of implementing the block encoding (often denoted T_A in complexity statements) can be expressed as a sparsity-dependent factor times polylog(N) overhead, rather than being treated as an opaque black-box assumption. This is the main reason sparse access is frequently singled out in the quantum linear-systems literature.

Once we have a block encoding of A/α (with $\alpha = O(s)$ when $\|A\|_{\max} \leq 1$, or more generally $\alpha = O(s\|A\|_{\max})$), the rest of the HHL pipeline becomes a coherent spectral transformation problem. In particular, QSVT (§ 11.6) provides a systematic way to implement an approximate reciprocal on the promised spectral region (e.g., $x \in [1/\kappa, 1]$ after normalization), thereby mapping $|b\rangle$ to a state proportional to $A^{-1}|b\rangle$ in a flagged subspace. In this sense, sparse access supplies a concrete route from an implicit matrix description to the QSVT input required by modern HHL-style solvers.

Finally, sparse block-encoding constructions typically introduce ancillas, and sometimes an explicit "good/bad" branching structure (for padding, boundary handling, or value loading). This is fully compatible with the block-encoding framework: the desired operator appears in a particular ancilla subspace and can be accessed by postselection. Moreover, because the success probability may be small when α or κ is large, this structure naturally motivates Oblivious Amplitude Amplification (OAA, § 9.3.2), which boosts the amplitude on the good subspace coherently, avoiding repeated measure-and-reprepare cycles.

Key Takeaways

1. Sparse access provides two primitives: an index oracle (locations of nonzeros) and a value oracle (loads the selected entry into an ancilla

amplitude).

2. A diffusion D_s over the sparsity index produces the normalization $1/s$ by interference.

3. The extracted block satisfies

$$\langle 0| \langle 0^m| \langle i| \, U_A \, |0\rangle \, |0^m\rangle \, |j\rangle = \frac{A_{ij}}{s},$$

so U_A is a block encoding of A/s (or $A/(s\|A\|_{\max})$ in general).

12.4 Applications and Limitations

This section summarizes representative application patterns and common limitations, reinforcing the output model from § 12.1.1. The emphasis is on how $|x\rangle$ is used inside larger workflows and on which assumptions typically determine whether the linear-systems primitive is beneficial.

12.4.1 Applications

HHL-type solvers are most natural when a larger task reduces to a well-conditioned (or effectively preconditioned) linear system, the input state $|b\rangle$ and a block encoding of A/α can be prepared efficiently, and the desired output is a small number of efficiently measurable functionals of the solution state $|x\rangle$—such as expectation values, overlaps, or sampling-based quantities—rather than the full classical solution vector.

1 Discretized Differential Equations and PDE-Style Linear Systems

Many boundary-value problems and elliptic partial differential equations discretize to sparse linear systems. The goal is often not to list the full discretized solution vector, but to compute functionals of the solution. Examples include averages, fluxes, energies, and other linear or quadratic forms. [43]

2 Least Squares and Regression as Subroutines

Least squares problems reduce to applying a pseudoinverse-like transformation to the singular values of a data matrix (see § 12.1.1). The resulting state can serve as an intermediate object in a larger pipeline. Downstream steps then extract a small number of derived quantities, rather than the full coefficient vector.

12.4.2 Limitations

A quantum linear-systems primitive does not automatically imply an end-to-end speedup. Any advantage is conditional on several requirements being satisfied simultaneously.

1 State Preparation and Data Access

Data loading can dominate. If preparing $|b\rangle$ requires loading an arbitrary classical vector of length N, this overhead can remove any advantage. Favorable regimes are those where $|b\rangle$ is produced by a short circuit or arises naturally within a quantum computation. Block encoding can also dominate. If constructing a block encoding of A is expensive, then the solver is expensive as well.

2   Conditioning and Preconditioning

Conditioning controls both cost and robustness. The condition number parameter κ enters the approximation complexity and the success-probability overhead. Ill-conditioned instances can therefore negate any advantage. Preconditioning, when available, aims to reduce the effective condition number before inversion.

3   Small-Output Goals

The output is not the full vector. HHL yields a quantum state $|x\rangle$, from which one can estimate certain functionals of x efficiently. Full readout is typically prohibitive. If the task requires all entries of x (or an explicit classical description), then measurement and reconstruction costs can eliminate any potential advantage.

4   Dequantization Risk In Low-Rank Regimes

A quantum linear-systems routine can be vulnerable to *dequantization* effects when the matrix A is low rank (or has very small *effective* rank). In such cases, classical randomized algorithms can sometimes reproduce comparable "small-output" quantities under access assumptions that mirror the quantum input model (for example, sampling or query access to rows/entries), weakening the case for a genuine quantum advantage. This issue typically does not arise in PDE-style discretizations, where the resulting sparse operators are generally high rank, but it has been identified as a concern in proposed applications that assume low-rank structure (such as matrix completion).

Chapter Takeaways

1. HHL is coherent matrix inversion on a promised spectral region. After scaling, we assume the relevant spectrum lies in an interval bounded away from 0, typically parameterized by κ.

2. Block encoding enables linear-algebra access. Provide a block encoding of A/α under a suitable input (access) model, together with an efficient procedure to prepare $|b\rangle$.

3. QSVT implements the reciprocal map in a flagged branch. Use QSVT to apply a bounded approximation to $1/x$ on the promised region, producing a branch proportional to $A^{-1}|b\rangle$.

4. OAA makes the procedure reliable. Use OAA to boost the flagged success probability to a constant so that the normalized solution state $|x\rangle$ can be prepared with high probability.

5. HHL is most compelling when κ is modest, state preparation for $|b\rangle$ and block encoding of A are efficient, and the goal is limited to expectation values or overlaps.

Problem Set 12

12.1 Let $|x\rangle = x/\|x\|$ where $x = A^{-1}b$.

   (a) Show that estimating x_j for a fixed index j to additive error η by repeated computational-basis measurements can require $\Omega(1/\eta^2)$ samples even if $|x\rangle$ is given for free.

   (b) Explain why reconstructing all N entries typically removes any speedup.

12.2 Let $g(x) = \gamma/x$ on $[1/\kappa, 1]$ and assume $|g(x)| \leq 1$ on this interval.

   (a) Find the largest admissible γ.

   (b) Suppose the good branch prepares a vector proportional to $g(A)\,|b\rangle$. Express the raw success probability p in terms of $\|g(A)\,|b\rangle\,\|$.

   (c) Give a worst-case upper bound on p in terms of κ (up to constants), and interpret it.

12.3 Let A be a (possibly rectangular) matrix. Consider the Hermitian embedding

$$H = \begin{pmatrix} 0 & A \\ A^\dagger & 0 \end{pmatrix}.$$

   (a) Show that the eigenvalues of H are $\pm\sigma_j(A)$, where $\sigma_j(A)$ are the singular values of A. (Hint: examine H^2.)

   (b) Explain how an inverse (or pseudoinverse) transformation for H corresponds to a singular-value transformation for A.

12.4 Assume the raw success probability is p.

   (a) Compute the expected number of repetitions under postselection.

   (b) Compute the number of uses of the flagged procedure required by amplitude amplification to reach constant success probability.

   (c) For $p = 10^{-3}$, compare the two numerically.

12.5 You implemented a linear-systems routine and obtained a small residual $\|Ax - b\|_2$, but your estimate of $\langle x|\,M\,|x\rangle$ is unstable across runs.

   (a) List three plausible causes (one algorithmic, one numerical/simulation-related, and one measurement-related).

   (b) For each cause, propose one diagnostic test you would run to isolate it.

12.6 Give a concrete example of a preconditioner P (classical or quantum) for which solving $PAx = Pb$ is easier than solving $Ax = b$. Explain how the effective condition number changes and what additional assumptions are required to implement or apply P in a quantum setting.

12.7 $*$ Let F_N denote the quantum Fourier transform on N basis states, with Fourier basis $|\tilde{k}\rangle := F_N\,|k\rangle$.

   (a) Show that the projector onto Fourier mode k_0 is $M = |\tilde{k}_0\rangle\langle\tilde{k}_0|$.

   (b) Show that $\langle x|\,M\,|x\rangle = |\langle\tilde{k}_0|x\rangle|^2$.

   (c) Give two other choices of efficiently measurable observables M and explain what property of x each choice extracts.

12.8 $*$ For a circulant matrix A, the Fourier basis diagonalizes A. (a) Derive an expression for the eigenvalues of the circulant tridiagonal matrix in terms of α, β, γ and the Fourier mode index k. (b) Using your expression, estimate the condition number parameter κ (after scaling) for $\alpha = 1$, $\beta = 2$, $\gamma = 3$, and $N = 256$. Comment on how this affects the difficulty of inversion.

12.9 $*$ Consider the $N \times N$ circulant tridiagonal matrix A from Eq. 12.4 with parameters $\alpha = 1$, $\beta = 2$, $\gamma = 3$, and $N = 256$. Let $|b\rangle = \frac{1}{\sqrt{N}} \sum_{j=0}^{N-1} |j\rangle$ (i.e., the normalized all-ones vector).

(a) Implement the sparse-matrix block encoding U_A using the D_s–O_C–O_A–$D_s^\dagger$ construction described in sec:HHL-block-encoding-sparse. You may use a quantum computing platform like Qiskit or Cirq. Verify numerically that the $(|0\rangle |0^m\rangle)$ block of U_A is proportional to $A/3$ by checking several matrix elements $\langle i| (\cdot) |j\rangle$.

(b) Using QSVT, implement an approximate inverse (or pseudoinverse with cutoff $|x| \geq 1/\kappa$) for A, and apply it to $|b\rangle$ to prepare a state proportional to $|x\rangle \approx A^{-1} |b\rangle$. Report (i) an estimate of $\langle x|M|x\rangle$ for a simple observable M of your choice (for example, $M = Z_\ell Z_{\ell'}$, or $M = |\widetilde{k}\rangle\langle \widetilde{k}|$ where $|\widetilde{k}\rangle = F_N^\dagger |k\rangle$ is a Fourier-basis state), and (ii) a numerical check of the residual $\|Ax - b\|_2$ using a classical post-processing step.

12.10 $*$ Let $T \in \mathbb{C}^{N \times N}$ be the tridiagonal Toeplitz matrix obtained from the circulant tridiagonal matrix in § 12.3.4 by setting the corner entries to zero. Concretely, T has constant diagonals

$$T_{i,i} = a, \qquad T_{i,i+1} = c \; (1 \leq i \leq N-1), \qquad T_{i,i-1} = b \; (2 \leq i \leq N),$$

and $T_{1,N} = T_{N,1} = 0$, with all other entries zero.

Using the same sparse-access block-encoding framework and the D_s–O_C–O_A–$D_s^\dagger$ scheme from § 12.3.4, construct a block encoding of T. Your construction should explicitly specify:

(a) the column-index oracle O_C (including how you handle the boundary rows $i = 1$ and $i = N$),

(b) the value oracle O_A, and

(c) the state-preparation operator D_s and the resulting normalization factor α.

13. Adiabatic and Variational Quantum Algorithms

Contents

This chapter presents two closely related approaches to quantum algorithms that are organized around *Hamiltonians* and *low-energy states*. In the adiabatic paradigm, computation is carried out by continuously deforming a Hamiltonian so that the system tracks its ground state, while in the variational paradigm one prepares a parameterized quantum state and classically optimizes its parameters to minimize an energy objective. These perspectives are especially important for near-term quantum computing, where hybrid quantum–classical loops, shallow circuits, and repeated measurement are often the dominant practical workflow.

This emphasis differs from many algorithms developed earlier in this book—such as the quantum Fourier transform, quantum phase estimation, and quantum signal processing/quantum singular value transformation—which are most naturally expressed as coherent, deep circuits and therefore align more directly with fault-tolerant quantum computing. By contrast, adiabatic- and variational-style methods can often be deployed in resource-limited settings, trading long coherent depth for programmability, sampling, and classical optimization.

We develop a unified picture that connects the main models and their modern implementations. We first introduce adiabatic quantum computing (AQC) as an ideal, closed-system model and explain how the minimum spectral gap shapes the runtime requirements for adiabatic tracking [10, 11, 51]. We then show how computational problems—particularly optimization tasks—are encoded into Ising or quadratic unconstrained binary optimization (QUBO) Hamiltonians, and how interpolation paths and schedules affect performance in both digital and annealing-style realizations [52, 53]. Finally, we present the variational quantum eigensolver (VQE) and the quantum approximate optimization algorithm (QAOA) as practical strategies for low-energy state preparation, emphasizing the relationship between them and the sense in which QAOA can be viewed as a discretized, variational counterpart of adiabatic evolution [54, 55]. Throughout, we treat these methods as complementary tools connected by a common language of Hamiltonians, energy landscapes, and measurement-driven hybrid workflows.

Overview

---

1. Adiabatic methods compute by *changing a Hamiltonian in time* so the system stays near a low-energy eigenstate.

2. Variational methods compute by *choosing a parameterized circuit* and using a classical optimizer to minimize an energy objective.

3. Both paradigms trade long coherent depth for repeated measurement and classical processing, making them central in near-term workflows.

13.1　＊Adiabatic Quantum Computing (AQC)

Adiabatic quantum computing (AQC), proposed by Farhi *et al.* [10], is a model of quantum computation rooted in the *adiabatic evolution* principle from quantum physics. In its simplest form, the principle says that if a quantum system is prepared in an eigenstate of a Hamiltonian and the Hamiltonian is changed slowly enough, then the system remains close to the corresponding instantaneous eigenstate during the evolution. In algorithmic terms, the word "adiabatic" refers to changing the Hamiltonian slowly relative to the inverse of the relevant spectral gap along the path.

In AQC we exploit this idea by choosing a *driver* Hamiltonian H_D with an easily prepared ground state, and a *problem* Hamiltonian H_P whose ground state encodes the desired solution. We start in the ground state of H_D and continuously deform the Hamiltonian toward H_P. If the evolution remains sufficiently adiabatic, the final state has large overlap with the ground state of H_P, and measurement yields the solution (or a high-quality candidate). Although the paradigm originates in physics, it applies broadly: many computational tasks can be phrased as preparing low-energy states, including combinatorial optimization, constraint satisfaction, and certain state-preparation primitives. (We will discuss optimization applications in detail in Chapter 20.)

AQC also connects two implementation regimes. On a universal, gate-based ("digital") quantum computer, adiabatic evolution is realized by *digitizing* the continuous-time dynamics, using Hamiltonian simulation to approximate the time-ordered evolution operator. On specialized analog hardware (quantum annealers), similar Hamiltonian deformations are implemented directly through programmable control parameters, and the device output is typically interpreted through sampling from low-energy configurations. The physical details differ, but the conceptual backbone is shared: computation by steering a quantum system toward the ground state of a carefully designed Hamiltonian. A fundamental milestone is that AQC is polynomially equivalent to the standard circuit model [11], while the closely related analog paradigm is commonly discussed under the name *quantum annealing* [53, 56].

Hamiltonian Quench vs. Adiabatic Evolution

A Hamiltonian quench is the opposite extreme of adiabatic evolution. Instead of changing the Hamiltonian slowly so that the state follows an instantaneous eigenstate, we change the Hamiltonian suddenly. A typical workflow is:

1. Prepare an initial state $|\psi_0\rangle$, often an eigenstate (or approximate eigenstate) of an "initial" Hamiltonian H_i.

2. At time $t = 0$, switch to a different Hamiltonian H_f (the *quench*).

3. Let the system evolve under H_f: $|\psi(t)\rangle = e^{-iH_f t}|\psi_0\rangle$.

Quenches are important because many physical questions are inherently non-equilibrium. They probe how disturbances propagate, how correlations build up or decay, and how quickly a system relaxes after a perturbation.

From an algorithmic viewpoint, quench dynamics is often simpler to implement than adiabatic schedules because the Hamiltonian is time-independent after the switch. One only needs a method to approximate $e^{-iH_f t}$ for selected times t (for example, by Trotter steps for a Pauli-sum Hamiltonian).

13.1.1   Problem Setting and Input/Output Models

At a high level, an adiabatic quantum algorithm solves a computational task by encoding the answer into the ground state of a *problem Hamiltonian* and then attempting to prepare that ground state through slow continuous-time evolution.

The basic ingredients are a family of n-qubit Hamiltonians $\{H(s)\}_{s\in[0,1]}$ together with a schedule $s(t)$ that increases from $s(0) = 0$ to $s(T) = 1$ over a total evolution time T.

1   Input Model

The input instance specifies (i) a problem Hamiltonian H_P whose ground state encodes the solution, (ii) a driver Hamiltonian H_D with an easily prepared ground state $|\psi_0\rangle$, and (iii) an interpolation path and schedule, namely $H(s)$ and $s(t)$. A common choice is a linear interpolation

$$H(s) = (1 - s)H_D + sH_P, \tag{13.1}$$

but more general paths are possible (and sometimes essential) in both theory and practice.

Access to the Hamiltonian.

In a gate-based setting, the input model also includes how the terms of $H(s)$ are accessed or implemented (for example, through a Hamiltonian simulation procedure). In an analog setting (quantum annealing), the input model is expressed in terms of programmable control coefficients for the Hamiltonian terms. In this chapter, we work at the algorithmic level and treat the triple $(H_D, H_P, s(t))$ as the defining input.

2   Output Model

What counts as "the output" depends on how the final state is used and what the computational goal is. We will use three closely related output models.

State-preparation output.

The strongest model is that the algorithm prepares a quantum state close to the ground state of H_P, i.e., $|\psi(T)\rangle \approx e^{i\phi} |\psi_{gs}\rangle$. This model is natural when the ground state is an input to further quantum processing, such as estimating observables.

Bitstring (sampling) output.

For many optimization encodings, the ground space of H_P is spanned by computational basis states $|x\rangle$ representing candidate solutions $x \in \{0,1\}^n$. Measuring in the computational basis yields a random bitstring X. Success can be defined as $\Pr[X \in \mathcal{S}_{opt}]$ for an optimal-solution set $\mathcal{S}_{opt}$, or more generally as producing samples biased toward low-energy strings.

Energy (objective-value) output.

Sometimes the goal is an energy estimate rather than an explicit solution string. One then treats the output as an estimate of $\langle H_P \rangle$ for the prepared state, or as an empirical best energy among sampled bitstrings x with energies $E(x) = \langle x| H_P |x\rangle$. This output model connects directly to variational methods later in the chapter.

These models share the same core design principle: construct H_P so that low-energy states correspond to good solutions, choose H_D so that its ground state is easy to prepare, and select an interpolation $(H(s), s(t))$ intended to keep the evolution in (or near) the instantaneous ground state. The next subsection makes this intuition quantitative by stating the adiabatic condition and identifying the key quantity that controls the required runtime: the minimum spectral gap along the path.

13.1.2　Adiabatic Evolution and the Adiabatic Condition

The core promise of AQC is that a system initialized in the ground state can track the instantaneous ground state as the Hamiltonian varies, provided the change is sufficiently slow. The adiabatic condition makes "sufficiently slow" quantitative by relating the runtime to spectral gaps and to coupling matrix elements between eigenstates.

1　Instantaneous Spectrum and Spectral Gap

Let $H(s)$ be a differentiable family of Hamiltonians on n qubits, with $s \in [0,1]$. Denote its eigenpairs by

$$H(s)\,|E_k(s)\rangle = E_k(s)\,|E_k(s)\rangle, \qquad E_0(s) \leq E_1(s) \leq \cdots . \tag{13.2}$$

The (instantaneous) spectral gap is

$$\Delta(s) = E_1(s) - E_0(s),$$

and the minimum gap along the path is

$$\Delta_{\min} = \min_{s\in[0,1]} \Delta(s).$$

Small gaps increase the likelihood of non-adiabatic transitions out of the ground state and therefore tend to force longer runtimes.

2　Time-Dependent Schrödinger Evolution

Choose a monotone schedule $s(t)$ with $s(0) = 0$ and $s(T) = 1$. The evolution is governed by the time-dependent Schrödinger equation

$$i\frac{d}{dt}\,|\psi(t)\rangle = H(s(t))\,|\psi(t)\rangle, \qquad |\psi(0)\rangle = |E_0(0)\rangle . \tag{13.3}$$

The goal is that $|\psi(T)\rangle$ has large overlap with $|E_0(1)\rangle$, the ground state of H_{P}.

3　✳ Adiabatic Theorem and a Representative Condition

Informally, the adiabatic theorem says that the evolution remains close to the instantaneous ground state provided the Hamiltonian changes slowly compared to the square of the inverse spectral gap. Many mathematically precise variants exist (with slightly different assumptions and prefactors), but a commonly used *representative sufficient condition* is

$$\max_{t\in[0,T]} \frac{\left| \langle E_1(s(t))|\, \frac{dH}{dt}\, |E_0(s(t))\rangle \right|}{\Delta(s(t))^2} \ll 1, \tag{13.4}$$

where $|E_0(s)\rangle$ and $|E_1(s)\rangle$ denote the instantaneous ground and first excited eigenstates of $H(s)$, and $\Delta(s) = E_1(s) - E_0(s)$ is the instantaneous gap.

Using $\frac{dH}{dt} = \dot{s}(t)\frac{dH}{ds}$, Eq. 13.4 becomes

$$\max_{t\in[0,T]} \frac{|\dot{s}(t)|\,\left| \langle E_1(s(t))|\, \frac{dH}{ds}\, |E_0(s(t))\rangle \right|}{\Delta(s(t))^2} \ll 1. \tag{13.5}$$

This inequality gives a concrete meaning to "slowly": if the gap becomes small, then $\dot{s}(t)$ must be reduced accordingly (unless the coupling matrix element is also small).

A Simple Runtime Estimate

For a linear schedule $s(t) = t/T$ (so $|\dot{s}(t)| = 1/T$), Eq. 13.5 suggests the scaling

$$T \gtrsim \max_{s \in [0,1]} \frac{\left| \langle E_1(s) | \frac{dH}{ds} | E_0(s) \rangle \right|}{\Delta(s)^2}. \tag{13.6}$$

This bound is generally not tight, but it captures the main idea: if $\Delta_{\min} = \min_{s \in [0,1]} \Delta(s)$ is small, then the runtime typically deteriorates at least on the order of $1/\Delta_{\min}^2$, up to problem-dependent factors arising from $\frac{dH}{ds}$.

> **Exercise 13.1** Consider $H(s) = (1 - s)Z + sX$ for $s \in [0, 1]$ acting on one qubit. Compute the instantaneous eigenvalues, the gap $\Delta(s)$, and the matrix element $\left| \langle E_1(s) | \frac{dH}{ds} | E_0(s) \rangle \right|$. Using Eq. 13.5, give an explicit scaling estimate for the runtime T under a linear schedule $s(t) = t/T$.

4   Operational Consequences

The bottleneck is near the minimum gap. Even if the gap is large for most values of s, a narrow region where $\Delta(s)$ becomes small can dominate the runtime, because the evolution must slow down enough to suppress transitions there.

The schedule is a design choice. Because the condition depends on $\dot{s}(t)$, one can (in principle) move quickly when the gap is large and slow down near small gaps. This motivates local-adiabatic schedule design, discussed next.

Finally, the adiabatic theorem is an ideal, closed-system statement. In open-system settings (such as many quantum annealers), thermal excitations and relaxation also play a role. Nevertheless, the gap-based picture remains a useful organizing principle for understanding why certain instances are easy or hard, and why interpolation and schedule choices can materially affect outcomes.

13.1.3   Interpolation Paths and Annealing Schedules

An adiabatic algorithm is not determined by H_{P} alone. Its performance can depend strongly on *how* we connect H_{D} to H_{P}. This design choice has two components: the interpolation path $H(s)$ (which Hamiltonians we traverse) and the schedule $s(t)$ (how fast we move along that path). The adiabatic condition shows why both matter: the runtime cost is concentrated near small gaps, and the schedule allocates time across the path.

1   Interpolation Paths

The default choice is the linear interpolation

$$H(s) = (1 - s)H_{\mathrm{D}} + sH_{\mathrm{P}}, \qquad s \in [0, 1],$$

but more general paths can introduce useful degrees of freedom.

Multi-parameter control.

In many settings one has a parameterized Hamiltonian

$$H(\boldsymbol{\lambda}) = \sum_{\ell} \lambda_\ell H_\ell,$$

with control parameters $\boldsymbol{\lambda} = (\lambda_1, \lambda_2, \dots)$. An interpolation path is then a curve $\boldsymbol{\lambda}(s)$ in this control space. Linear interpolation corresponds to the special case with two terms and $\boldsymbol{\lambda}(s) = ((1-s), s)$.

Intermediate (catalyst) terms.

One can add an intermediate Hamiltonian H_{I} that vanishes at the endpoints,

$$H(s) = A(s)H_{\mathrm{D}} + B(s)H_{\mathrm{P}} + C(s)H_{\mathrm{I}},$$

with boundary conditions such as

$$A(0) = 1, \ B(0) = 0, \ C(0) = 0, \qquad A(1) = 0, \ B(1) = 1, \ C(1) = 0.$$

The role of H_{I} is to reshape the spectrum in the middle of the path without changing the computational encoding at $s = 1$.

2 Annealing Schedules

The schedule is a monotone function $s(t)$ with $s(0) = 0$ and $s(T) = 1$.

Linear schedule.

The simplest choice is $s(t) = t/T$, which makes $\dot{s}(t) = 1/T$ constant. It is easy to analyze but often wastes time in regions where the gap is large.

Nonlinear schedule.

A nonlinear schedule changes speed along the path. Motivated by the adiabatic condition, it can move faster when $\Delta(s)$ is large and slow down when $\Delta(s)$ is small. In practice, schedule shaping is also constrained by control bandwidth and noise considerations.

3 Local Adiabatic Intuition

A useful conceptual guideline is to choose $\dot{s}(t)$ so that the adiabatic condition is satisfied roughly uniformly along the path:

$$\frac{|\dot{s}(t)| \, \left| \langle E_1(s(t))| \frac{dH}{ds} |E_0(s(t))\rangle \right|}{\Delta(s(t))^2} \approx \text{constant}.$$

This leads to the qualitative prescription

$$\dot{s}(t) \ \propto \ \frac{\Delta(s)^2}{\left| \langle E_1(s)| \frac{dH}{ds} |E_0(s)\rangle \right|}, \tag{13.7}$$

which slows the evolution near small gaps. The main message is not that one can implement this formula exactly, but that the schedule is a runtime-allocation rule: it concentrates time where adiabatic tracking is hardest.

Exercise 13.2 Starting from Eq. 13.5, show that choosing $\dot{s}(t)$ as in Eq. 13.7 keeps the adiabatic ratio in Eq. 13.5 approximately constant. If the proportionality constant is set by a target small parameter $\eta \ll 1$, express the resulting total runtime T as an integral over $s \in [0, 1]$.

Key Takeaways

1. A schedule $s(t)$ is best viewed as a *runtime-allocation rule*: it distributes total time T along the path $s \in [0, 1]$.

2. The local-adiabatic heuristic spends more time near difficult regions (typically where $\Delta(s)$ is small).

3. In practice, schedule shaping is constrained by control limitations and noise, so the goal is robust improvement rather than perfect adherence to a formula.

4   Endpoint Smoothness and Boundary Conditions

Abrupt changes in $H(s(t))$ can introduce excitations that are not captured well by the simplest adiabatic estimates. For this reason, schedules and envelope functions are often chosen to turn on and turn off smoothly, with small derivatives near the endpoints.

5   Discretization Viewpoint

Although AQC is naturally formulated in continuous time, it is useful—especially for gate-based implementations—to view the evolution as a product of short segments. Divide $[0, T]$ into r steps of duration $\delta t = T/r$, choose sample points $s_j = s(t_j)$, and approximate the time-ordered evolution operator from Eq. 13.3 by

$$U(T) \approx \prod_{j=1}^{r} \exp\bigl(-iH(s_j)\,\delta t \bigr), \qquad |\psi(T)\rangle = U(T)\,|\psi(0)\rangle .$$

This discretized picture clarifies two connections.

(a) *Analog vs. digital implementations.* In an analog device, $H(s(t))$ is implemented directly through controls. In a digital device, one approximates the continuous evolution by a sequence of unitaries, introducing Hamiltonian simulation (and possibly Trotterization) considerations.

(b) *A bridge to QAOA.* When $H(s)$ is built from a small set of Hamiltonians (often H_{D} and H_{P}), each short-time unitary resembles a product of exponentials of these components. QAOA replaces a fine discretization by a small number of alternating blocks whose angles are chosen variationally.

Key Takeaways

1. Continuous-time AQC can be approximated by a product of short time-evolution segments, linking AQC to gate-based implementations.

2. This discretized picture clarifies the analog–digital tradeoff: analog devices implement $H(s(t))$ directly, while digital devices simulate it with gates.

3. QAOA can be viewed as a coarse, variational discretization: it uses a small number of alternating blocks and optimizes their angles instead of using a fine time grid.

13.1.4   Encoding Optimization as a Hamiltonian for AQC

While adiabatic quantum computing is not limited to optimization, classical optimization problems provide a natural and widely used demonstration of the encoding idea. We map a classical objective function to a *problem Hamiltonian* H_P whose low-energy states represent good solutions, so that preparing (or sampling from) low-energy states becomes a computational method.

For many combinatorial problems, this encoding is most naturally expressed in either the Ising form (with spin variables $z_i \in \{\pm 1\}$) or the quadratic unconstrained binary optimization (QUBO) form (with binary variables $x_i \in \{0, 1\}$). These two representations are equivalent up to an affine change of variables and an additive constant. (We further explore the practical applications of these optimization techniques in Chapter 20.)

 Readers unfamiliar with QUBO and Max-Cut formulations may wish to consult the foundational introduction in QCI Book [2], Section 11.2: *QUBO, VQE, QAOA, and AQC.*

1   QUBO Formulation

A QUBO instance is specified by a real matrix $Q \in \mathbb{R}^{n \times n}$ (which we may take to be symmetric) and the objective

$$\min_{x \in \{0,1\}^n} f(x), \qquad f(x) = \sum_{1 \leq i \leq j \leq n} Q_{ij} x_i x_j. \tag{13.8}$$

Because $x_i^2 = x_i$ for $x_i \in \{0, 1\}$, diagonal terms act as linear terms. This upper-triangular convention avoids double counting and is equivalent to writing $f(x) = x^T Q x$ up to a harmless redefinition of off-diagonal coefficients. It is often convenient to write

$$f(x) = \sum_{i<j} Q_{ij} x_i x_j + \sum_i Q_{ii} x_i,$$

absorbing any constant shift into an irrelevant additive constant.

From Cost Function to Hamiltonian

To encode this objective into a Hamiltonian, we map each binary variable to a qubit observable via

$$x_i \longleftrightarrow \hat{x}_i := \frac{1 - Z_i}{2},$$

where Z_i is the Pauli-Z operator on qubit i. We interpret each computational basis state

$$|x\rangle = |x_1 x_2 \cdots x_n\rangle, \qquad x \in \{0, 1\}^n,$$

as the classical assignment x. The key point is that $\hat{x}_i$ is diagonal in the computational basis and has the correct eigenvalues:

$$\hat{x}_i |x\rangle = \frac{1 - (-1)^{x_i}}{2} |x\rangle = x_i |x\rangle.$$

Consequently, for any pair (i, j),

$$\hat{x}_i \hat{x}_j |x\rangle = x_i x_j |x\rangle.$$

Define the *problem Hamiltonian* by the operator substitution $x_i \mapsto \hat{x}_i$:

$$H_{\mathrm{P}} := f(\hat{x}) = \sum_{1 \leq i \leq j \leq n} Q_{ij} \hat{x}_i \hat{x}_j.$$

Because each $\hat{x}_i$ is diagonal, H_{P} is also diagonal in the computational basis, and the identities above imply

$$H_{\mathrm{P}} \ket{x} = f(x) \ket{x} \qquad \text{for all } x \in \{0,1\}^n.$$

Therefore, the computational basis states are eigenstates of H_{P}, and their eigenvalues are exactly the classical objective values.

Ground State $\leftrightarrow$ Optimal Solution

Let $f_{\min} = \min_{x \in \{0,1\}^n} f(x)$ and define the minimizer set

$$\mathcal{S}_{\min} = \{x \in \{0,1\}^n :\ f(x) = f_{\min}\}.$$

Since H_{P} is diagonal and has eigenvalues $\{f(x)\}$, its ground energy is $f_{\min}$ and its ground space is

$$\mathrm{span}\{\ket{x} : x \in \mathcal{S}_{\min}\}.$$

In particular, measuring any ground state of H_{P} in the computational basis returns an optimal assignment with probability 1. More generally, a state supported mostly on low-energy eigenspaces yields samples biased toward near-optimal solutions.

2   Ising Formulation and the QUBO–Ising Mapping

Expanding $\hat{x}_i = (1 - Z_i)/2$ shows that H_{P} can be written in Ising form as

$$H_{\mathrm{P}} = \sum_{i<j} J_{ij} Z_i Z_j + \sum_i h_i Z_i + c I, \tag{13.9}$$

for suitable real coefficients J_{ij}, h_i, and an additive constant c. The constant term cI shifts all energies equally and therefore does not affect the ground space or the set of minimizing assignments. This Hamiltonian is traditionally associated with the Ising spin model, hence its name.

The Ising form uses spin variables $z_i \in \{\pm 1\}$ and an energy function

$$E(z) = \sum_{i<j} J_{ij} z_i z_j + \sum_i h_i z_i, \qquad z \in \{\pm 1\}^n.$$

The QUBO and Ising representations are related by the affine change of variables

$$z_i = 1 - 2x_i, \qquad \text{equivalently} \qquad x_i = \frac{1 - z_i}{2}.$$

Under this map, minimizing $f(x)$ is equivalent to minimizing $E(z)$ up to a constant shift and a straightforward reparameterization of coefficients. In quantum language, this correspondence is exactly the operator substitution $z_i \leftrightarrow Z_i$ and $x_i \leftrightarrow (1 - Z_i)/2$.

3   Constraints via Penalty Terms

Many optimization problems are naturally expressed with constraints, for example

$$\min f(x) \quad \text{subject to} \quad g_k(x) = 0, \ \ k = 1, \ldots, m,$$

or inequality constraints such as $g_k(x) \leq 0$. A standard encoding strategy converts such constraints into energy penalties so that feasible assignments have lower energy than infeasible ones. For equality constraints, one often uses a quadratic penalty

$$f_{\text{pen}}(x) = f(x) + \sum_{k=1}^{m} \lambda_k \, g_k(x)^2,$$

where $\lambda_k > 0$ are penalty weights. If the penalties are chosen large enough, then any minimizer of $f_{\text{pen}}(x)$ is also a constrained minimizer of the original problem. Inequality constraints can be handled similarly, sometimes with slack variables that preserve a quadratic form.

Constraint Enforcement

From the Hamiltonian viewpoint, penalty terms enlarge the energy of invalid configurations while leaving feasible ones energetically preferred. Thus the low-energy sector of the encoded Hamiltonian is arranged to coincide with the feasible region of the original optimization problem. In this sense, penalties do more than merely "discourage" bad solutions: they reshape the energy landscape so that constraint satisfaction becomes part of the ground-state structure.

Ancilla-Based Reductions

Penalty methods are also useful for a second reason. Many natural problem encodings produce interaction terms involving three or more binary variables, whereas practical hardware often favors 2-local Hamiltonians. A common strategy is therefore to introduce *ancilla* variables together with additional penalty terms that enforce consistency relations between the original variables and the ancillas. In this way, a higher-body logical relation can be represented within a larger Hamiltonian containing only 1-local and 2-local terms.

For commuting terms in a common basis, one can use nonperturbative ancilla-based constructions to convert certain k-body interactions exactly into 2-body Hamiltonians [57]. This is conceptually useful because it shows that penalty constructions are not merely ad hoc modeling devices. They are part of a broader Hamiltonian-engineering toolkit that connects abstract optimization formulations to implementable low-locality Hamiltonians.

Choosing Penalty Weights

Penalty weights must be large enough to separate feasible from infeasible assignments in energy, but overly large coefficients can worsen conditioning and increase control sensitivity. This tradeoff is important in analog annealers, which have bounded coefficient ranges, and in digital implementations, where large coefficient ratios can increase simulation cost or amplify noise sensitivity. Ancilla-based reductions introduce an additional tradeoff: they can reduce locality, but at the cost of extra qubits and potentially less favorable energy scales.

4　Max-Cut as an AQC Task

Max-Cut is an archetypal QUBO/Ising optimization problem. Let $G = (V, E)$ be an undirected graph with $|V| = n$ vertices. A cut is specified by a bitstring $x \in \{0, 1\}^n$, where x_i indicates which side of the cut contains vertex i. The cut size is the number of edges whose endpoints lie on different sides:

$$\text{Cut}(x) = \sum_{(i,j)\in E} \big(x_i(1 - x_j) + (1 - x_i)x_j\big). \tag{13.10}$$

Equivalently, using spin variables $z_i \in \{\pm 1\}$ with $z_i = 1 - 2x_i$, we have

$$\text{Cut}(z) = \sum_{(i,j)\in E} \frac{1 - z_i z_j}{2}.$$

Max-Cut asks us to maximize $\text{Cut}(z)$ over $z \in \{\pm 1\}^n$. Defining the Ising energy

$$E(z) = -\text{Cut}(z) = \frac{1}{2} \sum_{(i,j)\in E} z_i z_j \; + \; (\text{constant}),$$

we may drop the constant and take the problem Hamiltonian

$$H_\text{P} = \frac{1}{2} \sum_{(i,j)\in E} Z_i Z_j. \tag{13.11}$$

For any computational basis state $|x\rangle$ (with $Z_i\,|x\rangle = z_i\,|x\rangle$), the energy is

$$\langle x|\, H_\text{P}\, |x\rangle = \frac{1}{2} \sum_{(i,j)\in E} z_i z_j, \qquad \text{so} \qquad \text{Cut}(x) = \frac{|E|}{2} - \langle x|\, H_\text{P}\, |x\rangle.$$

Thus minimizing H_P is equivalent to maximizing the cut size.

A standard driver Hamiltonian is the transverse field

$$H_\text{D} = \frac{1}{2} \sum_{i=1}^{n} X_i,$$

whose ground state is $|+\rangle^{\otimes n}$, where $|+\rangle = (|0\rangle + |1\rangle)/\sqrt{2}$. The simplest AQC interpolation is then

$$H(s) = (1 - s)H_\text{D} + sH_\text{P}, \qquad s(0) = 0, \; s(T) = 1.$$

Starting from $|\psi(0)\rangle = |+\rangle^{\otimes n}$, we evolve under $H(s(t))$ for time T and measure in the computational basis at the end.

Exercise 13.3 Let $G = (V, E)$ be an unweighted graph. Show that the MaxCut objective can be written as

$$\text{Cut}(z) = \sum_{(i,j)\in E} \frac{1 - (-1)^{z_i \oplus z_j}}{2},$$

and that maximizing $\text{Cut}(z)$ is equivalent to maximizing the expectation of the diagonal Hamiltonian

$$H_C = \sum_{(i,j)\in E} \frac{I - Z_i Z_j}{2}.$$

Explain briefly why H_C is diagonal in the computational basis.

Interpreting the Measurement

The measurement outcome is a bitstring $x \in \{0,1\}^n$, which defines a cut. If the evolution remains close to the ground state of H_P, then the output distribution is biased toward maximum cuts (and more generally toward large cuts). In an ideal closed-system AQC picture, increasing T improves the probability of sampling an optimal cut. In an open-system annealing picture, the distribution is influenced by temperature and relaxation, but the same Hamiltonian encoding provides the link between measured bitstrings and objective values.

A Small-Graph Instance

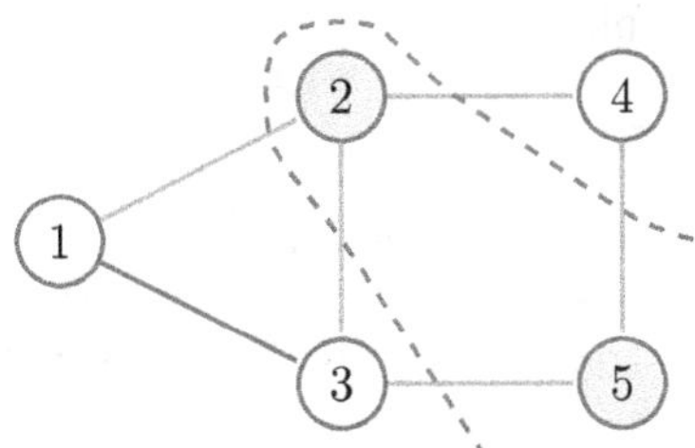

Figure 13.1: Max-Cut Example

For the five-vertex graph in Fig. 13.1, we have $V = \{1,2,3,4,5\}$ and

$$E = \{(1,2),(1,3),(2,3),(2,4),(3,5),(4,5)\}, \qquad |E| = 6.$$

The objective function is

$$\mathrm{Cut}(x) = \big(x_1(1-x_2) + (1-x_2)x_1\big) + \ldots + \big(x_4(1-x_5) + (1-x_5)x_4\big).$$

The cut shown by the dashed curve corresponds to the partition $\{1,3,4\}$ and $\{2,5\}$, which is represented by $|01001\rangle$ (big-endian notation). This cut has size 5; the only uncut edge is $(1,3)$. The ground space of H_P is degenerate and includes

$$|01001\rangle, \; |10110\rangle, \; |00110\rangle, \; |11001\rangle,$$

which all achieve the maximum cut size 5 and satisfy

$$\langle x | H_P | x \rangle = -2.$$

We will reuse this same Max-Cut encoding later when we discuss VQE and QAOA, so that the differences among adiabatic evolution, variational energy minimization, and alternating-operator circuits can be compared on a common example.

Key Takeaways

1. Optimization is encoded by a problem Hamiltonian whose low-energy (or high-energy) basis states correspond to good solutions.

2. The interpolation path chooses *which Hamiltonians* are traversed; the schedule chooses *how fast* each region is traversed.

> 3. Good performance depends on both the spectral landscape (especially gaps) and the practicality of implementing the chosen Hamiltonians.

13.1.5   Adiabatic State Preparation as a General Primitive

Optimization is a convenient entry point because it makes the Hamiltonian encoding idea concrete. However, adiabatic evolution supports a broader and more fundamental primitive: *adiabatic state preparation* (ASP), whose goal is to prepare a ground state (or, more generally, a low-energy eigenstate) of a target Hamiltonian. This primitive is central in areas such as quantum chemistry, condensed matter physics, and materials science, where the object of interest is not a bitstring solution but a quantum state whose properties (energies, correlations, and response functions) are physically meaningful.

1   Problem Statement

In its basic form, ASP takes as input a target Hamiltonian H_{T} on n qubits (or n effective qubits after encoding), together with an initial Hamiltonian H_0 whose ground state $|E_0(0)\rangle$ can be prepared efficiently. One then chooses an interpolation path

$$H(s) = (1 - s)H_0 + sH_{\mathrm{T}}, \qquad s \in [0, 1],$$

or a more general path $H(s)$ with the same endpoints. Starting from $|\psi(0)\rangle = |E_0(0)\rangle$, one evolves under the time-dependent Schrödinger equation

$$i\frac{d}{dt}|\psi(t)\rangle = H(s(t))|\psi(t)\rangle$$

for a runtime T, with the aim that $|\psi(T)\rangle$ is close to the ground state of H_{T}.

2   Why This Is More General Than Optimization

In combinatorial optimization encodings, the target Hamiltonian is typically diagonal in the computational basis and its ground states correspond to optimal bitstrings. In ASP, the target Hamiltonian may be a fully quantum operator with noncommuting terms, such as an electronic structure Hamiltonian after a fermion-to-qubit mapping, or a lattice model Hamiltonian from many-body physics. The algorithmic output is therefore a quantum state rather than a classical assignment.

Observable Estimation

Once a ground state $|\psi_{\mathrm{gs}}\rangle = |\psi(T)\rangle$ is prepared, one can estimate physical quantities by measuring observables O and forming $\langle\psi_{\mathrm{gs}}|O|\psi_{\mathrm{gs}}\rangle$. This is often the practical goal in simulation settings.

3   Adiabatic Gaps, Phase Transitions, and Hard Instances

The adiabatic condition from § 13.1.2 applies equally to ASP. The runtime is controlled by the spectral gap along the path and by coupling matrix elements between low-lying eigenstates. In many-body settings, the gap can become small near avoided crossings or quantum phase transitions. This provides both an opportunity and a limitation:

(a) ASP can be efficient when the chosen path avoids exponentially small gaps.

(b) ASP can become impractical when the path encounters regions where the gap shrinks rapidly with system size.

This viewpoint emphasizes that ASP is not merely a heuristic; it is a well-defined algorithmic primitive whose complexity is governed by spectral structure.

4　Connection to Hamiltonian Simulation and Eigenvalue Problems

ASP can be viewed as complementary to the Hamiltonian simulation and eigenvalue-estimation tools introduced earlier in this book.

Continuous-Time vs. Gate-Based Viewpoints

The evolution under $H(s(t))$ is a continuous-time process. On a gate-based quantum computer, one implements ASP by *digitizing* the evolution, approximating the time-ordered unitary

$$\mathcal{T} \exp\left(-i \int_0^T H(s(t))\, dt \right)$$

using Hamiltonian simulation techniques. In this sense, ASP is a particular application of Hamiltonian simulation where the goal is state preparation rather than direct time evolution under a fixed Hamiltonian.

From State Preparation to Eigenvalue Estimation

Preparing an approximate eigenstate is often a prerequisite for eigenvalue estimation. For example, if ASP yields a state with large overlap with the ground state of H_T, then phase-estimation-based methods (in a fault-tolerant setting) or variational energy estimation (in a near-term setting) can be used to extract the ground energy. Conversely, variational approaches such as VQE can be interpreted as alternative ground-state preparation strategies that replace adiabatic tracking with classical optimization. We will make this relationship explicit when we introduce VQE and its connection to digitized adiabatic paths.

Exercise 13.4 Assume along an adiabatic path that $\left| \langle E_1(s) | \frac{dH}{ds} | E_0(s) \rangle \right| \leq M$ for all $s \in [0, 1]$ and that $\Delta(s) \geq \Delta_{\min} > 0$. Use Eq. 13.6 to show $T \gtrsim M/\Delta_{\min}^2$. In one or two sentences, explain why this worst-case estimate can be pessimistic.

Key Takeaways

1. Adiabatic state preparation is a general primitive: it prepares a low-energy eigenstate by slowly deforming a Hamiltonian with a known ground state into a target Hamiltonian.

2. The quality of the prepared state is controlled primarily by the spectral gap along the path and by how the schedule allocates time near small-gap regions.

3. Once a ground state is prepared, it can be reused as an input for other tasks such as estimating observables, generating correlated samples, or serving as a starting point for variational refinement.

13.1.6 Universality and Limitations

1 Universality and Relation to the Circuit Model

AQC is more than a heuristic inspired by physics: in an ideal, closed-system setting it forms a *universal* model of quantum computation [11]. At a high level, this means that AQC and the gate-based circuit model can simulate one another with at most polynomial overhead in resources (such as system size and runtime). In particular:

(a) *Circuit* $\Rightarrow$ *AQC*. Any polynomial-size quantum circuit can be encoded into the ground state of a suitably constructed Hamiltonian, and an adiabatic evolution can be designed to prepare a state from which the circuit's output can be recovered with high probability.

(b) *AQC* $\Rightarrow$ *circuit*. Conversely, the continuous-time evolution generated by a time-dependent Hamiltonian can be approximated by a quantum circuit using Hamiltonian simulation techniques, so an AQC computation can be implemented on a universal gate-based machine.

Taken together, these facts justify treating AQC as a full-fledged computational model rather than a special-purpose optimization method.

The equivalence to the circuit model provides a useful pedagogical viewpoint. It clarifies that the central algorithmic ideas in AQC—encoding, interpolation paths, schedules, and gap-based runtime reasoning—are not tied to a particular hardware platform. They are alternative ways of organizing quantum computation that emphasize Hamiltonians, energy landscapes, and state preparation. In later sections, this viewpoint also helps explain why "digitized" versions of adiabatic evolution naturally lead to variational algorithms such as QAOA.

A Caution on Hardware-Restricted Models

Universality is a statement about *general* adiabatic computation under sufficiently flexible Hamiltonians. Practical devices may implement restricted families of Hamiltonians (for example, limited connectivity, limited coefficient ranges, or additional sign constraints). Such restrictions may prevent universality even if the ideal AQC model is universal.

2 Universality with Restricted Two-Local Hamiltonians

The universality of adiabatic quantum computation is usually stated for sufficiently general families of local Hamiltonians, and the standard equivalence result shows that ideal AQC is polynomially equivalent to the gate-based circuit model [11]. From a physical perspective, however, one would like to know whether universality survives when the allowed interactions are substantially restricted. This question matters because practical devices rarely implement arbitrary local Hamiltonians; instead, they offer only specific coupling types, limited connectivity, and constrained control parameters.

A notable result in this direction was given by Biamonte and Love, who showed that certain restricted 2-local Hamiltonian families are already sufficient for universal adiabatic quantum computation [58]. In particular, they showed that an Ising-type model can be made universal by adding tunable 2-local XX couplings, and they also established universality for models built from 1-local X and Z terms together with 2-local ZX interactions [58]. In this sense, universality can persist even under interaction sets that are much more structured than a fully general 2-local Hamiltonian.

This observation is conceptually important for two reasons. First, it narrows the gap between abstract universality proofs and Hamiltonians that are closer to experimentally meaningful spin models. Second, it shows that the computational power of the adiabatic model is not tied only to highly flexible, artificial Hamiltonian constructions. Rather, appropriately chosen restricted interaction families can already support universal computation.

At the same time, one should interpret this result carefully. It does not imply that every device implementing ZZ-type and XX-type couplings is automatically a practical universal adiabatic computer. Universality is an existence statement about what can be achieved with suitable encodings, controls, and problem constructions within a given interaction family. Questions of efficiency, calibration, connectivity, noise, and spectral-gap behavior remain separate and may still limit practical performance.

3   Practical Limitations and When Speedups Fail

Although AQC is universal in principle, it does not guarantee speedups in practice. The adiabatic condition highlights why: the runtime depends on spectral structure along the interpolation path, and real devices operate in imperfect, often open-system regimes. This subsection summarizes the most common bottlenecks and failure modes.

Gap Closing and Hard Instances

The adiabatic condition typically forces longer runtimes when the minimum gap $\Delta_{\min}$ becomes small. In optimization encodings, small gaps often arise near avoided crossings associated with competing low-energy configurations. Conceptually, one can view these as the quantum analogue of rugged energy landscapes: the system must navigate regions where the ground state changes rapidly as a function of s, making diabatic transitions likely unless the schedule slows down substantially.

First-Order Transitions and Exponentially Small Gaps

In many encodings, the most problematic behavior is a sharp rearrangement of the ground state that resembles a first-order phase transition in the large-n limit. In such cases the gap may shrink exponentially with problem size, and the adiabatic runtime bound becomes exponential. This is one reason AQC should be treated as an algorithmic framework: performance depends critically on the instance family, the encoding, and the chosen interpolation path.

Stoquastic Structure and Classical Simulability

Many annealing-style Hamiltonians used in practice are *stoquastic* in the computational basis (informally, they have non-positive off-diagonal matrix elements). Stoquasticity is important because it often makes certain families of Hamiltonians more amenable to classical simulation via quantum Monte Carlo methods, at least in equilibrium and in the absence of a sign problem. While stoquastic Hamiltonians can still be computationally challenging, this structure weakens generic claims of quantum advantage and helps explain why speedups are instance-dependent rather than automatic.

Open-System Effects: Temperature and Decoherence

Quantum annealers and noisy near-term devices do not implement perfectly coherent unitary evolution. Instead, the system interacts with an environment, so the state may thermalize, dephase, or relax. As a result:

(a) *Sampling vs. optimization.* The output is often better interpreted as samples biased toward low-energy configurations rather than a guaranteed optimizer. This can still be useful, but it changes the success criterion.

(b) *Energy scale matters.* Rescaling H_P does not change the ground state mathematically, but it changes the energy scale relative to temperature and noise, and therefore changes the observed distribution over bitstrings.

Control Errors and Limited Expressivity

Practical implementations face additional constraints: finite precision in couplers and fields, calibration drift, limited connectivity, and restrictions on available driver terms. These factors can distort the intended Hamiltonian, alter the gap profile, and degrade success even when the idealized encoding is sound. On digital devices, the analogous issues appear as Hamiltonian simulation error, gate noise, and finite sampling error in energy estimation.

When Adiabatic-Inspired Methods Still Help

These limitations do not make the paradigm obsolete. Instead, they motivate two common strategies that appear repeatedly in modern quantum algorithms:

(a) *Schedule and path engineering* (including pauses, nonlinear schedules, and intermediate terms) to avoid or soften bottlenecks associated with small gaps.

(b) *Variational alternatives* that replace strict adiabatic tracking by classical optimization over a parameterized family of states, as in VQE and QAOA, thereby trading long coherent evolution for repeated short circuits and measurement-driven feedback.

Key Takeaways

1. In principle, adiabatic quantum computing is computationally universal: with appropriate Hamiltonians, it can simulate gate-based quantum computation.

2. Universality does not imply practicality: performance is highly instance-dependent because runtime can scale poorly when the minimum gap is very small.

3. Practical limitations include control constraints (implementable Hamiltonian families), noise and decoherence (long coherent evolution), and the difficulty of diagnosing small-gap bottlenecks from experiments.

13.2 Variational Quantum Eigensolver (VQE)

The variational quantum eigensolver (VQE) is a hybrid quantum–classical approach to preparing low-energy eigenstates and estimating eigenvalues. Instead of relying on adiabatic tracking in continuous time, VQE searches over a parameterized family of quantum states and uses a classical optimizer to minimize an energy objective estimated from repeated measurements. This workflow is especially natural on near-term devices because it trades long coherent evolution for short circuits and sampling.

Overview

---

1. VQE estimates the ground energy by minimizing $\langle H \rangle$ over a parameterized family of states $|\psi(\boldsymbol{\theta})\rangle$ prepared by a quantum circuit.

2. The quantum device provides noisy estimates of energies (and possibly gradients) from repeated measurements, while a classical optimizer updates $\boldsymbol{\theta}$.

3. VQE succeeds when the ansatz is expressive enough *and* the optimization/measurement costs are manageable; there is no general convergence guarantee.

13.2.1 Variational Principle and Objective Functions

VQE is grounded in the Rayleigh–Ritz variational principle. Let H be a Hermitian operator on an n-qubit Hilbert space, and let E_0 denote its ground energy. For any normalized state $|\psi\rangle$,

$$\langle \psi | H | \psi \rangle \geq E_0,$$

with equality if and only if $|\psi\rangle$ is a ground state (or lies in the ground space when the ground energy is degenerate). This inequality turns ground-state preparation into an optimization task: minimize the energy expectation over a chosen family of trial states.

1   Parameterized State (Ansatz) Family

VQE chooses a parameterized ansatz state

$$|\psi(\boldsymbol{\theta})\rangle = U(\boldsymbol{\theta}) |\psi_{\text{ref}}\rangle ,$$

where the term *ansatz* refers to a trial quantum state $|\psi(\boldsymbol{\theta})\rangle$ described by a set of tunable parameters $\boldsymbol{\theta}$. Here $U(\boldsymbol{\theta})$ is a parameterized circuit, $|\psi_{\text{ref}}\rangle$ is a reference state (often $|0^n\rangle$ or a problem-motivated initial state), and $\boldsymbol{\theta} \in \mathbb{R}^p$ is a vector of parameters to be optimized. The variational objective is

$$E(\boldsymbol{\theta}) = \langle \psi(\boldsymbol{\theta}) | H | \psi(\boldsymbol{\theta}) \rangle , \qquad \boldsymbol{\theta}^\star \in \arg\min_{\boldsymbol{\theta}} E(\boldsymbol{\theta}).$$

If the ansatz family is expressive enough to approximate the true ground state, then the minimum value $E(\boldsymbol{\theta}^\star)$ approximates E_0, and $|\psi(\boldsymbol{\theta}^\star)\rangle$ approximates a ground state.

2   Common Objective Variants

The basic objective $E(\boldsymbol{\theta})$ is often supplemented with additional terms to reflect constraints or problem structure.

Constraints and Symmetries

If the physical problem has conserved quantities or known symmetries (such as particle number, total spin, or parity), one can enforce them by restricting the ansatz or by adding penalty terms. A typical penalty objective has the form

$$E_{\text{pen}}(\boldsymbol{\theta}) = \langle \psi(\boldsymbol{\theta})| H |\psi(\boldsymbol{\theta})\rangle + \sum_k \lambda_k \langle \psi(\boldsymbol{\theta})| C_k |\psi(\boldsymbol{\theta})\rangle \,,$$

where each C_k is a nonnegative operator that vanishes on the desired symmetry sector. This approach is conceptually similar to penalty methods used in QUBO encodings.

Excited States

VQE is most commonly used for ground states, but there are extensions that target excited states by enforcing orthogonality to previously found states or by optimizing within a low-dimensional subspace. In this chapter we emphasize ground-state VQE, since it is the shared core underlying many applications and provides the most direct comparison with AQC and QAOA.

13.2.2 Ansatz Design

The performance of VQE depends strongly on the choice of ansatz. An ansatz family should be expressive enough to represent low-energy states while remaining trainable under finite sampling noise and hardware imperfections. These requirements often pull in opposite directions.

1 Hardware-Efficient Ansatzes

Hardware-efficient ansatzes are built from layers of native gates that match the device connectivity. A typical pattern alternates single-qubit rotations and entangling gates in repeated blocks. Such ansatzes can be easy to implement and scale in a straightforward way with system size. However, they may require many layers to represent structured ground states, and sufficiently deep circuits can become difficult to optimize.

2 Problem-Inspired Ansatzes

Problem-inspired ansatzes incorporate knowledge of the Hamiltonian. In chemistry, for example, one often begins from a physically motivated reference state and applies structured excitations; in spin models, one may use circuits based on terms in H itself. These designs can reduce parameter count and improve interpretability, but they may rely on gates that are not native to a given device or require nontrivial compilation.

3 Expressibility vs. Trainability

A central tradeoff in ansatz design is expressibility vs. trainability. Very expressive ansatzes can represent many states but may suffer from flat optimization landscapes (the barren plateau phenomenon) or from sensitivity to noise and shot fluctuations. Conversely, very shallow or overly constrained ansatzes may be trainable but may not reach sufficiently low energy. In practice, one typically starts with a modest depth and increases complexity only when needed.

One important failure mode is the barren plateau phenomenon, which we discuss in § 13.2.5.

13.2.3 Classical Optimization Loop and Gradients

VQE alternates between quantum measurements and classical parameter updates. At iteration t, one prepares $|\psi(\boldsymbol{\theta}_t)\rangle$, estimates $E(\boldsymbol{\theta}_t)$, and then updates $\boldsymbol{\theta}_t \mapsto \boldsymbol{\theta}_{t+1}$ using a classical optimization rule.

1 Gradient-Based vs. Gradient-Free Updates

Two broad approaches are common:

(a) *Gradient-free methods* such as Nelder–Mead or COBYLA, which can be robust when gradients are noisy but may scale poorly with parameter dimension.

(b) *Gradient-based methods* such as stochastic gradient descent or quasi-Newton methods, which can converge faster when gradients can be estimated reliably.

Because VQE measurements are noisy, many practical workflows use stochastic variants of these optimizers.

2 Parameter-Shift Rule

For many parameterized gates of the form $e^{-i\theta G/2}$, derivatives of expectation values can be estimated by running the *same* circuit at shifted parameter values. This gives a practical gradient estimator using only expectation measurements.

Let
$$f(\theta) = \langle 0| U(\theta)^\dagger O\, U(\theta) |0\rangle,$$
and assume θ appears in a gate $V(\theta) = e^{-i\theta G/2}$ whose generator satisfies $G^2 = I$ (equivalently, eigenvalues ± 1). This covers common rotation gates such as $R_x(\theta)$, $R_y(\theta)$, and $R_z(\theta)$.

In this case,
$$e^{-i\theta G/2} = \cos\tfrac{\theta}{2} I - i\sin\tfrac{\theta}{2} G,$$
$$\frac{d}{d\theta} f(\theta) = \frac{1}{2}\left[f(\theta + \tfrac{\pi}{2}) - f(\theta - \tfrac{\pi}{2}) \right].$$

Operationally, we estimate $f(\theta + \pi/2)$ and $f(\theta - \pi/2)$ by measurement, then take their scaled difference.

Key Takeaways

1. The VQE loop is a noisy optimization problem: the objective values (and gradients, if used) are estimated with statistical uncertainty.

2. Optimization difficulty can come from expressivity/landscape issues (e.g., flat regions) as well as from noise and finite-shot fluctuations.

3. Practical workflows often rely on good initializations, problem structure, and lightweight diagnostics rather than on guarantees of global optimality.

Exercise 13.5 Suppose an optimizer returns parameters $\boldsymbol{\theta}_t$ and $\boldsymbol{\theta}_{t+1}$ that yield nearly identical energies within statistical error bars. Give two diagnostics (one quantum-side, one classical-side) that could help decide whether the optimization has truly converged or whether the result is limited by measurement noise.

Key Takeaways

1. VQE errors have multiple sources: device noise, finite-shot estimation, ansatz bias (restricted expressivity), and classical optimizer failure.

2. Mitigation patterns include measurement reduction, error-aware optimization, symmetry checks, and validation against small instances or classical baselines.

3. The most reliable practice is iterative debugging: identify the dominant bottleneck (sampling, noise, or optimization) and target it explicitly.

13.2.4 Example Workflows and Typical Use Cases

VQE is best viewed as a flexible workflow rather than a single fixed circuit. A typical implementation involves repeated cycles of state preparation, measurement, and classical optimization, together with problem-specific choices about ansatz structure and measurement strategy.

1 Typical Workflow

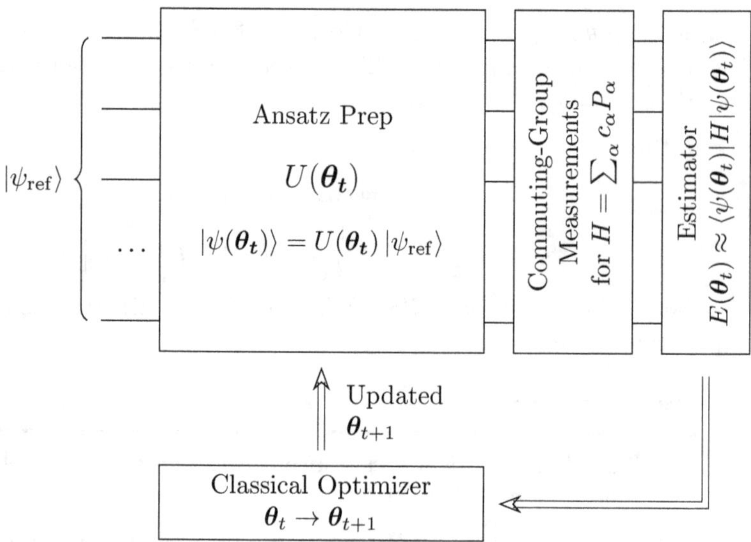

Figure 13.2: VQE Ansatz Preparation and Hybrid Optimization Loop

As illustrated in Fig. 13.2, a standard VQE workflow proceeds as follows:

(a) Choose a target Hamiltonian H and express it as a Pauli decomposition (see § 14.2).

(b) Choose an ansatz circuit $U(\boldsymbol{\theta})$ and a reference state $|\psi_{\text{ref}}\rangle$.

(c) Initialize parameters $\boldsymbol{\theta}_0$ (often using a physically motivated guess or a shallow warm start).

(d) Iterate: estimate $E(\boldsymbol{\theta}_t)$ from measurements and update $\boldsymbol{\theta}_t$ using a classical optimizer.

(e) Return $\boldsymbol{\theta}^\star$ and report the estimated energy and, when relevant, additional observables measured on $|\psi(\boldsymbol{\theta}^\star)\rangle$.

2   Common Use Cases

Two representative use cases are:

(a) *Ground-state properties of structured Hamiltonians,* such as small molecular or lattice-model instances, where one aims to estimate energies and observables rather than a classical bitstring.

(b) *Heuristic optimization for Ising-type objectives,* where H is diagonal and classical evaluation of energies is efficient, but finding the minimum-energy bitstring is combinatorially hard; a VQE-style workflow can be used as a sampling-based heuristic that concentrates probability mass on low-energy configurations.

The workflow above is a general procedure. A practical first step is to implement it on a simulator for small instances, and then apply the same loop on hardware as resources allow.

In the next section, we specialize this general variational framework to a particularly important structured ansatz: the quantum approximate optimization algorithm (QAOA). We will also connect its alternating-operator form to the discretized adiabatic viewpoint developed earlier in this chapter.

13.2.5   The Barren Plateau Problem and Its Mitigation

The barren plateau problem refers to the phenomenon that, for certain ansatz families and cost functions, typical gradients become exponentially small in the number of qubits. In this regime, the objective landscape becomes effectively flat at typical parameter values, so gradient-based training becomes statistically impractical: resolving a gradient signal above shot noise may require an exponential number of circuit evaluations.

Gradient-free methods are not immune either: when the landscape is exponentially flat, function-value differences between nearby parameter settings can also be exponentially small, so distinguishing genuine improvements above shot noise may still require an exponential number of measurements. In this sense, barren plateaus reflect a lack of usable training signal rather than a failure of gradients alone, and we will demonstrate this phenomenon through a simple random-circuit setting.

1   ✳ A Random-Circuit Demonstration

Setup

Consider an n-qubit VQE objective

$$E(\boldsymbol{\theta}) = \langle 0 | \, U(\boldsymbol{\theta})^\dagger H \, U(\boldsymbol{\theta}) \, | 0 \rangle \, ,$$

where H is a Hermitian cost operator (or the Hamiltonian). Focus on a single parameter θ_k that appears in a gate

$$V_k(\theta_k) = e^{-i\theta_k P/2} \, , \qquad P^2 = I \, ,$$

embedded inside the full circuit

$$U(\boldsymbol{\theta}) = U_2\, V_k(\theta_k)\, U_1,$$

where U_1 is the part *before* the parameterized gate and U_2 is the part *after* it.

Differentiate $E(\boldsymbol{\theta})$ using $\frac{d}{d\theta}V_k(\theta) = -\frac{i}{2}PV_k(\theta)$ and $\frac{d}{d\theta}V_k(\theta)^\dagger = \frac{i}{2}V_k(\theta)^\dagger P$:

$$\frac{\partial E}{\partial \theta_k} = \frac{i}{2}\,\langle 0|\, U_1^\dagger V_k(\theta_k)^\dagger \left[P,\, U_2^\dagger H U_2\right] V_k(\theta_k)U_1\, |0\rangle\,.$$

Define the commutator operator

$$K := \frac{i}{2}\left[P,\, U_2^\dagger H U_2\right],$$

which is Hermitian. Then the gradient is simply an expectation value:

$$\frac{\partial E}{\partial \theta_k} = \langle \psi|\, K\, |\psi\rangle\,, \qquad |\psi\rangle := V_k(\theta_k)U_1\, |0\rangle\,.$$

Random-Circuit Assumption

Suppose that for the relevant depth regime, the state $|\psi\rangle$ behaves like a *typical* (pseudo-)random state, e.g., because the prefix circuit U_1 is well approximated by a unitary 2-design [59].

Let $d = 2^n$ be the Hilbert-space dimension. For a Haar-random pure state $|\psi\rangle$ (and likewise for a 2-design), the first and second moments satisfy

$$\mathbb{E}\big[\langle \psi|\, A\, |\psi\rangle\big] = \frac{\mathrm{tr}(A)}{d}\,, \qquad \mathbb{E}\big[\langle \psi|\, A\, |\psi\rangle^2\big] = \frac{\mathrm{tr}(A^2) + \mathrm{tr}(A)^2}{d(d+1)}\,.$$

Applying this to $A = K$ gives

$$\mathbb{E}\left[\frac{\partial E}{\partial \theta_k}\right] = \frac{\mathrm{tr}(K)}{d}\,.$$

But $\mathrm{tr}([P, B]) = 0$ for any B, hence $\mathrm{tr}(K) = 0$, so the *mean gradient is zero*.

More importantly, the variance is

$$\mathrm{Var}\left(\frac{\partial E}{\partial \theta_k}\right) = \mathbb{E}\left[\langle \psi|\, K\, |\psi\rangle^2\right] - \mathbb{E}[\langle \psi|\, K\, |\psi\rangle]^2 = \frac{\mathrm{tr}(K^2)}{d(d+1)}\,.$$

If $\|K\|$ is bounded independently of n (for example, after a suitable normalization of H), then

$$\mathrm{tr}(K^2) \le d\,\|K\|^2,$$

and therefore

$$\mathrm{Var}\left(\frac{\partial E}{\partial \theta_k}\right) \le \frac{\|K\|^2}{d+1} = O(2^{-n})\,.$$

Thus the *typical gradient magnitude* scales as

$$\sqrt{\mathrm{Var}\left(\frac{\partial E}{\partial \theta_k}\right)} = O\!\left(2^{-\frac{n}{2}}\right),$$

which vanishes exponentially with the number of qubits. This is the barren plateau effect in its simplest quantitative form: once the ansatz behaves like a random circuit (relative to the cost), gradients concentrate near zero exponentially fast.

When This Argument Does and Does Not Apply

The exponential suppression above is most directly associated with *global* cost functions (observables that act nontrivially on $O(n)$ qubits) and with ansatz depths sufficient to approximate random states. If the cost is *local* (e.g., a sum of few-qubit terms) and the ansatz remains shallow or structured, gradients can scale more favorably.

2 Common Mitigation Strategies

Several mitigation strategies aim to keep the optimization landscape trainable by avoiding the "effectively random" regime, or by changing what is optimized.

- *Local or layerwise cost functions.* Replace a global cost by a sum of local costs, or train one layer (or block) at a time, so that early training signals do not vanish exponentially.

- *Problem-inspired or symmetry-preserving ansatzes.* Impose conserved quantities or physics structure to restrict the search to a relevant subspace, reducing the tendency to form highly random states.

- *Careful initialization.* Initialize near the identity (or near a known reference such as Hartree–Fock) so the circuit begins in a "close-to-classical" regime where gradients are measurable.

- *Progressive depth growth / parameter transfer.* Start from small depth and increase depth gradually, reusing parameters from shallower circuits as warm starts.

- *Adaptive ansatz construction.* Methods such as ADAPT-VQE build the circuit by adding only operators that produce a measurable gradient signal, often yielding shallower and more interpretable ansatzes.

- *Geometry-aware optimization.* Natural-gradient updates incorporate the information geometry of the variational manifold and can improve conditioning and convergence in practice [60].

- *Noise-aware training.* Since device noise can further flatten landscapes, combine trainability ideas above with error-mitigation and validation checks (symmetry checks, resampling stability, and small-instance benchmarking).

These methods do not replace fault-tolerant error correction, but they can improve performance in near-term regimes.

For original and review treatments of barren plateaus and mitigation ideas, see [61, 62].

3 Other Error Sources

Besides barren plateaus, VQE performance is limited by both statistical and systematic errors. Understanding these limitations is essential for interpreting results and for designing robust workflows.

Finite-Shot Noise

Expectation estimates are affected by sampling fluctuations. Finite-shot noise introduces stochasticity into the objective landscape and can mask small improvements, especially late in optimization. Shot allocation strategies and measurement grouping can reduce this cost, but they do not remove it entirely. (See § 14.2.)

Hardware Noise and Model Mismatch

Gate errors, decoherence, and measurement error shift the observed objective away from its ideal value. In addition, the implemented circuit may differ from the intended ansatz due to compilation and calibration drift. These effects can bias the optimizer and lead to an apparent energy minimum that does not correspond to a true low-energy state of H.

13.2.6   Universality in the Broader Variational Framework

1   The Broader Variational Model

Variational quantum algorithms are often introduced through concrete near-term examples such as the variational quantum eigensolver for ground-state estimation or parametrized circuits for optimization. In that practical setting, the emphasis is usually on ansatz design, measurement overhead, optimizer choice, noise sensitivity, and trainability, so the variational approach can appear mainly as a heuristic framework for approximate computation.

At a broader level, however, the variational paradigm can be viewed as a computational model in which a parametrized quantum process is repeatedly evaluated by measurements, assigned an objective value, and then updated through a classical optimization or decision loop. From this perspective, familiar algorithms such as VQE are instances of a larger hybrid measurement-and-feedback framework [63].

2   Universality Through Measurement and Feedback

The essential ingredients of this broader model are a parametrized quantum state-preparation or circuit procedure, measurements used to estimate one or more objective functions, and a classical routine that updates parameters or choices based on those measurement outcomes. The repeated measurement-and-update loop is therefore not merely a numerical convenience; it is part of the computational model itself.

This notion of universality is distinct from the universality of adiabatic quantum computation, which concerns Hamiltonian evolution and its polynomial equivalence to the circuit model [11] (see § 13.1.6). Here the claim is instead that, when formulated in this broader measurement-and-feedback setting, variational quantum computation also has, in principle, the expressive power of a universal model of quantum computation. Biamonte *et al.* showed that suitably designed objective functions can encode the behavior of an arbitrary target circuit [63]. The variational procedure is then guided, through repeated measurement and classical feedback, toward states and parameter settings whose measured properties reproduce the intended computation.

An intuitive way to view this is to think of the objective function as a certificate for the computation. It can be constructed so that low objective value is achieved only by states or parametrized processes that behave as though the target circuit had been carried out correctly. The variational loop therefore need not imitate the circuit gate by gate in a literal sense; instead, it searches for a state or process whose observable properties certify the same computational outcome.

3 Interpretation and Practical Limits

The significance of this result is conceptual rather than directly practical. It does not mean that an ordinary low-depth, problem-tailored VQE instance automatically functions as a practical universal quantum computer. It does not remove familiar difficulties such as sampling cost, optimizer instability, barren plateaus, or hardware noise, nor does it imply that variational algorithms are always the best way to implement a given quantum computation.

What the result does show is that variational methods should not be viewed only as ad hoc near-term heuristics for eigenvalue estimation or optimization. Rather, they occupy a broader place in the landscape of quantum algorithms: at one end are application-driven ansatzes designed for specific tasks, while at the other is a general hybrid model in which measurement-defined objectives and classical feed-forward can emulate arbitrary quantum circuit computation. In this sense, the variational paradigm stands alongside other universal models of quantum computation at the level of formal expressive power.

13.3 QAOA as a Structured Variational Algorithm

The quantum approximate optimization algorithm (QAOA), originally proposed by Farhi *et al.* [55], is a prominent variational algorithm for combinatorial optimization. It can be motivated from two complementary viewpoints. From the adiabatic perspective, QAOA is a coarse, gate-based discretization of an interpolation between a driver Hamiltonian and a problem Hamiltonian. From the variational perspective, QAOA is a structured ansatz family within the VQE framework, tailored to Ising- and QUBO-type objectives.

13.3.1 From Adiabatic Evolution to Alternating Operators

Consider the standard adiabatic interpolation

$$H(s) = (1 - s)H_B + sH_C, \qquad s \in [0, 1],$$

together with a schedule $s(t)$ over $t \in [0, T]$.

We adopt the standard QAOA notation $H_C \equiv H_P$ and $H_B \equiv H_D$. Here H_C is the *cost Hamiltonian*, which encodes the objective whose low-energy states represent good solutions, and H_B is the *mixer Hamiltonian*, which redistributes amplitude among computational basis states to explore the solution space.

On a gate-based device, we can approximate the continuous-time evolution by dividing the interval into p segments and replacing the time-ordered unitary by a product of short evolutions. If, within segment ℓ, the Hamiltonian is approximately constant and dominated by a linear combination of H_B and H_C, then a first-order product-formula approximation (see § 10.2) yields a unitary resembling

$$e^{-i\beta_\ell H_B} e^{-i\gamma_\ell H_C}$$

for some effective durations β_ℓ and γ_ℓ. QAOA takes the next conceptual step: rather than fixing these durations to match a specific schedule, it treats them as *variational parameters* that are optimized to minimize the expected cost. In this sense, QAOA preserves the Hamiltonian viewpoint of AQC while adopting the hybrid optimization workflow of VQE.

13.3.2 Standard QAOA Formulation

In the standard formulation, QAOA is defined by a pair of Hamiltonians on n qubits: a mixer Hamiltonian H_B and a cost Hamiltonian H_C. Typically, H_C is diagonal in the computational basis and represents an Ising or QUBO objective, while H_B is chosen to be simple and highly mixing, most commonly

$$H_B = \sum_{i=1}^{n} X_i.$$

(Up to an overall sign convention; only relative conventions matter.)

Exercise 13.6 Show that $|+\rangle^{\otimes n}$ is the unique ground state of $-H_B$. What is the ground energy?

1 The p-Layer Circuit

Let $\boldsymbol{\gamma} = (\gamma_1, \ldots, \gamma_p)$ and $\boldsymbol{\beta} = (\beta_1, \ldots, \beta_p)$ be real parameters. Define

$$U_C(\gamma) = e^{-i\gamma H_C}, \qquad U_B(\beta) = e^{-i\beta H_B}. \tag{13.12}$$

The depth-p QAOA unitary is

$$U_p(\boldsymbol{\gamma}, \boldsymbol{\beta}) = \prod_{\ell=1}^{p} U_B(\beta_\ell)\, U_C(\gamma_\ell), \tag{13.13}$$

where the product is ordered from $\ell = 1$ (rightmost) to $\ell = p$ (leftmost). Starting from an easily prepared initial state (also an eigenstate of H_B), usually

$$|+\rangle^{\otimes n}, \qquad |+\rangle = \frac{|0\rangle + |1\rangle}{\sqrt{2}},$$

QAOA prepares the trial state

$$|\psi_p(\boldsymbol{\gamma}, \boldsymbol{\beta})\rangle = U_p(\boldsymbol{\gamma}, \boldsymbol{\beta})\, |+\rangle^{\otimes n}. \tag{13.14}$$

Measuring $|\psi_p(\boldsymbol{\gamma}, \boldsymbol{\beta})\rangle$ in the computational basis yields a bitstring $x \in \{0,1\}^n$ distributed according to

$$\Pr[X = x] = |\langle x|\psi_p(\boldsymbol{\gamma}, \boldsymbol{\beta})\rangle|^2.$$

2 Objective Function

The standard objective is the expected cost (energy)

$$E_p(\boldsymbol{\gamma}, \boldsymbol{\beta}) = \langle \psi_p(\boldsymbol{\gamma}, \boldsymbol{\beta})|H_C|\psi_p(\boldsymbol{\gamma}, \boldsymbol{\beta})\rangle.$$

For minimization tasks, one seeks

$$(\boldsymbol{\gamma}^\star, \boldsymbol{\beta}^\star) \in \arg\min_{\boldsymbol{\gamma}, \boldsymbol{\beta}} E_p(\boldsymbol{\gamma}, \boldsymbol{\beta}),$$

and then samples bitstrings from $|\psi_p(\boldsymbol{\gamma}^\star, \boldsymbol{\beta}^\star)\rangle$. For maximization tasks such as Max-Cut, one may equivalently minimize $-H_C$ or maximize the expected cut value, depending on the chosen sign convention.

Exercise 13.7 Consider MaxCut on the single-edge graph with two vertices. With $H_C = \frac{I - Z_1 Z_2}{2}$ and $H_B = X_1 + X_2$, define the $p = 1$ QAOA state

$$|\psi(\gamma, \beta)\rangle = e^{-i\beta H_B} e^{-i\gamma H_C} |+\rangle^{\otimes 2}.$$

Compute $\langle H_C \rangle(\gamma, \beta)$ explicitly and find a pair (γ, β) that maximizes $\langle H_C \rangle$.

13.3.3 Algorithm Steps and Resource Discussion

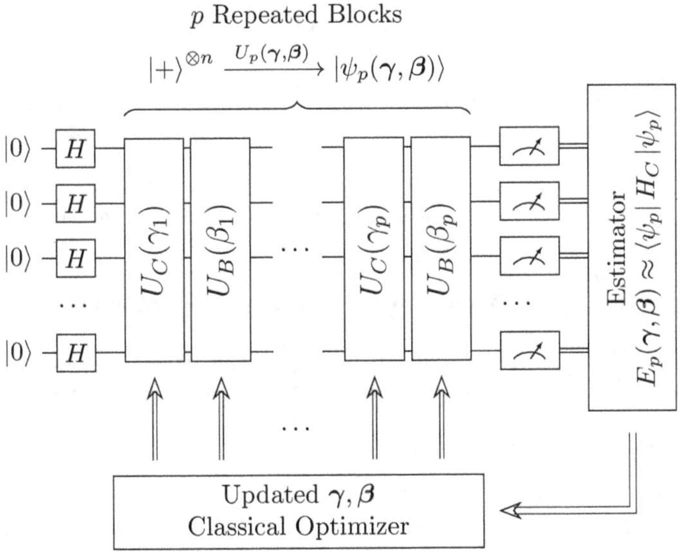

Figure 13.3: Depth-p QAOA Ansatz and Hybrid Optimization Loop

As illustrated in Fig. 13.3, QAOA follows the same hybrid loop as VQE, with the key difference that the ansatz circuit is fixed by the alternating-operator structure.

1 Algorithm Outline

A standard QAOA workflow is:

(a) Choose an encoding and construct the cost Hamiltonian H_C.

(b) Choose a mixer Hamiltonian H_B (often $\sum_i X_i$) and a depth p.

(c) Initialize parameters (γ, β).

(d) Iterate: estimate $E_p(\gamma, \beta)$ by measurement and update parameters using a classical optimizer.

(e) With optimized parameters, sample bitstrings and return the best observed solution(s) or summary statistics (e.g., best energy, empirical distribution).

2 Resources and the Role of Depth p

The depth parameter p controls the expressibility of the ansatz and the circuit cost. Larger p expands the family of reachable states and can improve the achievable objective value, but it also increases circuit depth and the number of parameters. The overall cost is typically dominated by repeated measurement for energy estimation and sampling, as in VQE.

From a conceptual standpoint, p plays a role analogous to a discretization level: small p corresponds to a very coarse digitization of an adiabatic path, while large p can better approximate continuous-time behavior—subject to hardware noise and optimization difficulty.

> **Key Takeaways**
>
> ---
>
> 1. QAOA alternates evolutions under a cost Hamiltonian and a mixer Hamiltonian; the angles are optimized classically from measurement data.
>
> 2. Increasing depth p typically increases expressivity but also increases circuit depth and can make optimization harder.
>
> 3. QAOA is best viewed as a structured variational method with problem-dependent performance, rather than a guaranteed convergent procedure.

13.3.4 Variants and Design Degrees of Freedom

Although the standard choice $H_B = \sum_i X_i$ is widely used, QAOA has several design degrees of freedom that can matter as much as depth [64].

1 Mixers for Constrained Optimization

When the feasible set is a strict subset of $\{0,1\}^n$, naive mixing can move probability mass into infeasible bitstrings. One strategy is to design a mixer Hamiltonian whose dynamics preserves feasibility, so the algorithm explores only the feasible subspace. Another strategy is to keep the standard mixer but encode constraints into H_C via penalty terms, trading feasibility preservation for simplicity; see, e.g., the constraint-mixer framework in [65].

2 Warm Starts and Problem-Informed Initialization

In some workflows, one initializes the algorithm using a classical heuristic so the variational search begins near a good region of parameter space. A common way to obtain such a heuristic is to solve a *continuous relaxation* of the discrete problem (for example, replacing $x \in \{0,1\}^n$ by $x \in [0,1]^n$), which is often easier because it admits mature convex-optimization or continuous-optimization methods.

A direct warm-start state of the form $\bigotimes_i \left(\sqrt{1 - x_i}\, |0\rangle + \sqrt{x_i}\, |1\rangle \right)$ is generally *not* an eigenstate of the standard mixer $H_B = \sum_i X_i$. Warm-start QAOA therefore modifies the mixer so that the chosen initial product state becomes a ground state (equivalently, an eigenstate with extremal eigenvalue) of the mixer Hamiltonian. One implementation uses a single-qubit, problem-informed mixer of the form

$$H_B(x) = \sum_i \left(\cos\theta_i\, X_i + \sin\theta_i\, Y_i \right),$$

with angles θ_i chosen from the relaxation solution x so that $|\psi_x\rangle$ is the ground state of $H_B(x)$. This construction "aligns" the mixer with the warm-start direction and can improve performance at small depth p and reduce optimizer burden under finite-shot noise [66].

3 Recursive QAOA

Recursive QAOA (RQAOA) augments QAOA with a classical *elimination* step that reduces the problem size. A typical pattern is: run a low-depth QAOA instance; estimate selected correlators (e.g., $\langle Z_i Z_j \rangle$ for Ising/QUBO cost Hamiltonians); use these to infer a high-confidence relation between variables (such as $z_i = \pm z_j$); *fix* one variable in terms of another (or fix a variable outright); simplify the Hamiltonian; and recurse on the reduced instance. In this way, the quantum circuit depth can remain small while the overall procedure becomes increasingly non-local through classical recursion. This hybrid strategy was proposed and analyzed in [67].

4 Problem-Structured Alternation

For problems with structure (for example, graph locality or layered constraints), one can use alternating operators that reflect this structure, such as separate parameter blocks for different edge sets or constraint families. These modifications preserve the QAOA philosophy while adapting the ansatz to the instance class; see [55] for the original framework and representative structured choices discussed in follow-up work.

13.3.5 Relationship Between QAOA and VQE

QAOA can be viewed as a special case of VQE in which the ansatz family is fixed to the alternating-operator form and the objective Hamiltonian is chosen to encode a combinatorial cost.

1 QAOA as a Structured VQE Ansatz

In VQE, we select a parameterized circuit family $U(\boldsymbol{\theta})$ and minimize $\langle \psi(\boldsymbol{\theta}) | H | \psi(\boldsymbol{\theta}) \rangle$. In QAOA, the parameter vector is $(\boldsymbol{\gamma}, \boldsymbol{\beta})$ and the circuit family is

$$U_p(\boldsymbol{\gamma}, \boldsymbol{\beta}) = \prod_{\ell=1}^{p} e^{-i\beta_\ell H_B} \, e^{-i\gamma_\ell H_C},$$

so QAOA is a VQE instance with a particular choice of ansatz and Hamiltonian structure.

2 Shared Challenges

Because both methods are variational, they share practical issues:

(a) *Finite-shot noise* introduces stochasticity into objective estimates and gradients.

(b) *Hardware noise* biases measured energies and can distort the optimization landscape.

(c) *Optimization difficulty* can arise from nonconvexity and from parameter scaling with depth.

The main difference is interpretability and structure: QAOA has a tighter connection to the underlying optimization Hamiltonians, while general VQE can target broader Hamiltonian families beyond diagonal Ising objectives.

3 When Each Viewpoint Is Useful

As a rule of thumb, QAOA is a natural first choice when the objective is a combinatorial problem with a clean Ising/QUBO encoding and when shallow, structured circuits are preferred. VQE is the more general framework when the target Hamiltonian is not diagonal (for example, in simulation settings) or when one wishes to explore ansatz families beyond alternating operators.

> **Exercise 13.8** Explain, in one paragraph, why QAOA can be viewed as a restricted variational family. In your explanation, identify (i) what is being optimized, (ii) what information is obtained from measurements, and (iii) why increasing p changes both expressivity and resource cost.

13.3.6 Worked Example: Max-Cut

We now connect QAOA to the Max-Cut encoding from § 13.1.4.4. Let $G = (V, E)$ be a graph on n vertices. The Ising Hamiltonian for Max-Cut (see Eq. 13.11) is

$$H_C = \frac{1}{2} \sum_{(i,j)\in E} Z_i Z_j.$$

Minimizing H_C is equivalent to maximizing the expected cut value (see Eq. 13.10). Choose the standard mixer Hamiltonian

$$H_B = \frac{1}{2} \sum_{i=1}^{n} X_i,$$

and the initial state $|+\rangle^{\otimes n}$.

The unitary block $U_C(\gamma)$ is

$$U_C(\gamma) = e^{-i\gamma H_C} = \prod_{(i,j)\in E} e^{-i\frac{\gamma}{2} Z_i Z_j} = \prod_{(i,j)\in E} \mathrm{ZZ}_{ij}(\gamma), \qquad (13.15)$$

where $\mathrm{ZZ}_{ij}(\gamma) \equiv e^{-i\frac{\gamma}{2} Z_i Z_j}$ is a ZZ-rotation gate acting on qubits i and j.

The unitary block $U_B(\beta)$ is

$$U_B(\beta) = e^{-i\beta H_B} = \prod_{i\in V} e^{-i\frac{\beta}{2} X_i} = \prod_{i\in V} R_{x_i}(\beta), \qquad (13.16)$$

where $R_{x_i}(\beta) \equiv e^{-i\frac{\beta}{2} X_i}$ is an X-rotation gate acting on qubit i.

1 Interpretation at Small Depth

At $p = 1$, QAOA already captures a nontrivial interaction between the objective and mixing dynamics: the phase-separation step $e^{-i\gamma H_C}$ imprints phases correlated with cut edges, and the mixing step $e^{-i\beta H_B}$ converts these phases into amplitude redistribution over bitstrings. Increasing p repeats this alternating process and can improve performance, while also increasing circuit depth and the difficulty of parameter optimization.

2 A Small-Graph Instance

For the five-vertex graph in Fig. 13.1, the building block for the QAOA ansatz comprising $U_C(\gamma)$ and $U_B(\beta)$ is shown in Fig. 13.4. With a sufficient number of QAOA layers (p) and optimization iterations, the algorithm can be driven toward the anticipated ground-state subspace, corresponding to a maximum cut of 5.

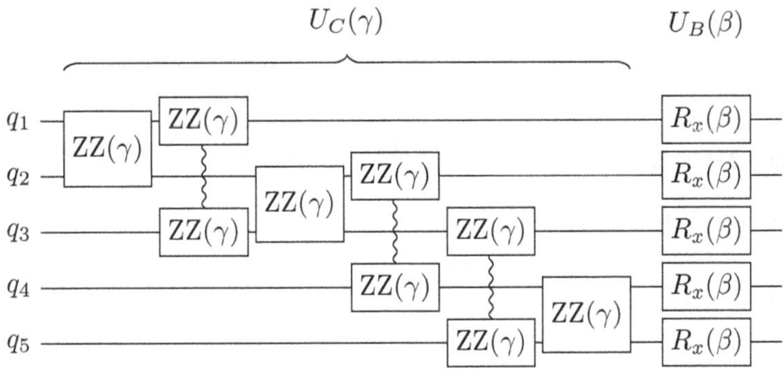

Figure 13.4: Example of a QAOA Ansatz Building Block

3　Connection Back to AQC

Comparing QAOA with AQC highlights the shared Hamiltonian backbone. In AQC, one aims to follow the instantaneous ground state of an interpolation from a driver Hamiltonian to a problem Hamiltonian. In QAOA, one replaces continuous-time tracking by a fixed number of alternating evolutions under a mixer and a cost Hamiltonian, and treats the corresponding durations as variational parameters. This viewpoint will be useful in the next section, where we discuss sampling-based quantum diagonalization (SQD) as another near-term strategy for extracting spectral information from measurement data.

13.4　* Sampling-based Quantum Diagonalization (SQD)

Sampling-based quantum diagonalization (SQD) is a hybrid quantum–classical approach for estimating low-lying eigenvalues and eigenstates of a Hamiltonian H by combining *quantum sampling* with *classical subspace diagonalization*. In contrast to variational algorithms such as VQE, SQD does not necessarily rely on iteratively updating a set of variational parameters; instead, it uses the quantum device primarily to *sample* candidate basis configurations and uses classical linear algebra to extract spectral information from the resulting subspace problem. Historically, this line of work was introduced under the name *quantum selected configuration interaction (QSCI)* by Kanno *et al.* [68], and the term SQD is now often used to emphasize the sampling-plus-diagonalization viewpoint.

The central idea is to use a quantum device to identify a small, "important" subset of basis configurations that carry significant weight in the target low-energy states, then project H onto the span of those configurations and solve the resulting eigenvalue problem classically.

Overview

1. SQD uses a quantum device to *sample* configurations that are likely to matter for low-energy states.

2. Those samples define a reduced subspace, and the main computation becomes a classical diagonalization of a projected eigenproblem.

3. The core tradeoff is shifting work from deep coherent circuits to sampling plus classical linear algebra on a carefully controlled subspace.

13.4.1   Subspace Projection as the Core Idea of SQD

1   Subspace Sampling and Generalized Diagonalization

To demonstrate the mechanics of subspace sampling, consider a system described by the Hamiltonian

$$H = \begin{bmatrix} 2.5 & 0 & 0 & 0.5 \\ 0 & 3 & -1 & 0 \\ 0 & -1 & 3 & 0 \\ 0.5 & 0 & 0 & 2.5 \end{bmatrix}.$$

The eigenvalues of H are $2, 2, 3, 4$. The smallest eigenvalue $(2$, the ground-state energy$)$ is degenerate, with a two-dimensional ground-state subspace $\mathcal{V}_0$. Suppose we sample vectors from $\mathcal{V}_0$ and obtain two (nonorthogonal) normalized vectors

$$|\phi_1\rangle = \frac{1}{2} \begin{bmatrix} 1 & 1 & 1 & -1 \end{bmatrix}^T, \qquad |\phi_2\rangle = \frac{1}{\sqrt{2}} \begin{bmatrix} 1 & 0 & 0 & -1 \end{bmatrix}^T.$$

Let $\mathcal{V} = \operatorname{span}\big(|\phi_1\rangle, |\phi_2\rangle\big)$. In general, when the basis is nonorthogonal, the restriction of H to $\mathcal{V}$ is expressed through the Rayleigh–Ritz matrices

$$\mathbf{H} = \begin{bmatrix} \langle\phi_1| H |\phi_1\rangle & \langle\phi_1| H |\phi_2\rangle \\ \langle\phi_2| H |\phi_1\rangle & \langle\phi_2| H |\phi_2\rangle \end{bmatrix}, \qquad \mathbf{S} = \begin{bmatrix} \langle\phi_1|\phi_1\rangle & \langle\phi_1|\phi_2\rangle \\ \langle\phi_2|\phi_1\rangle & \langle\phi_2|\phi_2\rangle \end{bmatrix}.$$

A direct calculation gives

$$\mathbf{H} = \begin{bmatrix} 2 & \sqrt{2} \\ \sqrt{2} & 2 \end{bmatrix}, \qquad \mathbf{S} = \begin{bmatrix} 1 & 1/\sqrt{2} \\ 1/\sqrt{2} & 1 \end{bmatrix}.$$

Because the basis is nonorthogonal, the reduced eigenproblem takes the generalized form

$$\mathbf{H}\alpha = \lambda \, \mathbf{S}\alpha.$$

In this example, one finds $\mathbf{H} = 2\mathbf{S}$, and therefore the generalized eigenvalue is $\lambda = 2$ with multiplicity 2, as expected for a two-dimensional ground space. A convenient choice of two linearly independent generalized eigenvectors is

$$\alpha_1 = \begin{bmatrix} 1 \\ 0 \end{bmatrix}, \qquad \alpha_2 = \begin{bmatrix} 0 \\ 1 \end{bmatrix},$$

corresponding to the physical vectors

$$|\psi_1\rangle = |\phi_1\rangle, \qquad |\psi_2\rangle = |\phi_2\rangle.$$

More generally, any nonzero coefficient vector $\alpha = (\alpha_1, \alpha_2)^T$ defines a ground-state vector

$$|\psi\rangle = c_1 |\phi_1\rangle + c_2 |\phi_2\rangle$$

with energy 2. (If the sampled basis were orthonormal, then $\mathbf{S} = I$ and the reduced problem would become an ordinary eigenvalue problem.)

> **Exercise 13.9** In the computational basis $\{|00\rangle, |01\rangle, |10\rangle, |11\rangle\}$, let
>
> $$H = \begin{pmatrix} 2 & 0 & 0 & 0 \\ 0 & 1 & \lambda & 0 \\ 0 & \lambda & 1 & 0 \\ 0 & 0 & 0 & 0 \end{pmatrix}.$$
>
> Restrict H to the subspace $\mathcal{S} = \mathrm{span}\{|01\rangle, |10\rangle\}$. In the orthonormal basis $\{|01\rangle, |10\rangle\}$, form the Rayleigh–Ritz matrix $\mathbf{H}$ (so $\mathbf{S} = I$) and diagonalize it. For which values of λ does $\mathcal{S}$ already contain the true ground state of H?

2　Bridging to SQD

The example above is deliberately small: it reduces diagonalization of a 4×4 matrix to that of a 2×2 reduced problem, and it assumes that we can sample vectors from the true ground-space $\mathcal{V}_0$. In realistic quantum many-body problems, neither assumption holds in such an explicit form.

Nevertheless, the same projection idea becomes powerful at scale. In many physics and chemistry settings, the ground state lives in an exponentially large Hilbert space, yet its weight is often concentrated on a comparatively small set of basis configurations. If we can identify a subspace that captures most of this weight, then projecting the Hamiltonian into that subspace can yield a dramatic reduction in dimension, potentially bringing the reduced eigenproblem within reach of classical linear algebra.

SQD follows this strategy. A quantum device is used to prepare and sample states with substantial overlap on the low-energy sector (for example, via a variational state-preparation procedure). The measurement outcomes reveal a problem-adapted set of configurations that define a candidate subspace. One then assembles the projected eigenproblem and performs classical diagonalization to extract low-lying energies and approximate eigenstates. By iterating this quantum sampling and classical diagonalization loop (often described as *configuration recovery* or *subspace refinement*), the method aims to converge toward a self-consistent approximation of the true ground state and nearby excited states.

13.4.2　Problem Setting and Algorithmic Template

SQD targets the low-energy eigensystem of an n-qubit Hamiltonian H,

$$H |\psi_k\rangle = E_k |\psi_k\rangle, \qquad E_0 \leq E_1 \leq \cdots,$$

where H is commonly available as a Pauli decomposition $H = \sum_\ell c_\ell P_\ell$. The goal is to estimate one or more low-lying energies $E_0, \ldots, E_{k_{\max}}$ and (when useful) to output approximate eigenstates in an interpretable basis.

1　Inputs

A representative SQD workflow assumes a Hamiltonian description (often a Pauli expansion), a state-preparation routine that yields one or more trial states with non-negligible overlap on the low-energy sector, a measurement basis that defines the configurations to be sampled (typically the computational basis or an encoding-induced variant), and practical choices such as the shot budget and selection thresholds.

2 Outputs

SQD outputs estimated energies $\widetilde{E}_k$ from a reduced eigenproblem, together with approximate eigenstates expressed in the chosen reduced basis (often a selected configuration basis). It is also natural to report diagnostics such as residual norms and stability under subspace enlargement.

3 Algorithmic Template

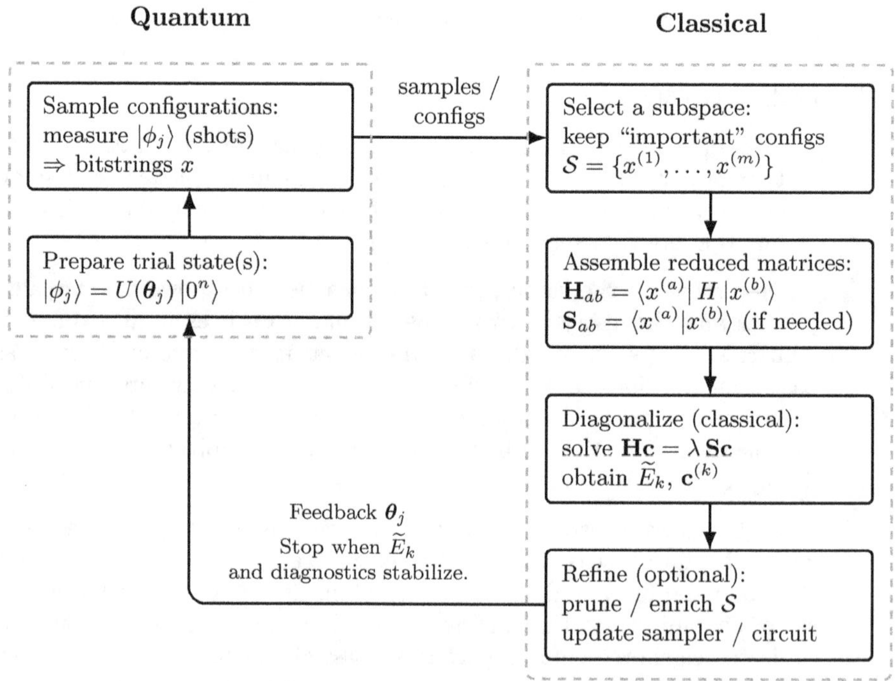

Figure 13.5: SQD Algorithmic Template

At a high level, SQD alternates between quantum sampling and classical diagonalization (see Fig. 13.5):

1. *Sample* one or more trial states on a quantum device and record measurement outcomes that define candidate configurations (or sampled vectors).

2. *Select and filter* a compact subspace intended to capture the low-energy sector, enforcing known constraints (e.g., symmetries) when available.

3. *Assemble* the reduced Hamiltonian on that subspace (and an overlap matrix if the spanning set is nonorthogonal).

4. *Diagonalize* the reduced (ordinary or generalized) eigenproblem to obtain $\widetilde{E}_k$ and approximate eigenvectors in the subspace.

5. *Refine* the sampler and/or enlarge the subspace using the classical solution, and repeat until energies and diagnostics stabilize.

This five-step view subsumes the common "self-consistent" SQD loop: quantum sampling proposes the subspace, classical diagonalization extracts the best low-energy information available within that subspace, and the resulting coefficients guide the next round of sampling.

13.4.3　✳ Sample-Defined Subspace and Projected Eigenproblem

Let $\{|x\rangle\}_{x\in\{0,1\}^n}$ denote the computational basis. Measuring a trial state $|\phi\rangle$ produces a multiset of bitstrings; from these outcomes one selects a configuration set

$$\mathcal{S} = \{x^{(1)}, x^{(2)}, \ldots, x^{(m)}\} \subseteq \{0,1\}^n,$$

and defines the sampled subspace

$$\mathcal{V} = \mathrm{span}\{|x^{(1)}\rangle, \ldots, |x^{(m)}\rangle\}.$$

When $\mathcal{S}$ contains distinct bitstrings, the induced basis is orthonormal and the reduced problem is an ordinary eigenvalue problem. More general SQD variants enrich the subspace using additional (not necessarily orthogonal) vectors; in that case one must solve a generalized eigenproblem.

1　Orthonormal Subspace Case

If $\{|x^{(a)}\rangle\}_{a=1}^m$ is orthonormal, the reduced Hamiltonian matrix is

$$(\mathbf{H})_{ab} = \langle x^{(a)}|\, H\, |x^{(b)}\rangle, \qquad a, b \in \{1, \ldots, m\},$$

and classical diagonalization yields

$$\mathbf{H}\boldsymbol{\alpha}^{(k)} = \widetilde{E}_k\, \boldsymbol{\alpha}^{(k)}.$$

The associated approximate eigenstate in the full Hilbert space is

$$|\widetilde{\psi}_k\rangle = \sum_{a=1}^m \alpha_a^{(k)} |x^{(a)}\rangle.$$

2　Nonorthogonal Enrichment and Rayleigh–Ritz Form

If the subspace is spanned by nonorthogonal vectors $\{|\varphi_a\rangle\}_{a=1}^m$, one forms the Rayleigh–Ritz matrices

$$(\mathbf{H})_{ab} = \langle \varphi_a|\, H\, |\varphi_b\rangle, \qquad (\mathbf{S})_{ab} = \langle \varphi_a|\varphi_b\rangle,$$

and solves the generalized eigenvalue problem

$$\mathbf{H}\boldsymbol{\alpha}^{(k)} = \widetilde{E}_k\, \mathbf{S}\boldsymbol{\alpha}^{(k)}.$$

This is the appropriate formulation whenever the sampled spanning set is not orthonormal, as in § 13.4.1.

3　Diagnostics

A compact diagnostic for each approximate eigenpair $(\widetilde{E}_k, |\widetilde{\psi}_k\rangle)$ is the residual norm

$$r_k = \big\| H\, |\widetilde{\psi}_k\rangle - \widetilde{E}_k\, |\widetilde{\psi}_k\rangle \big\|.$$

Small residuals and stability of $\widetilde{E}_k$ under modest subspace enlargement provide practical evidence that the sampled subspace captures the intended low-energy sector.

Exercise 13.10 Let S be an m-dimensional subspace with orthonormal basis $\{|v_1\rangle, \dots, |v_m\rangle\}$. Define the isometry

$$V = \begin{bmatrix} |v_1\rangle & \cdots & |v_m\rangle \end{bmatrix}, \qquad \text{so that} \qquad V^\dagger V = I_m,$$

and the Rayleigh–Ritz (restricted) matrix

$$\mathbf{H} = V^\dagger H V.$$

Show that minimizing $\langle \psi| H |\psi\rangle$ over all normalized $|\psi\rangle \in S$ reduces to minimizing $\mathbf{x}^\dagger \mathbf{H} \mathbf{x}$ over all $\mathbf{x} \in \mathbb{C}^m$ with $\|\mathbf{x}\|_2 = 1$, and conclude that the minimum equals the smallest eigenvalue of $\mathbf{H}$. Conclude further that this smallest eigenvalue is an upper bound on the true ground energy of H.

13.4.4 ∗Subspace Construction and Refinement

1 Matrix Elements and Pauli Structure

When $H = \sum_\ell c_\ell P_\ell$ is given as a sum of Pauli strings, matrix elements can be assembled term-by-term,

$$\langle x^{(a)}| H |x^{(b)}\rangle = \sum_\ell c_\ell \, \langle x^{(a)}| P_\ell |x^{(b)}\rangle.$$

Each Pauli string P_ℓ flips only the bits where it contains X or Y factors and contributes phases from Z or Y factors, so for a fixed $|x^{(b)}\rangle$ only a limited set of $|x^{(a)}\rangle$ yield nonzero $\langle x^{(a)}| P_\ell |x^{(b)}\rangle$. In practice this often makes the reduced matrix comparatively sparse, which can reduce both assembly time and diagonalization cost. (Some SQD variants also estimate certain matrix elements on quantum hardware, but the simplest and most common presentation treats the quantum device primarily as a sampler and performs the reduced assembly classically.)

2 Selection, Filtering, and Symmetry

The raw measurement outcomes are typically refined before diagonalization. The guiding principle is to keep configurations that are likely to contribute meaningfully to low-energy eigenstates, while controlling the subspace dimension m.

Frequency and Weight Thresholds

Let N_{shot} be the number of samples and let $f(x)$ be the observed frequency of configuration x, with empirical probability $\widehat{p}(x) = f(x)/N_{\text{shot}}$. Common selection rules include (i) keep the top-m configurations by $f(x)$, (ii) keep all configurations with $\widehat{p}(x) \geq p_{\min}$, or (iii) keep the smallest set whose cumulative mass $\sum_{x \in S} \widehat{p}(x)$ exceeds a target (e.g., 0.9). These rules provide transparent knobs that trade off subspace size against the risk of omitting important support.

Union Across Multiple Trial States

To reduce sampling bias, one often samples several trial states (for example, different initializations, different ansatz families, or states targeted toward excited sectors) and takes S as a union across runs. In chemistry settings, this is analogous to broadening the selected determinant set across multiple references, and it is especially useful when the spectrum has near-degeneracies or multiple competing configuration families.

Physics-Guided Pruning and Expansion

Selection is frequently paired with lightweight classical heuristics that either prune or expand $\mathcal{S}$:

- *Diagonal screening:* compute the diagonal energies $h(x) := \langle x| H |x\rangle$ and discard configurations with $h(x)$ far above the current best energy estimate, unless they are strongly connected to retained configurations.

- *Connectivity expansion:* for each retained configuration x, add configurations y that are reachable by applying dominant terms of H (e.g., bit flips induced by large-magnitude Pauli strings, or single/double excitations in second-quantized chemistry Hamiltonians). This mirrors the "selected CI" intuition: important configurations often appear together with their strongly coupled neighbors.

- *Coefficient-based pruning:* after diagonalization, keep only configurations whose coefficients satisfy $|\alpha_a^{(k)}| \geq \alpha_{\min}$ for one or more targeted states k, then resample and expand around this refined support.

Symmetry and Constraint Filtering

If an observable Q commutes with the Hamiltonian H, $[H, Q] = 0$, then the time evolution generated by H preserves each eigenspace of Q, and we say that Q is a conserved quantity (or that the symmetry generated by Q is conserved). Many Hamiltonians conserve quantities that define invariant sectors, and enforcing these constraints at the selection stage can substantially improve both accuracy and numerical conditioning.

Typical examples include particle-number conservation in electronic-structure problems (bitstrings with a fixed Hamming weight corresponding to a fixed number of electrons), parity symmetries in spin and fermionic encodings, and other model-specific constraints (e.g., fixed total magnetization sectors in lattice models). Operationally, symmetry filtering means restricting $\mathcal{S}$ to configurations that satisfy the constraint *exactly* (hard filtering), or else recording the constraint value for each configuration and discarding those that fall outside the intended sector (soft filtering). This reduces spurious basis elements and focuses the reduced diagonalization on the physically relevant block of H.

3 Iteration and Configuration Recovery

SQD is often run as an iterative loop: sample $\rightarrow$ select $\mathcal{S}$ $\rightarrow$ assemble $(\mathbf{H}, \mathbf{S})$ $\rightarrow$ diagonalize $\rightarrow$ update. The classical eigenvectors indicate which configurations carry the most weight in the current approximation, suggesting targeted subspace enrichment (for example, by connectivity expansion around large coefficients) or updating the state-preparation routine to increase overlap with the extracted low-energy states. Iterations stop when energies, residual norms, and selected observables stabilize under further subspace growth.

4 Classical Cost, Conditioning, and Failure Modes

The dominant classical post-processing cost is typically diagonalization, which scales as $O(m^3)$ in the worst case for dense matrices, so subspace control is essential. In nonorthogonal variants, the overlap matrix $\mathbf{S}$ can become ill-conditioned if sampled vectors are nearly linearly dependent, requiring pruning or regularization. SQD is most effective when low-energy eigenstates are concentrated on a modest set of configurations in the chosen basis and when the sampler can discover those configurations with feasible shot budgets; failure modes include sampling bias (important configurations have extremely small measurement probability), hardware noise that distorts the sample distribution, and uncontrolled subspace growth that increases classical cost or degrades conditioning. These considerations mirror the intuition behind selected configuration interaction methods; in quantum chemistry contexts, SQD is often discussed alongside quantum selected configuration interaction (QSCI).

> **Exercise 13.11** Assume SQD produces a nested sequence of subspaces $\mathcal{S}_1 \subset \mathcal{S}_2 \subset \cdots$ and Rayleigh–Ritz minima $E_m = \min_{|\psi\rangle \in \mathcal{S}_m,\ \||\psi\|=1} \langle\psi| H |\psi\rangle$. Prove that $\{E_m\}$ is nonincreasing and bounded below by the true ground energy E_0. Give one practical stopping criterion based on $E_{m-1} - E_m$.

> **Exercise 13.12** Identify two distinct failure modes for SQD: one due to insufficient sampling concentration and one due to an inadequate or poorly conditioned subspace basis. For each failure mode, propose one numerical diagnostic you could compute from classical data produced during the SQD workflow.

Key Takeaways

1. SQD is effective when low-energy eigenstates are concentrated on a modest set of configurations in the chosen basis.

2. Key failure modes include sampling bias (important configurations are rarely observed), noise-distorted samples, and uncontrolled subspace growth.

3. Classical post-processing can become dominant: dense diagonalization scales as $O(m^3)$, so pruning and conditioning control are essential.

13.5 * Unifying View and Further Reading

This and the preceding chapters presented several paradigms for extracting low-energy and optimization-relevant information from quantum systems. Table 13.1 summarizes these methods from a common viewpoint: *what the quantum device is asked to do* (coherent evolution, expectation estimation, or sampling) and *what the classical computer is asked to do* (post-processing, optimization, or reduced diagonalization).

A particularly useful discriminator is the role of measurement. In variational approaches, repeated measurements estimate $\langle\psi(\boldsymbol{\theta})| H |\psi(\boldsymbol{\theta})\rangle$ (or related objective functions) to drive a classical optimizer. In SQD, measurements primarily *identify a problem-adapted subspace* through sampling, after which classical diagonalization extracts low-lying spectral information. In contrast, phase-estimation methods shift most of the spectral extraction into a coherent quantum routine, at the cost of substantially deeper circuits and (typically) fault-tolerant resources.

Method	Quantum Role	Classical Role	Key Challenge

QPE	Coherent controlled evolutions and phase readout to estimate eigenvalues; provable convergence when the initial overlap is $\Omega(1/\operatorname{poly}(n))$.	Light post-processing of measurement outcomes.	Requires deep circuits and fault-tolerant processors.
QSVT	Implement signal-processing transformations of H to construct polynomial functions of H and related spectral filters.	Design polynomial approximants and analyze resource costs.	Typically requires block-encodings and deep fault-tolerant circuits.
VQE	Prepare $|\psi(\boldsymbol{\theta})\rangle$ and estimate $\langle\psi(\boldsymbol{\theta})|\,H\,|\psi(\boldsymbol{\theta})\rangle$ from measurements.	Optimize parameters $\boldsymbol{\theta}$ and process measurement data.	Measurement overhead and optimizer pathologies; no general convergence guarantee.
QAOA / VQAs	Prepare shallow structured ansatz states and estimate objective values from measurements.	Optimize parameters and select depth/schedules.	No general convergence guarantee and strongly problem-dependent performance.
AQC	Continuous-time evolution intended to track the ground state.	Choose schedules and parameters with modest post-processing.	Runtime depends on spectral gaps and requires long coherent evolution.
SQD	Prepare trial states and sample configurations (bitstrings) that define a reduced subspace; in the simplest form, no direct measurement of H is required.	Assemble a reduced eigenproblem on the sampled subspace and diagonalize to obtain low-lying spectrum.	Success depends on sampling concentration and subspace control; classical cost grows with m.

Table 13.1: A Unified Comparison of Key Quantum Algorithms

Key Takeaways

1. Across paradigms, the central design question is how to split work between quantum resources (depth vs. sampling) and classical resources (optimization vs. diagonalization).

2. Fault-tolerant methods emphasize deep coherent circuits with explicit

accuracy control, while near-term methods emphasize measurement-driven hybrid loops.

3. For a new application, the most informative first comparison is: circuit depth budget, measurement budget, and the dominant classical bottleneck.

✳ Exploration and References

The following references provide accessible entry points, broader context, and pointers to recent developments for the main algorithmic families discussed in this part.

Variational Algorithms (VQE and Related VQAs)

For a broad survey of variational quantum algorithms (including conceptual foundations and known pitfalls such as trainability and sampling cost), see the review by Cerezo *et al.* [69]. For a quantum-chemistry-focused and practice-oriented review of VQE (ansatz families, measurement strategies, and numerical considerations), see Tilly *et al.* [70]. For a wider NISQ-era perspective on hybrid algorithms and applications, see Bharti *et al.* [71].

QAOA and Structured VQAs

A useful entry point to QAOA performance, parameter structure, and implementation considerations is the detailed study by Zhou *et al.* [72].

Adiabatic Quantum Computing and Quantum Annealing

For a thorough review of adiabatic quantum computation (including the adiabatic condition, complexity considerations, and the relationship to annealing), see Albash and Lidar [73].

Fault-Tolerant Lens: Phase Estimation and QSVT

Quantum phase estimation is the canonical route to eigenvalues in the fault-tolerant setting, and it is often discussed together with Hamiltonian simulation. A modern unifying framework for many fault-tolerant linear-algebraic primitives is quantum singular value transformation (QSVT); see Gilyén *et al.* [41].

Sampling-Based Quantum Diagonalization (SQD)

SQD-style workflows are closely connected to selected configuration interaction ideas: quantum sampling identifies an important configuration set, after which a reduced eigenproblem is assembled and solved classically. For concrete SQD/QSCI algorithmic developments and chemistry-motivated demonstrations, see Kanno *et al.* [68] and Robledo-Moreno *et al.* [74]. For a recent theoretical treatment that formalizes SQD from an algorithmic/complexity viewpoint (and clarifies conditions under which the sampled subspace can succeed), see Piccinelli *et al.* [75].

Chapter Takeaways

1. Adiabatic quantum computing (AQC) prepares ground states by continuous-time evolution from a driver Hamiltonian H_D to a problem Hamiltonian H_P; the required runtime is governed by spectral gaps and, in the ideal model, assumes long coherent evolution.

2. Variational quantum eigensolver (VQE) is a hybrid loop: choose H and an ansatz $U(\theta)$, estimate $E(\theta)$ from measurements, and classically update θ; its practical cost is dominated by ansatz choice, measurement overhead, and optimization difficulty.

3. The quantum approximate optimization algorithm (QAOA) is a structured variational algorithm for Ising/QUBO objectives; it interpolates between circuit-based optimization and discretized adiabatic evolution through alternating "mixer" and "problem" generators.

4. Sampling-based quantum diagonalization (SQD) uses quantum sampling to build a problem-adapted low-dimensional subspace and then diagonalizes the reduced Hamiltonian classically, trading coherent depth for repeated sampling and classical postprocessing.

5. These approaches are especially relevant for near-term (NISQ) devices because they emphasize short circuits and measurement-driven robustness, allowing useful approximations even when fully fault-tolerant, high-depth algorithms are not yet practical.

Problem Set 13

13.1 Consider the MaxCut instance shown in Fig. 13.1. Let $G = (V, E)$ be the corresponding graph with $n = |V|$ vertices. Define the standard cost Hamiltonian

$$H_C = \sum_{(i,j)\in E} \frac{I - Z_i Z_j}{2},$$

so that for any computational basis state $|z\rangle$, the value $\langle z| H_C |z\rangle$ equals the cut size of z.

Design and carry out a small simulation study that compares multiple algorithmic approaches from this chapter on the same instance. You may use any software stack (e.g., Qiskit or PennyLane) and you may work entirely in statevector simulation.

(a) Compute the exact MaxCut optimum by exhaustive search over $z \in \{0,1\}^n$. Report the maximizing bitstring(s) and the optimal cut value.

(b) QAOA simulation. Implement QAOA for p from 2 to 10 with mixer $H_B = \sum_i X_i$. Optimize (γ, β) to maximize $\langle H_C \rangle$. Report the best achieved $\langle H_C \rangle$, the optimized parameters, and the most probable measured bitstring(s). Compare to the exact optimum.

(c) VQE-style simulation. Choose an ansatz family (e.g., a hardware-efficient layer pattern or a problem-inspired ansatz) and use the objective $-\langle H_C \rangle$ (or equivalently minimize $\langle |E| - H_C \rangle$). Optimize the parameters and report the best achieved expected cut value, together with the final measurement distribution. Compare your results to QAOA at the same approximate circuit depth.

(d) AQC-style simulation. Define an interpolation Hamiltonian

$$H(s) = (1-s)\left(-\sum_{i=1}^{n} X_i\right) + s\left(-H_C\right), \qquad s \in [0, 1],$$

so that the final ground state encodes the MaxCut optimum. Simulate
the Schrödinger evolution under a schedule $s(t)$ for total time T (you may
use a linear schedule and, optionally, a simple local-adiabatic schedule).
For each T, estimate the success probability of obtaining an optimal cut
bitstring upon measurement. Plot or tabulate success probability vs. T.

(e) SQD-style simulation. Using the same H_C, construct a reduced subspace
from samples as follows: run a low-depth circuit (e.g., a shallow QAOA
or random circuit) to generate a measurement distribution over bitstrings,
select the top m most frequent bitstrings, and let $\mathcal{S}$ be the subspace
spanned by $\{|z_1\rangle, \ldots, |z_m\rangle\}$. Form the projected matrix $(H_C)_{\mathcal{S}}$ and
diagonalize it classically. Report the best cut value observed within the
subspace and how it changes with m. Explain why $(H_C)_{\mathcal{S}}$ is trivial
to form in this basis, and discuss what changes if the objective were a
non-diagonal Hamiltonian.

(f) Summarize your findings in a short comparison table. Your table should
include (i) best achieved cut value (exact and expected), (ii) an estimate
of quantum resource usage (circuit depth or evolution time), and (iii) the
dominant classical workload (parameter optimization, time stepping, diagonalization, etc.). Based on this instance, state one practical advantage
and one limitation for each approach.

13.2 (QAOA)

(a) Give a concise conceptual explanation of why QAOA can be viewed as
a digitized version of adiabatic evolution, and clarify one key respect in
which the analogy can fail.

(b) Consider $H(s) = (1-s)H_M + sH_C$ and discretize the evolution into p
steps with a first-order Trotter approximation. Derive a mapping between
the step sizes and QAOA angles $\{\beta_\ell, \gamma_\ell\}_{\ell=1}^{p}$. Under a simple bounded-
commutator assumption, provide an error scaling estimate in terms of
the maximum step size.

13.3 (SQD)

(a) Explain why subspace methods such as SQD can, in favorable cases, avoid
direct measurement of a large Hamiltonian expectation value, and why
this can reduce measurement overhead.

(b) Consider a nested sequence of subspaces $\mathcal{S}_1 \subset \mathcal{S}_2 \subset \cdots$ with Rayleigh–
Ritz minima E_m. Prove that E_m is nonincreasing and bounded below
by the true ground energy. Then relate this monotonicity to a practical
stopping criterion and to a potential pitfall (premature stagnation).

13.4 (SQD and QSVT) Assume you have access to a small error-corrected device
with a limited number of logical qubits and moderately deep circuits.

(a) Propose a hybrid workflow that combines an SQD-style reduced subspace
construction with a QSVT-style spectral filter (conceptual design is
sufficient).

(b) Identify the main technical obstacles to making such a workflow rigorous and scalable (e.g., state preparation, block-encoding assumptions, resource estimates, and robustness to noise).

(c) Compare your proposed workflow to a more standard VQE pipeline on the same hardware regime, emphasizing where the quantum/classical burdens shift.

13.5 In SQD-style workflows, one often forms an approximate reduced eigenproblem from sampled configurations (bitstrings) and then diagonalizes classically.

(a) Propose a concrete criterion for deciding when to enlarge the sampled subspace (e.g., based on residual norms, stability of low-lying eigenpairs, or out-of-subspace weight).

(b) Provide a failure mode where the reduced problem becomes ill-conditioned, and propose at least one numerical stabilization technique (e.g., regularization, orthogonalization, truncation of small singular values).

(c) Let m be the subspace dimension. Give a rough scaling estimate for the classical cost of building and diagonalizing the reduced problem, and explain when this becomes the bottleneck.

13.6 ✳ Consider VQE with a hardware-efficient ansatz $U(\boldsymbol{\theta})$ and cost

$$E(\boldsymbol{\theta}) = \langle 0 | U(\boldsymbol{\theta})^\dagger H U(\boldsymbol{\theta}) | 0 \rangle .$$

(a) Explain why "barren plateaus" are a concern for certain ansatz families, and state one sufficient condition (informal is fine) under which gradients are expected to vanish exponentially in n.

(b) Propose two mitigation strategies that are compatible with the chapter's "shallow + hybrid" perspective, and describe one concrete diagnostic you would run to evaluate whether each mitigation is working.

13.7 ✳ Let $U(\theta) = e^{-i\theta P/2}$ with $P^2 = I$ and define $f(\theta) = \langle \psi | U(\theta)^\dagger H U(\theta) | \psi \rangle$.

(a) Prove a parameter-shift identity for $f'(\theta)$ and extend it to the multi-parameter setting $\boldsymbol{\theta} = (\theta_1, \ldots, \theta_L)$ when $U(\boldsymbol{\theta})$ is a product of such one-parameter Pauli rotations.

(b) Assume you want an ℓ_2-accurate gradient estimate $\|\widehat{\nabla f} - \nabla f\|_2 \leq \varepsilon$ with high probability. Give a shot-complexity estimate (up to polylog factors) in terms of L and ε under a simple bounded-variance model.

13.8 ✳ Consider an AQC algorithm with $H(s) = (1 - s)H_0 + sH_1$ and schedule $s(t)$.

(a) Starting from the representative condition Eq. 13.5, derive an expression for a local-adiabatic schedule that attempts to keep the adiabatic ratio bounded by a target $\eta \ll 1$.

(b) Give an example (realistic or toy) in which the matrix element $\left| \langle E_1(s) | \frac{dH}{ds} | E_0(s) \rangle \right|$ becomes small near the minimum gap, and explain how this changes the runtime estimate compared to a naive $1/\Delta_{\min}^2$ scaling.

13.9 Let $G = (V, E)$ be a graph and consider MaxCut with

$$H_C = \sum_{(i,j) \in E} \frac{I - Z_i Z_j}{2}, \qquad H_B = \sum_{i \in V} X_i.$$

For $p = 1$ QAOA on an n-vertex graph, the state is

$$|\psi(\gamma, \beta)\rangle = e^{-i\beta H_B} e^{-i\gamma H_C} |+\rangle^{\otimes n}.$$

(a) Show that $\langle H_C \rangle (\gamma, \beta)$ can be expressed as a sum over edges where each edge contribution depends only on a constant-size neighborhood of that edge.

(b) Specialize to a d-regular graph and explain why, at $p = 1$, all edges contribute the same amount by symmetry. Derive a simplified expression for $\langle H_C \rangle (\gamma, \beta)$ in terms of d, γ, β.

13.10 ✳ Consider the same QAOA setting as above. Let Π be any permutation of vertices that preserves the graph (an automorphism).

(a) Show that $\langle H_C \rangle (\gamma, \beta)$ is invariant under such automorphisms.

(b) Explain how this invariance can be used to reduce the dimension of a classical search or to simplify analytic evaluation on highly symmetric graphs.

13.11 ✳ A common practical question in QAOA is whether increasing depth p meaningfully increases expressivity.

(a) Give a concrete example of a graph family (e.g., cycles, complete graphs, or bipartite graphs) where $p = 1$ is provably limited or empirically limited, and explain the mechanism behind the limitation.

(b) Propose a principled way to compare p vs. $p + 1$ that does not rely only on the final optimized objective value (e.g., parameter landscape diagnostics, concentration of measure, or circuit-lightcone considerations).

13.12 ✳ Let H be a Hermitian matrix and let $\mathcal{S}$ be a subspace with orthonormal basis columns of V. Define $H_{\mathcal{S}} = V^{\dagger} H V$.

(a) Prove the Rayleigh–Ritz variational principle in this setting: the smallest eigenvalue of $H_{\mathcal{S}}$ upper-bounds the true ground energy of H.

(b) Suppose $\mathcal{S}$ is spanned by states obtained from sampling a probability distribution over computational basis strings. Explain how sampling bias can lead to a systematic (not merely statistical) error in the Rayleigh–Ritz estimate.

14. Measurement Primitives

Contents

Many quantum algorithms aim to prepare a state from which one extracts a small number of scalars, such as $\langle O \rangle$, an event probability, or an overlap, rather than the full state description. This "small-output" viewpoint is already central in quantum linear systems: the HHL family is best understood as a state-preparation primitive whose value depends on how efficiently one can *read out* the downstream quantity of interest (see § 12.1.1). In practice, the readout layer is not an afterthought. It often dominates the end-to-end cost, and it strongly shapes which algorithms are realistic in near-term settings.

This chapter gathers a set of reusable measurement and estimation primitives that appear throughout the book. We begin by organizing common output targets and by quantifying the basic shot-noise scaling that underlies sampling-based estimators. We then develop three widely used families of primitives.

1. Estimating Pauli observables and Hamiltonian expectation values, including Pauli decomposition, commuting-group measurement, and practical shot-allocation ideas.

2. Interference-based primitives that trade shots for controlled operations, focusing on the Hadamard-test family for quantities of the form $\langle\psi|\,U\,|\psi\rangle$ and the swap test for overlaps and purities.

3. Classical shadows and randomized measurement, which amortize measurement effort when many observables must be estimated from repeated preparations of the same state.

The goal is to provide a "toolbox" that later chapters can cite directly, so that discussions of variational algorithms, quantum simulation, and linear-systems applications can focus on modeling choices and resource drivers rather than re-deriving readout techniques.

14.1 Readout Targets and Measurement Costs

Readout is often the dominant "last mile" in quantum algorithms: once a state is prepared, we still need to estimate the quantity of interest from repeated measurements. This section clarifies what we mean by algorithmic output and summarizes the basic scaling laws that govern measurement cost.

14.1.1 What Counts as Output

1 Observables, Events, and Overlaps

In the circuit model, a quantum computation ends with measurement outcomes. In algorithm design, however, it is often clearer to describe the *output* as the mathematical quantity that those outcomes estimate.

The simplest and most common target is an expectation value. If the algorithm prepares a state $|\psi\rangle$ and we wish to evaluate an observable O, we write

$$\langle O\rangle \equiv \langle\psi|\,O\,|\psi\rangle. \tag{14.1}$$

Here O can be a Pauli string, a Hamiltonian H, a projector onto a subspace, or any efficiently measurable operator in the chosen access model.

A second common output type is an *event probability*. Given a projector Π that encodes an event of interest (for example, "the ancilla is 1" or "the measured bitstring lies in a set S"), the desired quantity is

$$p = \langle\psi|\,\Pi\,|\psi\rangle. \tag{14.2}$$

Sampling-based readout estimates p by repeating state preparation and measurement. Many algorithmic success probabilities and acceptance tests can be expressed in this form.

A third output type is an *overlap* or *fidelity*. Given another efficiently preparable state $|\phi\rangle$, one may seek the inner product $\langle\phi|\psi\rangle$ or the fidelity

$$F(\phi,\psi) = |\langle\phi|\psi\rangle|^2. \tag{14.3}$$

Overlap targets appear in benchmarking, subspace methods, quantum kernels, and other "compare-two-states" tasks. Their most direct estimators are often interference-based rather than simple basis measurements.

These categories are not disjoint: an overlap can be written as an expectation value of a suitable observable on an extended system, and an event probability is the expectation value of a projector. The main point is that algorithmic output is typically a *small set of numbers* with a specified accuracy requirement, and the measurement primitive is chosen to minimize the total cost of producing those numbers.

14.1.2 Shot Noise, Estimators, and Shot Complexity

Any measurement-based output is statistical: it is computed from a finite number of repeated runs (shots). In this subsection we fix notation for the *true* expectation value, the *sample* (empirical) mean, and the sampling uncertainty that results from finite N.

 For the requisite background on probability and statistics, see QC Math [1], Chapter 14: *Fundamentals of Probability*.

1 Notation for True Mean, Samples, and Variance

Assume the algorithm prepares a state $|\psi\rangle$ and we wish to estimate the expectation value of an observable O. The *true mean* is

$$\langle O \rangle \equiv \langle \psi|O|\psi \rangle .$$

A single shot produces a random outcome, which we denote by

$$O_j \equiv \langle O \rangle_j ,$$

or simply O when it is unambiguous from context. The *single-shot variance* is

$$\mathrm{Var}(O) = \langle O^2 \rangle - \langle O \rangle^2 . \tag{14.4}$$

Given N independent shots, the *sample mean* (empirical expectation value) is

$$\overline{\langle O \rangle} = \frac{1}{N} \sum_{j=1}^{N} O_j. \tag{14.5}$$

By the law of large numbers, the empirical mean converges to the true mean as the number of shots grows:

$$\overline{\langle O \rangle} \to \langle O \rangle \quad \text{as} \quad N \to \infty.$$

Moreover, the sample mean is unbiased for every N, meaning that its expected value (over repeated experiments with N shots) equals the true mean:

$$\mathbb{E}\left[\overline{\langle O \rangle}\right] = \langle O \rangle .$$

2 Variance of the Sample Mean and the SEM

If the N shots are independent and identically distributed, then the variance of the sample mean satisfies

$$\mathrm{Var}\left(\overline{\langle O \rangle}\right) = \mathrm{Var}\left(\frac{1}{N} \sum_{j=1}^{N} O_j\right) = \frac{1}{N^2} \mathrm{Var}\left(\sum_{j=1}^{N} O_j\right).$$

Independence implies that variances add (intuitively, the cross terms average out), so $\operatorname{Var}\left(\sum_{j=1}^{N} O_j\right) = \sum_{j=1}^{N} \operatorname{Var}(O_j)$, and therefore

$$\operatorname{Var}\left(\overline{\langle O\rangle}\right) = \frac{1}{N^2} \sum_{j=1}^{N} \operatorname{Var}(O_j) = \frac{\operatorname{Var}(O)}{N}. \tag{14.6}$$

The corresponding *standard error of the mean* (SEM) is

$$\operatorname{SEM}\left(\overline{\langle O\rangle}\right) \equiv \sqrt{\operatorname{Var}\left(\overline{\langle O\rangle}\right)} = \sqrt{\frac{\operatorname{Var}(O)}{N}}. \tag{14.7}$$

The SEM, proportional to $1/\sqrt{N}$, is a convenient proxy for the typical sampling error scale: to reduce the typical error by a factor of 10, one needs about 100 times as many shots.

3   Confidence Intervals and Shot Complexity

For large N, the central limit theorem implies that the sample mean is approximately normally distributed:

$$\overline{\langle O\rangle} \approx \mathcal{N}\left(\mu = \langle O\rangle,\ \sigma = \operatorname{SEM}\left(\overline{\langle O\rangle}\right)\right). \tag{14.8}$$

Using the normal approximation, a two-sided confidence interval at confidence level $1 - \delta$ takes the form

$$\overline{\langle O\rangle} \approx \langle O\rangle \pm z_\delta \operatorname{SEM}\left(\overline{\langle O\rangle}\right), \tag{14.9}$$

where z_δ is the standard-normal quantile (e.g., $z_{0.32} \approx 1$ for 68% and $z_{0.05} \approx 1.96$ for 95%).

Consequently, to achieve additive error at most ϵ with confidence $1 - \delta$, it suffices (up to the quality of the normal approximation) to choose

$$N \gtrsim \frac{z_\delta^2 \operatorname{Var}(O)}{\epsilon^2}. \tag{14.10}$$

When $\operatorname{Var}(O)$ is not known in advance, one may estimate it from a small pilot sample or use a conservative upper bound based on the spectrum of O (for Pauli outcomes, $\operatorname{Var}(O) \leq 1$).

4   Single Qubit Measurements

Now suppose $|\psi\rangle = \alpha\,|0\rangle + \beta\,|1\rangle$ with $|\alpha|^2 + |\beta|^2 = 1$. We compare two closely related observables: the Pauli-Z operator and the projector onto $|0\rangle$,

$$Z = |0\rangle\langle 0| - |1\rangle\langle 1|, \qquad \Pi \equiv \frac{I + Z}{2} = |0\rangle\langle 0|. \tag{14.11}$$

A computational-basis measurement produces a classical bit $b \in \{0, 1\}$. Interpreting the same readout as a ± 1 variable yields a Z measurement, while interpreting it as a $\{0, 1\}$ indicator yields a Π measurement. Table 14.1 summarizes the associated statistics.

Quantity	$O = Z$	$O = \Pi$
Outcomes	$-1, 1$	$0, 1$
Probabilities	$\Pr(+1) = \|\alpha\|^2$, $\Pr(-1) = \|\beta\|^2$	$\Pr(1) = \|\alpha\|^2$, $\Pr(0) = \|\beta\|^2$
True mean	$\langle Z \rangle = \|\alpha\|^2 - \|\beta\|^2 = 2\|\alpha\|^2 - 1$	$\langle \Pi \rangle = \|\alpha\|^2 = \dfrac{1 + \langle Z \rangle}{2}$
Second moment	$\langle Z^2 \rangle = 1$	$\langle \Pi^2 \rangle = \langle \Pi \rangle = \|\alpha\|^2$
Single-shot variance	$\mathrm{Var}(Z) = 1 - \langle Z \rangle^2 = 4\|\alpha\beta\|^2$	$\mathrm{Var}(\Pi) = \langle \Pi \rangle - \langle \Pi \rangle^2 = \|\alpha\|^2\|\beta\|^2$
Sample mean (empirical)	$\overline{\langle Z \rangle} = \dfrac{1}{N} \sum_{j=1}^{N} z_j,\ z_j \in \{\pm 1\}$	$\overline{\langle \Pi \rangle} = \dfrac{1}{N} \sum_{j=1}^{N} \pi_j,\ \pi_j \in \{0, 1\}$
Variance of sample mean	$\dfrac{\mathrm{Var}(Z)}{N} = \dfrac{4\|\alpha\beta\|^2}{N}$	$\dfrac{\mathrm{Var}(\Pi)}{N} = \dfrac{\|\alpha\|^2\|\beta\|^2}{N}$
SEM	$\sqrt{\dfrac{\mathrm{Var}(Z)}{N}} = \dfrac{2\|\alpha\beta\|}{\sqrt{N}}$	$\sqrt{\dfrac{\mathrm{Var}(\Pi)}{N}} = \dfrac{\|\alpha\|\|\beta\|}{\sqrt{N}}$

Table 14.1: Statistics for Z and Π Measurements on $|\psi\rangle = \alpha|0\rangle + \beta|1\rangle$

Exercise 14.1 Let O be a Hermitian observable with spectrum contained in $[a, b]$. One shot produces an outcome $O_j \in [a, b]$ and the empirical mean is $\overline{\langle O \rangle} = \frac{1}{N} \sum_{j=1}^{N} O_j$.

(a) Show that $\mathrm{Var}(O) \leq \frac{(b-a)^2}{4}$.

(b) Using the normal approximation, derive a sufficient condition on N so that $|\overline{\langle O \rangle} - \langle O \rangle| \leq \epsilon$ with confidence $1 - \delta$.

Key Takeaways

1. Readout is statistical: each circuit execution produces a random outcome, so precision is achieved by repetition.

2. For an observable O, the empirical mean $\overline{\langle O \rangle}$ is unbiased: $\mathbb{E}[\overline{\langle O \rangle}] = \langle O \rangle$.

3. The variance of the sample mean scales as $\mathrm{Var}(\overline{\langle O \rangle}) = \mathrm{Var}(O)/N$, so the typical sampling error (SEM) scales as $1/\sqrt{N}$.

4. A normal-approximation confidence interval gives $N \gtrsim z_\delta^2\,\mathrm{Var}(O)/\epsilon^2$ to achieve additive error ϵ with confidence $1 - \delta$.

14.1.3 A Decision Guide for Measurement Primitives

Measurement primitives can be viewed as different answers to the same question: given a state-preparation procedure for $|\psi\rangle$, how can we estimate the desired scalar(s) with minimal total cost? The choice depends on the number and type of targets, and on whether controlled operations are available.

1 Few Targets vs. Many Targets

If only a few expectation values are needed, direct estimation of those specific observables is usually simplest. In Pauli-based settings, this often means measuring a small set of Pauli strings, possibly with commuting-group measurement to reduce the number of distinct measurement settings.

If one needs *many* expectation values of different observables on the same underlying state (for example, many correlators or many Hamiltonian terms), it can be advantageous to use a scheme that amortizes measurement effort across targets. Classical shadows and related randomized measurement methods are designed for this regime: one collects a single dataset of randomized measurements and then post-processes it to estimate many observables.

2 Diagonal vs. Non-Diagonal Targets

If the target observable is diagonal in the computational basis—equivalently, if the output is a classical function of the measured bitstring—then readout can sometimes be done by straightforward sampling in the computational basis. This is common in optimization problems where the cost function is diagonal in Z and can be evaluated classically on sampled bitstrings.

For non-diagonal observables, one typically needs basis changes (for example, single-qubit rotations for local Paulis) or more elaborate coherent procedures. The diagonal/non-diagonal distinction is often the difference between "sample-and-evaluate" readout and genuine quantum measurement design.

3 Destructive Sampling vs. Coherent Tests

Most near-term readout is *destructive*: one measures the state and uses the resulting samples to build an estimator. Coherent tests follow a different pattern. They introduce an ancilla and use interference to convert the desired quantity into an ancilla measurement statistic, at the price of implementing controlled operations.

The Hadamard-test family estimates quantities of the form $\langle \psi | U | \psi \rangle$ (and hence many expectation values and overlaps when U is chosen appropriately). The swap test estimates overlaps and purities by coherently comparing two states.

These primitives can be attractive when controlled operations are available and when the target is naturally expressed as an interference quantity. In near-term settings, however, their depth and control requirements may outweigh their benefits, and sampling-based methods or randomized measurements may be preferable.

14.2 Estimating Pauli Observables and Hamiltonians

A large fraction of near-term quantum algorithms reduce readout to estimating expectation values of Pauli observables. This happens explicitly in the variational quantum eigensolver (VQE), where one minimizes $\langle H \rangle$ over a parameterized family of states, but the same measurement pattern appears whenever an observable can be expanded as a linear combination of Pauli strings:

$$H = \sum_{k=1}^{m} c_k P_k, \tag{14.12}$$

with real coefficients c_k and Pauli strings P_k. The target output is

$$\langle H \rangle = \sum_{k=1}^{m} c_k \langle P_k \rangle. \tag{14.13}$$

A basic strategy is to estimate each $\langle P_k \rangle$ (possibly after grouping commuting terms) and then combine the estimates classically. The overall cost depends on two choices: how many distinct measurement settings are used, and how shots are allocated among settings and terms.

A useful rule of thumb is that terms with larger $|c_k|$ and larger intrinsic variance deserve more measurement effort. Grouping can reduce the number of settings, while shot allocation trades precision among terms to minimize total shots for a target error. We develop these ideas more systematically in this section.

 This section builds on QC Math [1], Chapter 12: *Pauli Matrices, Strings, and Groups*. See also § 10.3.1.

14.2.1 Pauli Decomposition and Term-Wise Estimation

1 From a Pauli Expansion to Shot Budgets

Assume the algorithm prepares $|\psi\rangle$ and we wish to estimate $\langle H \rangle = \langle \psi | H | \psi \rangle$ for a Hermitian operator H. In many applications, H is specified (or approximated) by a Pauli expansion, so that $\langle H \rangle$ becomes a weighted sum of Pauli-string expectation values. We now quantify how sampling error propagates through this sum.

Fix a shot allocation $(N_1, \ldots, N_m)$. Here m is the number of Pauli strings in the expansion of H, and N_k is the number of measurement shots allocated to estimating $\langle P_k \rangle$ (so $N_{\text{tot}} = \sum_{k=1}^{m} N_k$). For each Pauli string P_k, measure P_k in its eigenbasis N_k times, obtaining outcomes

$$(P_k)_1, \ldots, (P_k)_{N_k} \in \{\pm 1\}.$$

The empirical mean for term k is

$$\overline{\langle P_k \rangle} = \frac{1}{N_k} \sum_{j=1}^{N_k} (P_k)_j, \tag{14.14}$$

which estimates the true mean $\langle P_k \rangle = \langle \psi | P_k | \psi \rangle$. The corresponding empirical estimator of $\langle H \rangle$ is the same linear combination with empirical means,

$$\overline{\langle H \rangle} = \sum_{k=1}^{m} c_k \overline{\langle P_k \rangle}. \tag{14.15}$$

This estimator is unbiased: $\mathbb{E}[\overline{\langle H \rangle}] = \langle H \rangle$. If the shot pools used for different terms are independent, then the variance adds across terms and we obtain

$$\text{Var}\left(\overline{\langle H \rangle}\right) = \sum_{k=1}^{m} c_k^2 \, \text{Var}\left(\overline{\langle P_k \rangle}\right) = \sum_{k=1}^{m} c_k^2 \, \frac{\text{Var}(P_k)}{N_k}. \tag{14.16}$$

For Pauli outcomes, $\mathrm{Var}(P_k) = 1 - \langle P_k \rangle^2 \leq 1$, so a convenient worst-case bound is

$$\mathrm{Var}\left(\overline{\langle H \rangle}\right) \leq \sum_{k=1}^{m} \frac{c_k^2}{N_k}, \qquad \mathrm{SEM}\left(\overline{\langle H \rangle}\right) \leq \sqrt{\sum_{k=1}^{m} \frac{c_k^2}{N_k}}. \tag{14.17}$$

These expressions make the main resource drivers explicit: large coefficients $|c_k|$ amplify sampling uncertainty, and increasing N_k reduces it.

> **Exercise 14.2** Fix a Pauli string P with outcomes ± 1, measured N times on $|\psi\rangle$.
>
> (a) Express $\mathrm{Var}(P)$ in terms of $\langle P \rangle$.
>
> (b) Give a worst-case upper bound on $\mathrm{SEM}(\overline{\langle P \rangle})$ that does not depend on $\langle P \rangle$.
>
> (c) Interpret the worst case in terms of the distribution of outcomes.

> **Exercise 14.3** Let $H = \sum_{k=1}^{m} c_k P_k$ and suppose each P_k is measured independently with N_k shots.
>
> (a) Show that $\mathrm{Var}(\overline{\langle H \rangle}) = \sum_{k=1}^{m} c_k^2 \mathrm{Var}(P_k)/N_k$ under independent shot pools.
>
> (b) Derive the worst-case bound $\mathrm{Var}(\overline{\langle H \rangle}) \leq \sum_k c_k^2/N_k$.

Key Takeaways

1. If $H = \sum_{k=1}^{m} c_k P_k$, then $\langle H \rangle = \sum_k c_k \langle P_k \rangle$ can be estimated by measuring Pauli strings and combining results classically.

2. With independent shot pools, $\mathrm{Var}(\overline{\langle H \rangle}) = \sum_k c_k^2 \mathrm{Var}(P_k)/N_k$, and for Pauli outcomes $\mathrm{Var}(P_k) \leq 1$.

3. Large coefficients $|c_k|$ and high-variance terms dominate the measurement cost.

14.2.2 Commuting Groups and Measurement Reduction

Many Pauli strings can be estimated from the same measurement data if they are jointly measurable. Grouping terms into jointly measurable sets can reduce the number of distinct measurement settings, often substantially. In practice, constructing good groups is a heuristic optimization problem, and different grouping strategies trade fewer settings against deeper basis-change circuits and different statistical behavior.

The central idea is that if a set of Pauli strings can be measured in the same basis, then a single circuit execution yields samples for all of them. This motivates partitioning $\{P_k\}_{k=1}^{m}$ into groups that are jointly measurable.

Two settings are common in practice. Qubit-wise commutativity groups are easy to measure with only single-qubit rotations, while fully commuting groups can be larger but may require entangling (often Clifford) measurement circuits. The practical tradeoff is fewer measurement settings vs. more complex per-setting circuits, with shot allocation and error budgeting still necessary in either approach.

1 Qubit-Wise Commutativity (QWC)

Qubit-wise commutativity is a *sufficient* condition for joint measurement using only single-qubit basis rotations. Write two Pauli strings as

$$P = \bigotimes_{q=1}^{n} \sigma_q, \qquad Q = \bigotimes_{q=1}^{n} \tau_q, \qquad \sigma_q, \tau_q \in \{I, X, Y, Z\}.$$

They are qubit-wise commuting if, for every qubit q, either $\sigma_q = I$, $\tau_q = I$, or $\sigma_q = \tau_q$. In that case, there exists a *single product basis* in which both P and Q (and any QWC group built from them) are diagonal. Operationally, for each qubit choose the measurement axis Z, X, or Y according to the (common) non-identity Pauli appearing in the group on that qubit, and then apply a local basis change

$$U_{\text{basis}} = \bigotimes_{q=1}^{n} U_q \tag{14.18}$$

followed by computational-basis measurement, where U_q maps the chosen axis to Z (e.g., H for X-measurement, and $S^\dagger H$ for Y-measurement). A single shot then provides ± 1 outcomes that can be combined to estimate every Pauli string in the group.

A Concrete Example

Consider the three Pauli strings on two qubits

$$P_1 = Z_1 X_2, \qquad P_2 = Z_1, \qquad P_3 = X_2.$$

These operators are qubit-wise commuting and therefore jointly measurable using only single-qubit basis rotations. A single measurement setting is obtained by measuring qubit 1 in the Z basis and qubit 2 in the X basis (that is, apply a Hadamard gate on qubit 2 and then measure both qubits in the computational basis). Each shot produces a bitstring $(b_1, b_2) \in \{0, 1\}^2$, from which we infer the ± 1 outcomes

$$(Z_1)_j = (-1)^{b_1}, \qquad (X_2)_j = (-1)^{b_2}, \qquad (Z_1 X_2)_j = (Z_1)_j (X_2)_j.$$

Thus, the *same* shot record simultaneously contributes to the empirical means $\overline{\langle Z_1 X_2 \rangle}$, $\overline{\langle Z_1 \rangle}$, and $\overline{\langle X_2 \rangle}$.

2 Full Commutativity (FC)

Full commutativity requires $PQ = QP$ as matrices, but it does *not* guarantee a shared tensor-product eigenbasis. For Pauli strings, P and Q commute if the number of qubits on which σ_q and τ_q are both non-identity and different (hence anticommute) is even. Such a commuting set is simultaneously diagonalizable in principle, but the joint eigenbasis may be *entangled*. Measuring an FC group in one setting can therefore require an *entangling* basis change that maps the commuting algebra to Z-type operators before measurement. Concretely, one seeks a Clifford circuit U such that each operator in the group is transformed to a product of Z operators,

$$U P_j U^\dagger \in \{\pm Z\text{-type strings}\}, \tag{14.19}$$

and then measures in the computational basis. This diagonalization approach can reduce the number of settings compared with QWC grouping, but it introduces extra circuit depth and noise sensitivity, so the net benefit depends on hardware and error rates.

> **Exercise 14.4** You are given Pauli strings $\{Z_1Z_2,\ Z_2Z_3,\ X_1X_2,\ X_2X_3,\ Z_1,\ Z_3\}$. Partition them into qubit-wise commuting groups, and state a measurement basis (product of single-qubit bases) for each group.

Key Takeaways

---

1. Grouping reduces the number of distinct measurement settings by estimating multiple Pauli strings from the same shot record.

2. Qubit-wise commutativity (QWC) guarantees joint measurability using only single-qubit basis rotations.

3. Full commutativity (FC) allows larger groups in principle but may require entangling (often Clifford) basis changes, increasing circuit depth and noise sensitivity.

4. Fewer settings does not eliminate shot noise; one still needs shot allocation and error budgeting at the group or term level.

14.2.3 Shot Allocation Heuristics

Once the measurement settings are fixed (term-wise or grouped), the next question is how to distribute a finite shot budget. Shot allocation is a central practical knob in VQE and related workflows, because per-iteration measurement can dominate runtime.

The variance expressions above suggest the guiding principle: allocate more shots to terms (or groups) that contribute most to the uncertainty in $\overline{\langle H \rangle}$. In term-wise estimation, the dominant contributions are typically associated with large coefficients $|c_k|$ and with high-variance terms, where $\langle P_k \rangle \approx 0$.

Common heuristics include:

- *Coefficient-based allocation.* Allocate shots proportional to $|c_k|$ or c_k^2. This is simple and often effective when the coefficient distribution is highly nonuniform.

- *Variance-aware allocation.* Use a small pilot sample to estimate $\mathrm{Var}(P_k) \approx 1 - \overline{\langle P_k \rangle}^2$, then allocate shots based on both $|c_k|$ and the estimated variability. This adapts to the current state $|\psi\rangle$ and can reduce wasted shots on near-deterministic terms.

- *Group-level allocation.* When measuring commuting groups, allocate shots at the level of settings. One can score each group by coefficient magnitudes and empirical variability, then distribute N_ℓ across groups accordingly.

- *Importance sampling over terms.* Instead of measuring all terms every iteration, randomly sample a subset of terms using a distribution biased toward large $|c_k|$ (and possibly large estimated variance), and form an unbiased estimator of $\langle H \rangle$. This reduces per-iteration cost but introduces additional stochasticity into the optimization loop.

In many practical workflows, a staged approach works well: start with coarse coefficient-based budgeting early in the optimization (when parameter updates are large), then switch to variance-aware allocation near convergence, when higher precision is needed to reliably compare candidate updates.

> **Key Takeaways**
>
> ---
>
> 1. Shot allocation is a variance-minimization problem: spend shots where they most reduce $\mathrm{Var}(\overline{\langle H \rangle})$.
>
> 2. A reliable heuristic is to favor terms (or groups) with large $|c_k|$ and large estimated variance.
>
> 3. In practice, a staged strategy often works well: coarse coefficient-based allocation early, then variance-aware allocation near convergence.

14.3 Interference-Based Estimation Primitives

Sampling in a fixed measurement basis is the default readout model in many near-term workflows. A different family of primitives uses *interference*: we couple the state of interest to an ancilla and arrange a controlled operation so that the desired quantity appears as a bias in the ancilla measurement statistics. These methods can be viewed as "coherent estimators." They are especially natural when the target is an amplitude, an inner product, or an expectation value of a unitary. The price is that they typically require controlled operations (controlled-U or controlled-SWAP), which can be deep or hardware-demanding.

14.3.1 Hadamard Test: Estimating $\langle \psi | U | \psi \rangle$

Let $|\psi\rangle$ be a state that we can prepare efficiently, and let U be a unitary that we can implement efficiently. The Hadamard test estimates the complex number $\langle \psi | U | \psi \rangle$ by converting it into the expectation value of a single-qubit Pauli observable on an ancilla.

1 Real and Imaginary Parts

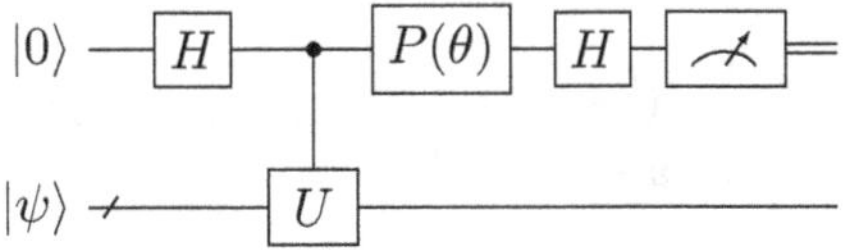

Figure 14.1: Hadamard Test Circuit

Prepare an ancilla qubit in $|0\rangle$ and apply a Hadamard gate, then apply controlled-U with the ancilla as the control and the system as the target, and finally apply another Hadamard on the ancilla before measuring it in the computational basis, as illustrated in Fig. 14.1. Let Z_a denote the Pauli-Z observable on the ancilla, and let $P(\theta)$ be a single-qubit phase rotation on the ancilla (with $\theta = 0$ in Fig. 14.1). A standard calculation (see § 7.3.1) shows that

$$\langle Z_a \rangle = \mathrm{Re}(\langle \psi | U | \psi \rangle), \qquad \text{with } \theta = 0. \tag{14.20}$$

Equivalently, if we measure the ancilla bit $b \in \{0, 1\}$ and map it to the ± 1 outcome $(-1)^b$, then the sample mean estimates $\mathrm{Re}(\langle \psi | U | \psi \rangle)$.

To extract the imaginary part, one inserts a phase on the ancilla (take $\theta = \pm\frac{\pi}{2}$ in Fig. 14.1, or equivalently prepare the ancilla in $|+i\rangle = \frac{1}{\sqrt{2}}(|0\rangle + i|1\rangle)$). The ancilla Z expectation then satisfies

$$\langle Z_a \rangle = \mathrm{Im}(\langle \psi | U | \psi \rangle), \qquad \text{with } \theta = \pm\frac{\pi}{2}, \tag{14.21}$$

up to a sign convention determined by where the phase gate is placed. In short, with minor modifications, the Hadamard test estimates either $\mathrm{Re}(\langle \psi | U | \psi \rangle)$ or $\mathrm{Im}(\langle \psi | U | \psi \rangle)$ using repeated measurements of a single ancilla qubit.

A useful special case occurs when U is both unitary and Hermitian (for example, a Pauli string P). Then $\langle \psi | U | \psi \rangle \in \mathbb{R}$ and the Hadamard test directly estimates $\langle U \rangle = \langle \psi | U | \psi \rangle$ without the $\theta = \frac{\pi}{2}$ setting. In that sense, the Hadamard test can be viewed as a coherent alternative to basis-rotation Pauli measurement.

2 When Controlled-U Is Reasonable

The Hadamard test trades measurement effort for coherent control. The shot-noise scaling is still governed by averaging ±1 outcomes on the ancilla, so additive error ϵ typically costs $\Theta(1/\epsilon^2)$ shots, as in § 14.1.2. The difference is that each shot requires implementing controlled-U.

In fault-tolerant settings, controlled operations are often available as a basic capability, and one can sometimes implement controlled-U with modest overhead if U is already synthesized in a reversible form. In that regime, the Hadamard test is a natural primitive for estimating amplitudes, inner products, and expectation values of unitary observables.

In near-term (NISQ) settings, however, controlled-U can be the dominant cost. If U is deep, has large two-qubit gate count, or requires non-native connectivity, the controlled version may introduce substantial overhead and noise sensitivity. As a result, the Hadamard test is most attractive on NISQ hardware when U is short and structured (for example, small Pauli strings, shallow Clifford circuits, or modest-depth problem-specific unitaries). Otherwise, destructive sampling methods (basis-rotation Pauli measurement, randomized measurement, or classical shadows) may yield better accuracy per unit runtime.

Exercise 14.5 Let the system register be prepared in $|\psi\rangle$ and consider the Hadamard test for a unitary U.

(a) Show that the ancilla measurement statistic can be written as an expectation value on the joint space of the ancilla and system registers:

$$\langle Z_a \rangle = \langle 0 | \langle \psi | A | 0 \rangle | \psi \rangle,$$

for a suitable Hermitian operator A acting on the ancilla⊗system space.

(b) Specialize to the case $U = P$ for a Pauli string P and compare the Hadamard-test readout to direct Pauli measurement, meaning: apply single-qubit basis rotations that diagonalize P and then measure the system register in the computational basis to obtain ±1 outcomes whose sample mean estimates $\langle P \rangle$.

Key Takeaways

1. The Hadamard test estimates $\langle\psi|\,U\,|\psi\rangle$ by converting it into an ancilla bias: $\langle Z_a \rangle = \mathrm{Re}(\langle\psi|\,U\,|\psi\rangle)$ (and, with a phase tweak, the imaginary part).

2. Shot complexity still scales as $\Theta(1/\epsilon^2)$ for additive error ϵ, but each shot requires controlled-U.

3. The Hadamard test is most attractive when controlled-U is not too costly (often in fault-tolerant settings or for shallow/structured U).

14.3.2 Swap Test: Estimating Overlaps and Purity

The swap test estimates how close two quantum states are by converting their overlap into an ancilla measurement statistic. It is widely used as a building block for overlap estimation, fidelity tests, and purity estimation.

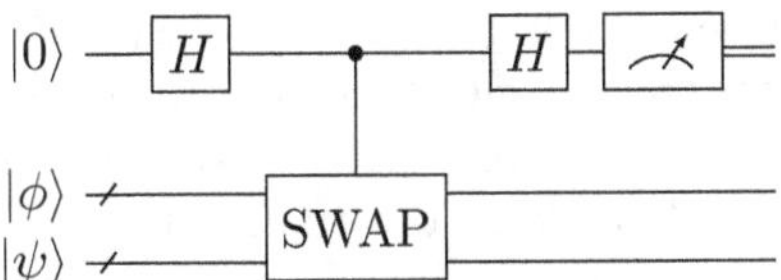

Figure 14.2: Swap Test Circuit

1 Overlap as a Measurement Target

Suppose we can prepare two states $|\psi\rangle$ and $|\phi\rangle$ on equal-size registers, as shown in Fig. 14.2. The swap test uses an ancilla to perform a controlled-SWAP (Fredkin) gate that swaps the two registers conditioned on the ancilla. Measuring the ancilla in the computational basis *after* the final Hadamard (equivalently, measuring the ancilla in the X basis *without* that final Hadamard) yields a Bernoulli outcome whose bias encodes the squared overlap:

$$\Pr(0) = \frac{1 + |\langle\phi|\psi\rangle|^2}{2}, \qquad \Pr(1) = \frac{1 - |\langle\phi|\psi\rangle|^2}{2}. \tag{14.22}$$

Thus,

$$|\langle\phi|\psi\rangle|^2 = 2\Pr(0) - 1 = \langle Z_a \rangle, \tag{14.23}$$

where Z_a is the ancilla Pauli-Z measured after the final Hadamard (with the usual ± 1 mapping). The fidelity $F(\phi,\psi) = |\langle\phi|\psi\rangle|^2$ is therefore accessible from repeated swap-test runs.

As a derivation, set $U = \mathrm{SWAP}$ and $|\Psi\rangle = |\phi\rangle\,|\psi\rangle$ in the Hadamard-test identity $\langle Z_a \rangle = \mathrm{Re}(\langle\Psi|\,U\,|\Psi\rangle)$. Then

$$\begin{aligned}
\langle Z_a \rangle &= \mathrm{Re}(\langle\phi|\,\langle\psi|\,\mathrm{SWAP}\,|\phi\rangle\,|\psi\rangle) \\
&= \mathrm{Re}(\langle\phi|\,\langle\psi|\,|\psi\rangle\,|\phi\rangle) \\
&= \mathrm{Re}(\langle\phi|\psi\rangle\,\langle\psi|\phi\rangle) \\
&= |\langle\phi|\psi\rangle|^2.
\end{aligned}$$

2 Purity as an Overlap

The same idea extends to mixed states. Let ρ be a density operator and consider two identical copies $\rho \otimes \rho$. The swap operator SWAP satisfies

$$\mathrm{tr}(\mathrm{SWAP}\,(\rho \otimes \rho)) = \mathrm{tr}\,\rho^2, \qquad (14.24)$$

which is the *purity* of ρ. Operationally, a swap-test-style measurement of the SWAP observable on two copies of the state estimates $\mathrm{tr}(\rho^2)$. This quantity appears in entanglement diagnostics, thermalization tests, and randomized-benchmarking variants.

3 When Controlled-SWAP Is Reasonable

Like the Hadamard test, the swap test produces ± 1 (or equivalently Bernoulli) outcomes, so the sampling cost for additive error ϵ is again $\Theta(1/\epsilon^2)$ shots. The distinguishing feature is the gate-level cost: one must implement a controlled-SWAP across two registers, which is typically decomposed into multiple two-qubit gates per swapped qubit pair.

In fault-tolerant settings, controlled-SWAP can be implemented reliably and the swap test becomes a clean primitive for overlaps and purities, assuming the ability to prepare two copies of the required states. In near-term settings, the controlled-SWAP depth and the requirement of two clean state preparations can make the primitive expensive. The swap test is therefore most attractive when the registers are small, when the states are easy to prepare, or when overlap information is essential and cannot be obtained more cheaply by problem-specific methods. In other cases, one may prefer alternatives such as direct destructive overlap estimators for special families of states, randomized measurement approaches, or shadow-based estimators that avoid a global controlled-SWAP.

> **Exercise 14.6** Let $|\phi\rangle$ and $|\psi\rangle$ be pure states prepared on equal-size registers, and let O be an observable on the same register. Assume O is unitary (for example, a Pauli string), so that $O\,|\phi\rangle$ is a valid state. Based on the Hadamard test and swap test circuits, propose a circuit that coherently estimates $|\langle\psi|O|\phi\rangle|^2$.
>
> *Hint:* reduce the task to an overlap between $|\psi\rangle$ and $O\,|\phi\rangle$, then use a swap test.

Key Takeaways

1. The swap test estimates fidelity: $|\langle\phi|\psi\rangle|^2 = 2\Pr(0) - 1 = \langle Z_a\rangle$.

2. It can be viewed as a Hadamard test with $U = \mathrm{SWAP}$ acting on $|\phi\rangle\,|\psi\rangle$.

3. With two copies, the same idea estimates purity $(\mathrm{tr}\,\rho^2)$.

4. The main cost is controlled-SWAP depth plus the need to prepare two registers (or two copies).

14.4 ✳ Classical Shadows and Randomized Measurements

A central challenge in quantum information science is the efficient extraction of property data from a many-body quantum state described by its density operator

ρ. Traditional strategies such as Pauli grouping are optimized for a fixed, *a priori* known set of observables, but can become costly when the number of targets grows or when many targets do not commute. In many modern applications—from variational algorithms to the characterization of many-body correlations—one wants a *large collection* of expectation values from the same prepared state. Related motivations also appear in the earlier "shadow tomography" line of work, which asks how to learn many properties of an unknown state from limited copies without reconstructing ρ explicitly [76].

Randomized measurement methods offer a different viewpoint. Instead of tailoring measurements to specific operators, one collects a task-agnostic classical record—a "classical shadow"—generated by randomized rotations and computational-basis measurements. This record can then be reused to estimate many observables by classical post-processing, amortizing the experimental cost across the entire set of diagnostics. The modern classical-shadow framework and its guarantees were developed by Huang, Kueng, and Preskill [77], and it connects naturally to a broader toolbox of randomized measurement techniques and applications surveyed in [78, 79].

 This section assumes familiarity with density operators and completely positive trace-preserving (CPTP) channels. For background, see QCI Book [2], Section 12.1: *Mixed States and Density Operators* and Section 12.4: *Quantum Channels and CPTP Maps.*

14.4.1   The Randomized Measurement Protocol

The classical shadow framework separates *data acquisition* from *querying*. The data acquisition phase applies randomized unitaries to ρ and measures in the computational basis, producing a sequence of pairs (U, b), as illustrated in Fig. 14.3.

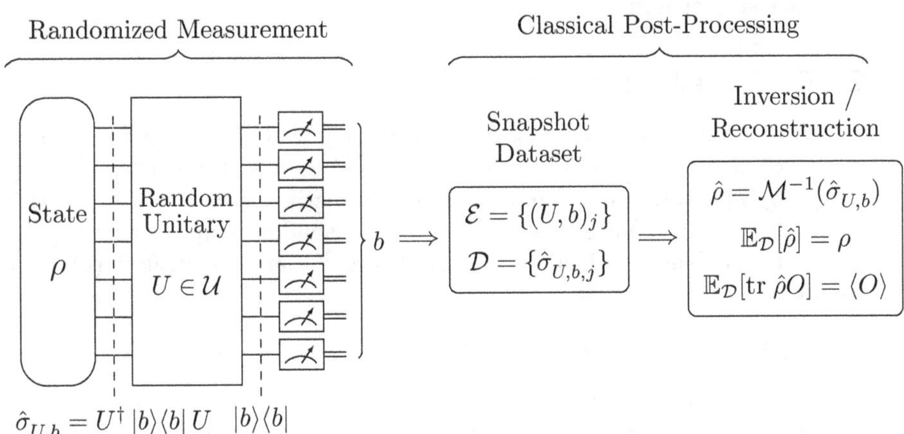

Figure 14.3: Classical Shadow Framework

1   Unitary Sampling

The protocol begins by sampling a unitary U from a fixed ensemble $\mathcal{U}$. The choice of $\mathcal{U}$ determines which observable families can be estimated efficiently.

- *Global Clifford ensemble.* U is sampled from the n-qubit Clifford group $\mathcal{C}_n$. This choice is well suited for certain global properties (e.g., fidelities with stabilizer-like targets) and for analyses that benefit from strong symmetry.

- *Local Pauli (local Clifford) ensemble.* $U = \bigotimes_{i=1}^{n} U_i$, where each U_i is sampled independently from a single-qubit Clifford set that maps Z-basis measurement to an effective X, Y, or Z measurement. This is equivalent to measuring each qubit in a randomly chosen Pauli basis and is well suited for k-local observables and correlation functions.

2   Measurement and Bitstrings

After applying U, we measure in the computational basis $\{|b\rangle : b \in \{0,1\}^n\}$. The measurement yields a bitstring b with probability

$$P(b) = \mathrm{tr}\big(U \rho U^\dagger |b\rangle\langle b|\big). \tag{14.25}$$

A single experimental shot thus produces the classical record (U, b).

3   The Data Snapshot

For each experimental realization, define the rotated projector

$$\hat{\sigma}_{U,b} \equiv U^\dagger |b\rangle\langle b| \, U. \tag{14.26}$$

This $\hat{\sigma}_{U,b}$ is the raw "snapshot" associated with the record (U, b). Collecting N shots produces the experimental record and the snapshot dataset

$$\mathcal{E} = \{(U,b)_1, (U,b)_2, \ldots, (U,b)_N\}, \qquad \mathcal{D} = \{\hat{\sigma}_{U,b,1}, \hat{\sigma}_{U,b,2}, \ldots, \hat{\sigma}_{U,b,N}\}. \tag{14.27}$$

An individual snapshot is generally not an unbiased estimator of ρ, but the ensemble of snapshots becomes useful once we account for the measurement bias through a reconstruction map.

14.4.2   The Classical Shadow Framework

The classical shadow framework is a randomized measurement-and-reconstruction method that produces a compact classical dataset from repeated preparations of ρ. By applying random unitaries, measuring in the computational basis, and then applying an explicit linear inverse (the "shadow map") in post-processing, one obtains unbiased estimators of ρ and hence of many quantities $\mathrm{tr}(O\rho)$ from the same dataset [76, 77].

1   The Measurement Map $\mathcal{M}$

The central mathematical object in the framework is the measurement map $\mathcal{M}$, which maps ρ to the statistical average of the snapshots produced by the randomized measurement protocol:

$$\mathcal{M}(\rho) = \mathbb{E}[\hat{\sigma}_{U,b}] = \mathbb{E}_{U \sim \mathcal{U}} \sum_{b \in \{0,1\}^n} \langle b| U \rho U^\dagger |b\rangle \, U^\dagger |b\rangle\langle b| \, U. \tag{14.28}$$

This map is a CPTP quantum channel. For highly symmetric ensembles (including unitary 2-designs such as the Clifford group), $\mathcal{M}$ is a depolarizing-type channel on operators, which makes $\mathcal{M}^{-1}$ simple to write down.

2 Inversion and Reconstruction

If $\mathcal{M}$ is invertible (or invertible on the operator subspace relevant to the intended observables), the *classical shadow* associated with a single snapshot is defined by

$$\hat{\rho}_{U,b} \equiv \mathcal{M}^{-1}(\hat{\sigma}_{U,b}) = \mathcal{M}^{-1}(U^{\dagger}\,|b\rangle\langle b|\,U). \tag{14.29}$$

By construction,

$$\mathbb{E}[\hat{\rho}_{U,b}] = \mathcal{M}^{-1}(\mathbb{E}[\hat{\sigma}_{U,b}]) = \mathcal{M}^{-1}(\mathcal{M}(\rho)) = \rho, \tag{14.30}$$

so $\hat{\rho}_{U,b}$ is an unbiased single-shot estimator of ρ. The reconstruction map $\mathcal{M}^{-1}$ is a *classical post-processing* rule and is not required to be CPTP.

For common ensembles, the inverse map has a simple closed form. Let $d = 2^n$.

- *Global Clifford (global 2-design).* The snapshot inversion takes the form

$$\mathcal{M}^{-1}(X) = (d+1)X - I. \tag{14.31}$$

- *Local Pauli / local Clifford (product ensemble).* The channel factorizes across qubits, and the inversion is applied locally. For a single-qubit snapshot X, one uses

$$\mathcal{M}_1^{-1}(X) = 3X - I, \tag{14.32}$$

  and for n qubits the full inverse is the tensor product of these local inverses acting on the corresponding per-qubit snapshot structure.

3 Median-of-Means Estimation

To obtain high-confidence error guarantees that are robust to heavy tails, one often uses a *median-of-means* (MoM) estimator. Split the N reconstructed snapshots into K batches of equal size $B = \lfloor N/K \rfloor$. For an observable O, compute the batch means

$$\mu_k = \frac{1}{B} \sum_{j \in \mathrm{Batch}_k} \mathrm{tr}(O\hat{\rho}_j), \qquad k = 1, \ldots, K, \tag{14.33}$$

and output the median $\mathrm{median}(\mu_1, \ldots, \mu_K)$. This improves concentration and yields failure probability that decreases exponentially in K under standard assumptions. See [77] for concrete bounds.

4 Example: Single-Qubit Randomized Measurement

To illustrate the mechanics, consider the single-qubit case shown in Fig. 14.4.

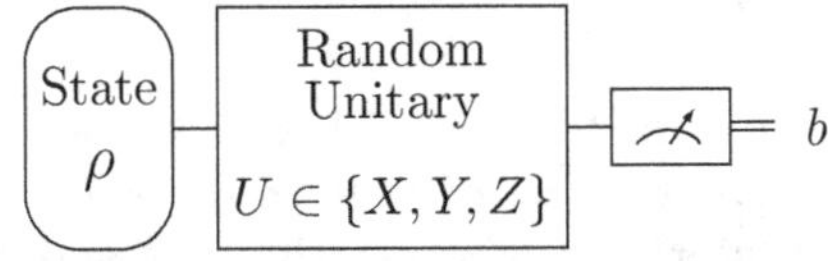

Figure 14.4: Single-Qubit Shadow Estimation

Basis Selection

We apply a random single-qubit Clifford U and then measure in the computational basis. Equivalently, we measure in a randomly chosen Pauli basis from $\{X, Y, Z\}$, each chosen with probability $1/3$:

$$U^\dagger Z U \in \{X, Y, Z\}.$$

The measurement outcome corresponds to one of the eigenstates

$$|\sigma\rangle \in \{|0\rangle, |1\rangle, |+\rangle, |-\rangle, |+_i\rangle, |-_i\rangle\}.$$

The Snapshot Distribution

Each shot yields the rank-one projector $\hat\sigma = |\sigma\rangle\langle\sigma|$. Its distribution is governed by two factors:

- *Probability by design:* the basis choice (X, Y, or Z) occurs with probability $1/3$;

- *Born rule:* given a chosen basis, the eigenstate outcome occurs with probability $\mathrm{tr}(\hat\sigma\rho)$.

Thus, for any particular eigenstate $\hat\sigma$ in the measurement set, the total probability is $P(\hat\sigma) = \frac{1}{3}\mathrm{tr}(\hat\sigma\rho)$.

The Channel and Its Inverse

A direct average of raw snapshots yields a depolarizing-type channel. In the single-qubit case,

$$\mathbb{E}_{\mathcal{D}}[\hat\sigma] = \mathcal{M}(\rho) = \frac{1}{3}\rho + \frac{1}{3}I. \tag{14.34}$$

To invert this map, define

$$\hat\rho = 3\hat\sigma - I \tag{14.35}$$

for any snapshot $\hat\sigma$. Then $\mathbb{E}_{\mathcal{D}}[\hat\rho] = \rho$, and averaging (or using MoM) yields an unbiased estimator for ρ and hence for $\langle O\rangle = \mathrm{tr}(O\rho)$.

Exercise 14.7 In the single-qubit local-Pauli shadow protocol, each snapshot is $\hat\sigma = |\sigma\rangle\langle\sigma|$ with $\sigma \in \{0, 1, +, -, +_i, -_i\}$ and the reconstruction is $\hat\rho = 3\hat\sigma - I$.

(a) Verify directly that $\mathbb{E}[\hat\rho] = \rho$ by expanding ρ in the Bloch form $\rho = \frac{1}{2}(I + r_x X + r_y Y + r_z Z)$.

(b) Use your computation to identify the induced measurement channel $\mathcal{M}$ on the Bloch vector.

Exercise 14.8 Investigate the single-qubit example above if the measurement ensemble is (i) $\{X, Z\}$ chosen uniformly, or (ii) $\{I, X, Y, Z\}$ chosen uniformly. For each case, determine whether the corresponding measurement map $\mathcal{M}$ is invertible and, if so, write down $\mathcal{M}^{-1}$.

Symmetry and Sampling in Quantum State Reconstruction

Tomographical Completeness

A set of measurements is *tomographically complete* if the information gathered is sufficient to uniquely identify any arbitrary quantum state ρ. If a measurement scheme is not tomographically complete, there exist "blind spots"—distinct quantum states that would produce identical measurement statistics, making the reconstruction map $\mathcal{M}^{-1}$ impossible to define.

For example, in the single-qubit case, tomographical completeness requires measurement along all three linearly independent axes (typically X, Y, and Z).

Unitary t-Designs

A unitary t-design is a finite ensemble of unitaries $\{U_i\}$ that mimics the low-order moments of the Haar measure up to order t. In shadow estimation, 2-design-like symmetry often makes $\mathcal{M}$ a depolarizing-type channel and enables clean variance bounds for many observables [80].

- *1-design:* A set that replicates the first moment $\mathbb{E}[U \rho U^\dagger]$.

- *2-design:* A set that also replicates second moments, enabling analytic variance control for estimators constructed from randomized measurements.

The Clifford Group

The Clifford group consists of unitaries that map Pauli operators to Pauli operators under conjugation (up to phase), for example $HXH^\dagger = Z$. It is a central example of a unitary 2-design and plays a key role in classical-shadow constructions. In *global Clifford shadows*, one samples U from the full n-qubit Clifford group and treats the system as a single block. In *local Clifford shadows* (random local Pauli bases), one samples U as a product of single-qubit Cliffords, which is equivalent to choosing a measurement basis from $\{X, Y, Z\}$ independently for each qubit.

Key Takeaways

---

1. Randomized measurement produces snapshots $\hat{\sigma}_{U,b} = U^\dagger \, |b\rangle\langle b| \, U$; their average is $\mathcal{M}(\rho) = \mathbb{E}[\hat{\sigma}_{U,b}]$.

2. $\mathcal{M}$ is a CPTP measurement channel; the reconstruction map $\mathcal{M}^{-1}$ is classical post-processing and need not be CPTP.

3. The reconstructed snapshot $\hat{\rho}_{U,b} = \mathcal{M}^{-1}(\hat{\sigma}_{U,b})$ is unbiased: $\mathbb{E}[\hat{\rho}_{U,b}] = \rho$.

4. Once a shadow dataset is collected, many observables can be estimated from the same data via $\mathrm{tr}(O \, \hat{\rho}_{\mathrm{avg}})$, often with median-of-means for high-confidence guarantees.

14.4.3 Sampling Complexity and Shadow Sizes

The utility of classical shadows is determined by the number of snapshots N required to estimate a set of L observables $\{O_\ell\}_{\ell=1}^{L}$ to within additive error ϵ with failure probability at most δ. A representative bound from the classical-shadow literature has the form

$$N \gtrsim \frac{\log(2L/\delta)}{\epsilon^2} \max_{1 \le \ell \le L} \|O_\ell\|_{\mathrm{shadow}}^2, \tag{14.36}$$

where $\| \cdot \|_{\text{shadow}}$ is an ensemble-dependent *shadow norm* that captures the effective variance of the estimator. Precise constants and the definition of $\| \cdot \|_{\text{shadow}}$ depend on the reconstruction rule and on whether median-of-means is used; see [77].

1   The Shadow Norm and Locality

The shadow norm reflects the compatibility between the observable family and the measurement ensemble.

- *Global Clifford shadows.* For many global targets of interest, the relevant shadow norm can be bounded independently of n (for example, in fidelity-type tasks), which is one reason global ensembles can be powerful for certain global properties.

- *Local Pauli shadows.* For a k-local observable, the shadow norm typically scales like 3^k (up to operator-norm factors), which is exponential in locality k but independent of the total system size n.

2   Comparison to Pauli Grouping

Pauli grouping partitions a given list of Pauli strings into commuting sets and measures each set in a dedicated basis. If many targets do not commute, the number of settings can be large. Classical shadows, in contrast, can estimate many non-commuting observables from a single randomized dataset, with sample complexity that depends only logarithmically on the number of targets (through $\log L$) but polynomially on accuracy and on the shadow norm.

3   Data Reusability

Because snapshots are stored classically, they can be reused: if a new observable O' is identified after the experiment, one can estimate $\text{tr}(O'\hat{\rho}_j)$ from the existing dataset, provided O' has a manageable shadow norm under the chosen ensemble.

Key Takeaways

1. Classical shadows target *many* observables at once: the number of shots can scale only logarithmically with the number of targets L (through $\log L$), rather than linearly.

2. The main precision drivers are the desired error ϵ (typically $1/\epsilon^2$) and an ensemble-dependent shadow norm that captures estimator variance for the observable family.

3. Local random Pauli bases are hardware-friendly and effective for k-local observables; more global ensembles can help with global properties but may require deeper circuits.

14.4.4   Advanced Techniques and Variations

While the standard classical shadow framework provides a robust foundation, several extensions reduce sampling overhead or adapt the protocol to near-term constraints; see, for example, the review perspective in [78, 79].

1 Derandomization and Biased Shadows

If there is prior knowledge about the observables of interest, purely random sampling can be suboptimal.

- *Derandomized shadows.* One can select measurement bases by a greedy or structured rule designed to reduce the predicted variance for a specific target set, while retaining the many-target benefits [81].

- *Biased (weighted) shadows.* By changing the sampling distribution over bases (e.g., favoring Z on qubits where Hamiltonian terms are mostly Z-type), one can reduce the shadow norm for the intended targets [82].

2 Shallow Shadows and Hardware Efficiency

Deep Clifford circuits are often prohibitive on NISQ devices. Shallow ensembles interpolate between local Pauli measurements (very shallow) and more global randomization (deeper), aiming to capture mid-range correlations while keeping circuit depth manageable [83]. Noise-aware variants replace $\mathcal{M}^{-1}$ with an approximate inverse calibrated to the device [84, 85].

3 Machine Learning and Shadow Data

The classical record (snapshots and measurement choices) can serve as input features for classical learning pipelines. In many-body settings, shadow datasets have been used to support tasks such as phase classification or property prediction, often leveraging the fact that the same dataset can be queried for many derived observables [86, 87, 88].

Problem Set 14

14.1 *Qubit-Wise Commutativity Check.* Consider the Pauli strings on three qubits:

$$P_1 = X_1 Z_2, \quad P_2 = X_1 Z_2 Z_3, \quad P_3 = Z_2, \quad P_4 = Y_1 Z_2, \quad P_5 = Z_1 Z_3.$$

(a) Identify a maximal qubit-wise commuting (QWC) group that contains P_1.

(b) Give the corresponding single-qubit measurement bases (one basis choice per qubit).

(c) Explain why P_4 cannot be placed in the same QWC group as P_1.

14.2 *Shot Complexity for Multiple Targets.* Suppose you want to estimate $\langle O_\ell \rangle = \mathrm{tr}(O_\ell \rho)$ for $\ell = 1, \ldots, L$ to additive error at most ϵ with confidence $1 - \delta$, using independent estimators for each target.

(a) Assuming each single-shot outcome is bounded in $[-1, 1]$, derive a sufficient number of shots per observable using a normal-approximation or a concentration bound.

(b) Compare the scaling in L under independent estimation versus a strategy that shares data across observables (e.g., classical shadows), and explain the regime where the difference is most significant.

14.3 $*$ *Optimal Shot Allocation for a Pauli Sum (Lagrange Multiplier).* Let $H = \sum_{k=1}^{m} c_k P_k$ and suppose you can choose N_k to minimize $\mathrm{Var}(\overline{\langle H \rangle})$ subject to a fixed total-shot budget $\sum_{k=1}^{m} N_k = N_{\mathrm{tot}}$.

(a) Using the approximation $\mathrm{Var}(P_k) \approx 1 - \langle P_k \rangle^2$, set up and solve the continuous optimization problem for the variance-minimizing N_k.

(b) Give the resulting scaling rule in terms of $|c_k|$ and an estimated standard deviation of P_k.

(c) Discuss why practical implementations use pilot estimates and regularization (e.g., floors on N_k).

14.4 *QWC vs Full Commutativity.* Construct two Pauli strings on $n \geq 2$ qubits that commute but are not qubit-wise commuting.

(a) Verify that they commute using the "even number of local anticommutes" rule.

(b) Show they fail the QWC criterion.

(c) Discuss what additional circuit resources might be required to jointly measure them in one setting.

14.5 *Hadamard Test for a Non-Hermitian Target.* Let U be a unitary and consider estimating $\langle \psi | U | \psi \rangle$.

(a) Derive how a phase choice on the ancilla yields $\mathrm{Re}(\langle \psi | U | \psi \rangle)$
and $\mathrm{Im}(\langle \psi | U | \psi \rangle)$.

(b) Show how to combine these estimates into an estimate of $| \langle \psi | U | \psi \rangle |^2$, and compare this to a swap-test-style overlap target.

14.6 *Swap Test for Purity.* Let ρ be an n-qubit mixed state and consider two copies $\rho \otimes \rho$.

(a) Prove $\mathrm{tr}(\rho^2) = \mathrm{tr}(\mathrm{SWAP}(\rho \otimes \rho))$.

(b) Describe a swap-test-style circuit that estimates $\mathrm{tr}(\rho^2)$ and identify the measurement statistic that yields it.

(c) Evaluate the purity for (i) a pure state, (ii) the maximally mixed state $I/2^n$, and (iii) a classical mixture $p |0\rangle\langle 0| + (1 - p) |1\rangle\langle 1|$ (single-qubit case).

14.7 *Single-Qubit Classical Shadow: Full Deduction.* For the single-qubit local-Pauli shadow protocol:

(a) Starting from $\rho = \frac{1}{2}(I + \boldsymbol{r} \cdot \boldsymbol{\sigma})$, compute $\mathcal{M}(\rho) = \mathbb{E}[\hat{\sigma}]$ explicitly and show it is depolarizing.

(b) Derive the reconstruction rule $\hat{\rho} = 3\hat{\sigma} - I$ by inverting $\mathcal{M}$.

(c) Using your expression, compute $\mathrm{Var}(\mathrm{tr}(O\hat{\rho}))$ for $O = X$ and interpret the result.

14.8 *Derandomized and Biased Shadows.* Consider estimating $\langle H \rangle$ where H is dominated by Z-type terms (e.g., diagonal cost Hamiltonians in optimization).

(a) Propose a biased measurement schedule that favors Z measurements on qubits appearing frequently in Z-heavy terms.

(b) Define a simple metric that predicts which basis choices reduce variance for the intended observable family.

(c) Test your schedule numerically against uniform local-Pauli sampling on a small instance.

14.9 ✳ *Shadow Estimation vs Grouping.* Pick an n-qubit Hamiltonian $H = \sum_k c_k P_k$ (e.g., transverse-field Ising on $n = 6$ qubits). Fix a state-preparation circuit (e.g., a shallow hardware-efficient ansatz).

(a) Implement (i) term-wise estimation, (ii) QWC grouping, and (iii) local-Pauli classical shadows.

(b) For a fixed shot budget, compare the empirical mean-squared error of the energy estimator across methods.

(c) Identify regimes (in m, coefficient spread, and target precision) where each method is preferable.

14.10 ✳ *Grouping as a Graph Problem.* Given a set of Pauli strings $\{P_k\}_{k=1}^{m}$, define a graph with vertices k and an edge between k and ℓ if P_k and P_ℓ are *not* jointly measurable under your chosen measurement model.

(a) Explain why finding the minimum number of measurement settings can be formulated as a graph coloring problem.

(b) For QWC grouping on n qubits, provide an explicit criterion for when two Pauli strings are adjacent.

(c) Implement a simple greedy coloring routine and test it on a small Hamiltonian (e.g., a 4-qubit Heisenberg chain) to compare the number of settings with term-wise measurement.

Supporting Materials

Bibliography

Books

[1] Peter Y. Lee, James M. Yu, and Ran Cheng. *Mathematical Foundations for Quantum Computing: A Scaffolding Approach*. 1st. Polaris QCI Publishing, 2025. ISBN: 978-1-961880-09-2 (cited on pages 3, 6, 92, 95, 104, 123, 126, 147, 160, 188, 225, 311, 315, 472, 637).

[2] Peter Y. Lee, Huiwen Ji, and Ran Cheng. *Quantum Computing and Information: A Scaffolding Approach*. 2nd. Polaris QCI Publishing, 2024. ISBN: 978-1-961880-06-1 (cited on pages 25, 26, 32, 37, 46, 49, 51, 66, 75, 92, 102, 271, 323, 630, 635, 637, 647).

[3] Michael A Nielsen and Isaac L Chuang. *Quantum Computation and Quantum Information: 10th Anniversary Edition*. Cambridge University Press, 2010. ISBN: 978-1-107-00217-3 (cited on pages 105, 658).

Articles and Other Resources

[4] David Deutsch. Quantum theory, the Church–Turing principle and the universal quantum computer. *Proceedings of the Royal Society of London. A. Mathematical and Physical Sciences* 400.1818 (1985) (cited on pages 22, 38).

[5] Ethan Bernstein and Umesh Vazirani. Quantum complexity theory. *SIAM Journal on Computing* 26.5 (1997). DOI: 10.1137/S0097539794262061 (cited on page 22).

[6] Cameron Cianci. The Quantum-Extended Church-Turing Thesis in Quantum Field Theory. *Preprint* arXiv: 2309.09000. (2023) (cited on page 22).

[7] Richard P Feynman. Simulating physics with computers. *International Journal of Theoretical Physics* 21.6 (1982) (cited on pages 23, 39, 603).

[8] Hsin-Yuan Huang et al. The vast world of quantum advantage. *Preprint* arXiv: 2508.05720. (2025) (cited on pages 27, 83, 613, 615, 618–620, 663).

[9] Olivia Lanes et al. A Framework for Quantum Advantage. *Preprint* arXiv: 2506.20658. (2025) (cited on pages 27, 83, 602, 605, 613, 618, 620, 663).

[10] Edward Farhi et al. Quantum Computation by Adiabatic Evolution. *Preprint* arXiv: quant-ph/0001106. (2000) (cited on pages 38, 264).

[11] Dorit Aharonov et al. Adiabatic quantum computation is equivalent to standard quantum computation. *SIAM Journal on Computing* 37.1 (2007) (cited on pages 38, 264, 265, 278, 288).

[12] John Preskill. Beyond NISQ: The MegaQuOp Machine. *Preprint* arXiv: 2502.17368. (2025) (cited on page 73).

[13] Rajeev Acharya et al. Quantum error correction below the surface code threshold. *Nature* 638.8052 (Dec. 2024). ISSN: 1476-4687. DOI: 10.1038/s41 586-024-08449-y (cited on pages 75, 659).

[14] Aosai Zhang et al. Demonstrating quantum error mitigation on logical qubits. *Nature Communications* 17.1021 (Dec. 2025). DOI: 10.1038/s41467-025-67 768-4 (cited on page 76).

[15] Samuel C. Smith, Benjamin J. Brown, and Stephen D. Bartlett. Mitigating errors in logical qubits. *Communications Physics* 7.386 (Nov. 2024). DOI: 10.1038/s42005-024-01883-4 (cited on page 76).

[16] Ryan Babbush et al. The Grand Challenge of Quantum Applications. *Preprint* arXiv: 2511.09124. (2025) (cited on pages 83, 508–510, 602, 605, 606, 613, 615, 618, 620, 663).

[17] Robert B Griffiths and Chi-Sheng Niu. Semiclassical Fourier transform for quantum computation. *Physical Review Letters* 76.17 (1996) (cited on page 107).

[18] A. Yu. Kitaev. Quantum measurements and the Abelian Stabilizer Problem. *Preprint* arXiv: quant-ph/9511026. (1995) (cited on pages 111, 123, 125).

[19] Richard Cleve et al. Quantum Algorithms Revisited. *Proceedings of the Royal Society of London. Series A: Mathematical, Physical and Engineering Sciences* 454.1969 (1998). DOI: 10.1098/rspa.1998.0164 (cited on page 111).

[20] Peter W. Shor. Algorithms for Quantum Computation: Discrete Logarithms and Factoring. (1994). DOI: 10.1109/SFCS.1994.365700 (cited on pages 111, 133).

[21] Krysta M. Svore, Matthew B. Hastings, and Michael Freedman. Faster Phase Estimation. *Preprint* arXiv: 1304.0741. (2013) (cited on page 125).

[22] Kentaro Yamamoto et al. Demonstrating Bayesian quantum phase estimation with quantum error detection. *Physical Review Research* 6.1 (Feb. 2024). ISSN: 2643-1564. DOI: 10.1103/physrevresearch.6.013221 (cited on page 126).

[23] Thomas Häner, Martin Roetteler, and Krysta M. Svore. Factoring using 2n+2 qubits with Toffoli based modular multiplication. *Preprint* arXiv: 1611.07995. *Quantum Information and Computation* 18.7–8 (2018) (cited on pages 146, 147).

[24] Craig Gidney and Martin Ekerå. How to factor 2048-bit RSA integers in 8 hours using 20 million noisy qubits. *Preprint* arXiv: 1905.09749. *Quantum* 5 (2021). DOI: 10.22331/q-2021-04-15-433 (cited on page 147).

[25] Madelyn Cain et al. Shor's algorithm is possible with as few as 10,000 reconfigurable atomic qubits. *Preprint* arXiv: 2603.28627. *arXiv preprint arXiv:2603.28627* (2026). DOI: 10.48550/arXiv.2603.28627 (cited on page 147).

[26] Lov K. Grover. A Fast Quantum Mechanical Algorithm for Database Search. (1996). DOI: 10.1145/237814.237866 (cited on page 159).

[27] Gilles Brassard et al. Quantum Amplitude Amplification and Estimation. *Preprint* arXiv: quant-ph/0005055. Contemporary Mathematics 305 (2002). Edited by Jr. Lomonaco Samuel J. DOI: 10.1090/conm/305/05215 (cited on pages 159, 169).

[28] Gilles Brassard, Peter Høyer, and Alain Tapp. Quantum Counting. *Lecture Notes in Computer Science* 1443 (1998). DOI: 10.1007/BFb0055105 (cited on page 159).

[29] Theodore J. Yoder, Guang Hao Low, and Isaac L. Chuang. Fixed-Point Quantum Search with an Optimal Number of Queries. *Physical Review Letters* 113.21 (2014). DOI: 10.1103/PhysRevLett.113.210501 (cited on pages 159, 174).

[30] Andrew M. Childs and Nathan Wiebe. Hamiltonian Simulation Using Linear Combinations of Unitaries. *Quantum Information and Computation* 12.11–12 (2012) (cited on pages 159, 197, 198).

[31] Dominic W. Berry et al. Simulating Hamiltonian Dynamics with a Truncated Taylor Series. *Physical Review Letters* 114.9 (2015). DOI: 10.1103/PhysRevLett.114.090502 (cited on pages 159, 196, 197, 205).

[32] Charles H. Bennett et al. Strengths and Weaknesses of Quantum Computing. *SIAM Journal on Computing* 26.5 (1997). DOI: 10.1137/S0097539796300933 (cited on page 167).

[33] Christof Zalka. Grover's Quantum Searching Algorithm Is Optimal. *Physical Review A* 60.4 (1999). DOI: 10.1103/PhysRevA.60.2746 (cited on page 167).

[34] Dmitry Grinko et al. Iterative Quantum Amplitude Estimation. *Preprint* arXiv: 1912.05559. *npj Quantum Information* 7 (2021). DOI: 10.1038/s41534-021-00379-1 (cited on page 170).

[35] Yohichi Suzuki et al. Amplitude Estimation without Phase Estimation. *Preprint* arXiv: 1904.10246. *Quantum Information Processing* 19 (2020). DOI: 10.1007/s11128-019-2565-2 (cited on page 170).

[36] Guang Hao Low and Isaac L. Chuang. Optimal Hamiltonian Simulation by Quantum Signal Processing. *Preprint* arXiv: 1606.02685. *Physical Review Letters* 118.1 (Jan. 2017). DOI: 10.1103/PhysRevLett.118.010501 (cited on pages 174, 177, 196, 197, 205, 214).

[37] Dominic W. Berry et al. Efficient Quantum Algorithms for Simulating Sparse Hamiltonians. *Communications in Mathematical Physics* 270 (2007). DOI: 10.1007/s00220-006-0150-x (cited on page 196).

[38] Guang Hao Low and Isaac L. Chuang. Hamiltonian Simulation by Qubitization. *Preprint* arXiv: 1610.06546. *Quantum* 3 (July 2019). DOI: 10.22331/q-2019-07-12-163 (cited on pages 196, 197, 214, 218).

[39] Andrew M. Childs and Yulong Su. Nearly Optimal Lattice Simulation by Product Formulas. *Physical Review Letters* 123.17 (2019). DOI: 10.1103/PhysRevLett.123.170504 (cited on page 196).

[40] Diyi Liu et al. Block encoding with low gate count for second-quantized Hamiltonians. *Preprint* arXiv: 2510.08644. (2025) (cited on pages 196, 210, 253).

[41] András Gilyén et al. Quantum singular value transformation and beyond: exponential improvements for quantum matrix arithmetics. *Preprint* arXiv: 1806.01838. (June 2019). DOI: 10.1145/3313276.3316366 (cited on pages 197, 218, 225, 232, 304).

[42] Christopher Kang and Yuan Su. Quantum matrix arithmetics with Hamiltonian evolution. *Preprint* arXiv: 2510.06316. (2026) (cited on pages 210, 236, 253, 663).

[43] Ahmad M. Alkadri et al. A Quantum Algorithm for the Finite Element Method. *Preprint* arXiv: 2510.18150. (2025) (cited on pages 210, 236, 253, 259, 663).

[44] QSPPACK Developers. QSPPACK: A Quantum Signal Processing Package. *https://qsppack.gitbook.io* (2023) (cited on pages 213, 217).

[45] Seyed Mohammad Motlagh and Nathan Wiebe. Generalized Quantum Signal Processing. *Preprint* arXiv: 2403.02805. *arXiv preprint* (2024) (cited on page 216).

[46] Samuel E. Skelton. Qubit-optimal quantum phase estimation of block-encoded Hamiltonians. *Preprint* arXiv: 2509.04246. (Sept. 2025) (cited on page 218).

[47] Kiichiro Toyoizumi, Naoki Yamamoto, and Kazuo Hoshino. Hamiltonian Simulation Using the Quantum Singular-Value Transformation: Complexity Analysis and Application to the Linearized Vlasov–Poisson Equation. *Physical Review A* 109.1 (Jan. 2024). DOI: 10.1103/PhysRevA.109.012430 (cited on page 235).

[48] Aram W. Harrow, Avinatan Hassidim, and Seth Lloyd. Quantum Algorithm for Linear Systems of Equations. *Physical Review Letters* 103.15 (2009). DOI: 10.1103/PhysRevLett.103.150502 (cited on page 243).

[49] Daan Camps et al. Explicit Quantum Circuits for Block Encodings of Certain Sparse Matrices. *Preprint* arXiv: 2203.10236. (2022) (cited on page 254).

[50] Daan Camps et al. Explicit Quantum Circuits for Block Encodings of Certain Sparse Matrices. *SIAM Journal on Matrix Analysis and Applications* (2024). DOI: 10.1137/22M1484298 (cited on page 254).

[51] Sabine Jansen, Mary-Beth Ruskai, and Ruedi Seiler. Bounds for the Adiabatic Approximation with Applications to Quantum Computation. *Journal of Mathematical Physics* 48.10 (2007). DOI: 10.1063/1.2798381 (cited on page 264).

[52] Andrew Lucas. Ising formulations of many NP problems. *Frontiers in Physics* 2 (2014). DOI: 10.3389/fphy.2014.00005 (cited on page 264).

[53] Tadashi Kadowaki and Hidetoshi Nishimori. Quantum Annealing in the Transverse Ising Model. *Physical Review E* 58.5 (1998). DOI: 10.1103/PhysRevE.58.5355 (cited on pages 264, 265).

[54] Alberto Peruzzo et al. A Variational Eigenvalue Solver on a Photonic Quantum Processor. *Nature Communications* 5 (2014). DOI: 10.1038/ncomms5213 (cited on page 264).

[55] Edward Farhi, Jeffrey Goldstone, and Sam Gutmann. A Quantum Approximate Optimization Algorithm. *Preprint* arXiv: 1411.4028. *arXiv e-prints* (2014) (cited on pages 264, 289, 293).

[56] Atanu Rajak et al. Quantum annealing: an overview. *Philosophical Transactions of the Royal Society A: Mathematical, Physical and Engineering Sciences* 381.2241 (2023). DOI: 10.1098/rsta.2021.0417 (cited on page 265).

[57] J. D. Biamonte. Non-Perturbative k-Body to Two-Body Commuting Conversion Hamiltonians and Embedding Problem Instances into Ising Spins. *Preprint* arXiv: 0801.3800. *Physical Review A* 77.5 (2008). DOI: 10.1103/PhysRevA.77.052331 (cited on page 273).

[58] Jacob D. Biamonte and Peter J. Love. Realizable Hamiltonians for Universal Adiabatic Quantum Computers. *Preprint* arXiv: 0704.1287. *Physical Review A* 78.1 (2008). DOI: 10.1103/PhysRevA.78.012352 (cited on page 278).

[59] Fernando G. S. L. Brandão, Aram W. Harrow, and Michał Horodecki. Local Random Quantum Circuits are Approximate Polynomial-Designs. *Communications in Mathematical Physics* 346.2 (2016). DOI: 10.1007/s00220-016-2706-8 (cited on page 286).

[60] Naoki Yamamoto. On the Natural Gradient for Variational Quantum Eigensolver. *Preprint* arXiv: 1909.05074. (Sept. 2019). DOI: 10.48550/arXiv.1909.05074 (cited on page 287).

[61] Jarrod R. McClean et al. Barren Plateaus in Quantum Neural Network Training Landscapes. *Nature Communications* 9 (2018). DOI: 10.1038/s41467-018-07090-4 (cited on page 287).

[62] M. Cerezo et al. Cost Function Dependent Barren Plateaus in Shallow Quantum Neural Networks. *Nature Communications* 12 (2021). DOI: 10.1038/s41467-021-21728-w (cited on page 287).

[63] Jacob Biamonte. Universal Variational Quantum Computation. *Preprint* arXiv: 1903.04500. *Physical Review A* 103.3 (2021). DOI: 10.1103/PhysRevA.103.L030401 (cited on page 288).

[64] Kostas Blekos et al. A Review on Quantum Approximate Optimization Algorithm and its Variants. *Physics Reports* (2024). DOI: 10.1016/j.physrep.2024.03.002 (cited on pages 292, 505, 508–510, 512, 663).

[65] Stuart Hadfield et al. From the Quantum Approximate Optimization Algorithm to a Quantum Alternating Operator Ansatz. *Algorithms* 12.2 (2019). DOI: 10.3390/a12020034 (cited on page 292).

[66] Daniel J. Egger, Jakub Mareček, and Stefan Woerner. Warm-starting quantum optimization. *Quantum* 5 (2021). DOI: 10.22331/q-2021-06-17-479 (cited on page 292).

[67] Sergey Bravyi et al. Hybrid quantum-classical algorithms for approximate graph coloring. *Preprint* arXiv: 2011.13420. *arXiv e-prints* (2020) (cited on page 293).

[68] Shota Kanno et al. Quantum selected configuration interaction: classical diagonalization of Hamiltonians in subspaces formed by measurements of variational states. *Preprint* arXiv: 2302.11320. (2023) (cited on pages 295, 304).

[69] M. Cerezo et al. Variational quantum algorithms. *Nature Reviews Physics* 3 (2021). DOI: 10.1038/s42254-021-00348-9 (cited on pages 304, 663).

[70] Jules Tilly et al. The Variational Quantum Eigensolver: A review of methods and best practices. *Physics Reports* 986 (2022). DOI: 10.1016/j.physrep.2022.08.003 (cited on page 304).

[71] Kishor Bharti et al. Noisy intermediate-scale quantum algorithms. *Reviews of Modern Physics* 94 (2022). DOI: 10.1103/RevModPhys.94.015004 (cited on pages 304, 663).

[72] Leo Zhou et al. Quantum Approximate Optimization Algorithm: Performance, Mechanism, and Implementation on Near-Term Devices. *Physical Review X* 10 (2020). DOI: 10.1103/PhysRevX.10.021067 (cited on page 304).

[73] Tameem Albash and Daniel A. Lidar. Adiabatic quantum computation. *Reviews of Modern Physics* 90 (2018). DOI: 10.1103/RevModPhys.90.01500 2 (cited on page 304).

[74] Javier Robledo-Moreno et al. Chemistry beyond the scale of exact diagonalization on a quantum-centric supercomputer. *Preprint* arXiv: 2405.05068. (2024) (cited on page 304).

[75] Lorenzo Piccinelli et al. A simple and provable algorithm for sampling-based quantum diagonalization. *Preprint* arXiv: 2506.09611. (2025) (cited on page 304).

[76] Scott Aaronson. Shadow Tomography of Quantum States. (2018). DOI: 10.1 145/3188745.3188802 (cited on pages 323, 324, 548, 551).

[77] Hsin-Yuan Huang, Richard Kueng, and John Preskill. Predicting Many Properties of a Quantum System from Very Few Measurements. *Nature Physics* 16 (2020). DOI: 10.1038/s41567-020-0932-7 (cited on pages 323–325, 328, 548, 551).

[78] Andreas Elben et al. The randomized measurement toolbox. *Nature Reviews Physics* (2022) (cited on pages 323, 328).

[79] Piotr Cieśliński et al. Analysing quantum systems with randomised measurements. *Physics Reports* (2024) (cited on pages 323, 328).

[80] Christoph Dankert et al. Exact and Approximate Unitary 2-Designs: Constructions and Applications. *Physical Review A* 80.1 (2009). DOI: 10.1103 /PhysRevA.80.012304 (cited on page 327).

[81] Hsin-Yuan Huang, Richard Kueng, and John Preskill. Efficient Estimation of Pauli Observables by Derandomization. *Physical Review Letters* 127.3 (2021). DOI: 10.1103/PhysRevLett.127.030503 (cited on page 329).

[82] Charles Hadfield et al. Measurements of Quantum Hamiltonians with Locally-Biased Classical Shadows. *Communications in Mathematical Physics* (2022). DOI: 10.1007/s00220-022-04343-8 (cited on page 329).

[83] H.-Y. Hu et al. Demonstration of robust and efficient quantum property estimation with shallow shadows. *Nature Communications* (2025). DOI: 10.1038/s41467-025-57349-w (cited on page 329).

[84] Dax Enshan Koh and Sabee Grewal. Classical Shadows with Noise. *Quantum* (2022). DOI: 10.22331/q-2022-08-16-776 (cited on page 329).

[85] R. Cioli et al. Approximate inverse measurement channel for shallow shadows. *Preprint* arXiv: 2407.11813. *arXiv* (2024) (cited on page 329).

[86] Hsin-Yuan Huang et al. Provably efficient machine learning for quantum many-body problems. *Science* (2022). DOI: 10.1126/science.abk3333 (cited on page 329).

[87] S. Jerbi et al. Shadows of quantum machine learning. *Nature Communications* (2024). DOI: 10.1038/s41467-024-49877-8 (cited on page 329).

[88] V. Wei et al. Neural-shadow quantum state tomography. *Physical Review Research* (2024) (cited on page 329).

[89] Emanuel Gull et al. Continuous-Time Monte Carlo Methods for Quantum Impurity Models. *Reviews of Modern Physics* 83.2 (2011). DOI: 10.1103/Re vModPhys.83.349 (cited on page 439).

[90] Hui Shao, Anders W. Sandvik, and Kevin S. D. Beach. Progress on Stochastic Analytic Continuation of Imaginary-Time Correlation Functions Computed by Quantum Monte Carlo Simulations. *Physics Reports* 1003 (2023). DOI: 10.1016/j.physrep.2022.11.003 (cited on page 439).

[91] Oles Shtanko et al. Uncovering Local Integrability in Quantum Many-Body Dynamics. *Preprint* arXiv: 2307.07552. (2023) (cited on pages 447, 612).

[92] Benedikt Fauseweh. Quantum many-body simulations on digital quantum computers: State-of-the-art and future challenges. *Nature Communications* 15.1 (2024). DOI: 10.1038/s41467-024-46402-9 (cited on pages 447, 663).

[93] Mari Carmen Bañuls et al. Tensor Network Algorithms: A Route Map. *Annual Review of Condensed Matter Physics* 14 (2023) (cited on pages 477, 478).

[94] J. Ignacio Cirac et al. Matrix product states and projected entangled pair states. *Reviews of Modern Physics* 93 (2021) (cited on pages 477, 478).

[95] Y.-Y. Shi, L.-M. Duan, and G. Vidal. Classical simulation of quantum many-body systems with a tree tensor network. *Physical Review A* 74 (2006) (cited on page 477).

[96] G. Vidal. Entanglement Renormalization. *Physical Review Letters* 99 (2007) (cited on page 478).

[97] Steven R. White. Density matrix formulation for quantum renormalization groups. *Physical Review Letters* 69.19 (1992). DOI: 10.1103/PhysRevLett.69.2863 (cited on page 478).

[98] Ulrich Schollwöck. The density-matrix renormalization group in the age of matrix product states. *Annals of Physics* 326.1 (2011). DOI: 10.1016/j.aop.2010.09.012 (cited on page 478).

[99] Román Orús. A Practical Introduction to Tensor Networks: Matrix Product States and Projected Entangled Pair States. *Annals of Physics* 349 (2014) (cited on page 478).

[100] Guifré Vidal. Efficient Simulation of One-Dimensional Quantum Many-Body Systems. *Physical Review Letters* 93 (2004). DOI: 10.1103/PhysRevLett.93.040502 (cited on page 478).

[101] A. J. Daley et al. Time-dependent density-matrix renormalization-group using adaptive effective Hilbert spaces. *Journal of Statistical Mechanics: Theory and Experiment* 2004.04 (2004). DOI: 10.1088/1742-5468/2004/04/P04005 (cited on page 478).

[102] Marco Cerezo et al. Variational quantum algorithms. *Nature Reviews Physics* 3.9 (2021). DOI: 10.1038/s42254-021-00348-9 (cited on pages 505, 507–510).

[103] Alexey Bochkarev et al. Quantum computing for discrete optimization: A highlight of three technologies. *Preprint* arXiv: 2409.01373. *European Journal of Operational Research* (2025). DOI: 10.1016/j.ejor.2025.07.063 (cited on pages 505, 507–510).

[104] Dylan Herman et al. Quantum computing for finance. *Preprint* arXiv: 2307.11230. *Nature Reviews Physics* 5.8 (2023). DOI: 10.1038/s42254-023-00603-1 (cited on pages 505, 510).

[105] Giovanni Buonaiuto et al. Best practices for portfolio optimization by quantum computing, experimented on real quantum devices. *Scientific Reports* 13 (2023). DOI: 10.1038/s41598-023-45392-w (cited on pages 505, 510).

[106] Panagiotis Kl. Barkoutsos et al. Improving Variational Quantum Optimization using CVaR. *Quantum* 4 (2020). DOI: 10.22331/q-2020-04-20-256 (cited on pages 505, 508, 510, 512).

[107] Sean J. Weinberg et al. Supply chain logistics with quantum and classical annealing algorithms. *Preprint* arXiv: 2205.04435. *Scientific Reports* 13 (2023). DOI: 10.1038/s41598-023-31765-8 (cited on pages 506, 508, 510, 512).

[108] Santiago Arroyo et al. Quantum Computing in Logistics and Supply Chain Optimization. *Preprint* arXiv: 2402.17520. *arXiv* (2024) (cited on page 506).

[109] Alexey Pyrkov et al. Quantum computing for near-term applications in generative chemistry and drug discovery. *Drug Discovery Today* 28.8 (2023). DOI: 10.1016/j.drudis.2023.103675 (cited on pages 506, 507).

[110] P. Das and S. Ray. A brief review on quantum computing based drug design. *Wiley Interdisciplinary Reviews: Data Mining and Knowledge Discovery* 14.6 (2024). DOI: 10.1002/widm.1553 (cited on pages 506, 507).

[111] Maximillian Zinner et al. Quantum computing's potential for drug discovery: Early stage industry dynamics. *Drug Discovery Today* 26.7 (2021). DOI: 10.1016/j.drudis.2021.06.003 (cited on page 506).

[112] John Jumper et al. Highly accurate protein structure prediction with AlphaFold. *Nature* 596.7873 (2021). DOI: 10.1038/s41586-021-03819-2 (cited on page 507).

[113] Kathryn Tunyasuvunakool et al. Highly accurate protein structure prediction for the human proteome. *Nature* 596.7873 (2021). DOI: 10.1038/s41586-021-03828-1 (cited on page 507).

[114] Martín Larocca et al. A Review of Barren Plateaus in Variational Quantum Computing. *Preprint* arXiv: 2405.00781. *arXiv* (2024) (cited on page 507).

[115] Y. Wang et al. A Comprehensive Review of Quantum Machine Learning. *Preprint* arXiv: 2401.11351. *arXiv* (2024) (cited on page 507).

[116] Pablo Bermejo et al. Variational quantum and quantum-inspired clustering. *Scientific Reports* 13 (2023). DOI: 10.1038/s41598-023-39771-6 (cited on page 507).

[117] Jules Tilly et al. The Variational Quantum Eigensolver: a review of methods and best practices. *Preprint* arXiv: 2111.05176. *Physics Reports* (2022). DOI: 10.1016/j.physrep.2022.08.003 (cited on pages 508, 509).

[118] Yu Du et al. New advances for quantum-inspired optimization. *International Transactions in Operational Research* (2023). DOI: 10.1111/itor.13420 (cited on pages 509, 512).

[119] Seth Lloyd, Masoud Mohseni, and Patrick Rebentrost. Quantum Principal Component Analysis. *Nature Physics* 10.9 (2014). DOI: 10.1038/nphys3029 (cited on page 546).

[120] Ewin Tang. Quantum Principal Component Analysis Only Achieves an Exponential Speedup Because of Its State Preparation Assumptions. *Physical Review Letters* 127 (2021). DOI: 10.1103/PhysRevLett.127.060503 (cited on page 546).

[121] Shijie Wang and Yong Liu. Quantum machine learning: a review. *Reports on Progress in Physics* 87.1 (2024). DOI: 10.1088/1361-6633/ad0d5a (cited on pages 550, 551).

[122]   Sandeep Gandhi and et al. A Comprehensive Review of Quantum Machine Learning. *Preprint* arXiv: 2401.11351. (2024) (cited on pages 550, 551).

[123]   M. Cerezo et al. Variational quantum algorithms. *Nature Reviews Physics* 3 (2021). DOI: 10.1038/s42254-021-00348-9 (cited on pages 550, 551).

[124]   Fei Tian et al. Quantum Generative Neural Networks: A Survey and Introduction. *IEEE Transactions on Pattern Analysis and Machine Intelligence* (2023) (cited on page 550).

[125]   N. Brustle et al. Quantum and Classical Algorithms for Nonlinear Unitary Dynamics via Carleman Linearization. *Quantum* (2025) (cited on pages 592, 596).

[126]   Hari Krovi. Improved Quantum Algorithms for Linear and Nonlinear Differential Equations. (2023) (cited on pages 592, 595, 596).

[127]   D. Shi et al. Koopman Spectral Linearization vs. Carleman Linearization: A Computational Comparison Study. *Mathematics* 12.14 (2024). DOI: 10.3390/math12142156 (cited on pages 592, 596).

[128]   Mike B. Giles. Multilevel Monte Carlo Path Simulation. *Operations Research* 56.3 (2008). DOI: 10.1287/opre.1070.0496 (cited on pages 593, 594, 596).

[129]   Desmond J. Higham. An Algorithmic Introduction to Numerical Simulation of Stochastic Differential Equations. *SIAM Review* 43.3 (2001). DOI: 10.1137/S0036144500378302 (cited on page 593).

[130]   Gilles Brassard et al. Quantum Amplitude Amplification and Estimation. 305 (2002) (cited on pages 594, 596).

[131]   S. Jaques. QRAM: A Survey and Critique. *Quantum* (2025) (cited on page 595).

[132]   Nai-Hui Chia et al. Sampling-Based Sublinear Low-Rank Matrix Arithmetic Framework for Dequantizing Quantum Machine Learning. *Proceedings of STOC 2020* (2020) (cited on pages 595, 596).

[133]   Dominic W. Berry. Quantum Algorithm for Linear Differential Equations with Exponentially Improved Dependence on Precision. *Communications in Mathematical Physics* 356.3 (2017). DOI: 10.1007/s00220-017-3002-y (cited on pages 595, 596).

[134]   M. Chaudhary et al. A Practical Quantum Solver for Multidimensional Partial Differential Equations. *Proceedings of ACM* (2025) (cited on page 595).

[135]   A. Novikau et al. Quantum Singular Value Transformation for Solving Partial Differential Equations. (2025) (cited on page 595).

[136]   Tao Xin et al. Quantum Algorithm for Solving Linear Differential Equations: Theory and Experiment. *Physical Review A* 101.3 (2020). DOI: 10.1103/PhysRevA.101.032307 (cited on page 596).

[137]   Scott Aaronson and Yuxuan Zhang. On verifiable quantum advantage with peaked circuit sampling. *Preprint* arXiv: 2404.14493. (2024) (cited on pages 601, 602, 605, 610, 614, 617, 618, 620).

[138]   Seth Lloyd. Universal Quantum Simulators. *Science* 273.5278 (1996). DOI: 10.1126/science.273.5278.1073 (cited on page 603).

[139]   Frank Arute et al. Quantum supremacy using a programmable superconducting processor. *Nature* 574.7779 (2019) (cited on pages 603, 617, 620).

[140] Google Quantum AI and Collaborators. Observation of constructive interference at the edge of quantum ergodicity. *Nature* 646 (2025). DOI: 10.1038/s 41586-025-09526-6 (cited on pages 603, 604, 607, 617, 620).

[141] Hrant Gharibyan et al. Heuristic Quantum Advantage with Peaked Circuits. *Preprint* arXiv: 2510.25838. (2025) (cited on pages 605, 610, 617, 620).

[142] Minzhao Liu et al. Certified randomness amplification by dynamically probing remote random quantum states. *Preprint* arXiv: 2511.03686. (2025) (cited on pages 605, 611, 618–620).

[143] Stephen P. Jordan et al. Optimization by decoded quantum interferometry. *Nature* 646 (2025). DOI: 10.1038/s41586-025-09527-5 (cited on pages 608, 609).

[144] Youngseok Kim et al. Evidence for the utility of quantum computing before fault tolerance. *Nature* 618.7965 (June 2023). ISSN: 1476-4687. DOI: 10.1038 /s41586-023-06096-3 (cited on page 612).

[145] Hongye Yu, Yusheng Zhao, and Tzu-Chieh Wei. Simulating large-size quantum spin chains on cloud-based superconducting quantum computers. *Physical Review Research* 5.1 (Mar. 2023). ISSN: 2643-1564. DOI: 10.1103/physrevre search.5.013183 (cited on page 612).

[146] Toshiki Yasuda et al. Quantum reservoir computing with repeated measurements on superconducting devices. *Preprint* arXiv: 2310.06706. (2023) (cited on page 612).

[147] Edward H. Chen et al. Realizing the Nishimori transition across the error threshold for constant-depth quantum circuits. *Preprint* arXiv: 2309.02863. (2023) (cited on page 612).

[148] Roland C. Farrell et al. Scalable Circuits for Preparing Ground States on Digital Quantum Computers: The Schwinger Model Vacuum on 100 Qubits. *Preprint* arXiv: 2308.04481. (2024) (cited on page 612).

[149] Elisa Bäumer et al. Efficient Long-Range Entanglement Using Dynamic Circuits. *PRX Quantum* 5.3 (Aug. 2024). ISSN: 2691-3399. DOI: 10.1103/pr xquantum.5.030339 (cited on page 612).

[150] Jacob Biamonte et al. Quantum machine learning. *Nature* 549 (2017) (cited on pages 619, 663).

[151] Srinivasan Arunachalam and Ronald de Wolf. A Survey of Quantum Learning Theory. *Preprint* arXiv: 1701.06806. (2017) (cited on page 619).

[152] Hsin-Yuan Huang et al. Quantum advantage in learning from experiments. *Science* 376.6598 (2022). DOI: 10.1126/science.abn7293 (cited on page 619).

[153] Valentin Gebhart et al. Learning quantum systems. *Nature Reviews Physics* 5.3 (2023). DOI: 10.1038/s42254-022-00552-1 (cited on page 619).

[154] Christian L. Degen, Friedemann Reinhard, and Paola Cappellaro. Quantum sensing. *Reviews of Modern Physics* 89 (2017). DOI: 10.1103/RevModPhys .89.035002 (cited on page 619).

[155] Luca Pezzè et al. Quantum metrology with nonclassical states of atomic ensembles. *Reviews of Modern Physics* 90 (2018). DOI: 10.1103/RevModPhy s.90.035005 (cited on page 619).

[156] Vittorio Giovannetti, Seth Lloyd, and Lorenzo Maccone. Advances in quantum metrology. *Nature Photonics* 5 (2011). DOI: 10.1038/nphoton.2011.35 (cited on page 619).

[157] Nabeel Aslam et al. Quantum sensors for biomedical applications. *Nature Reviews Physics* 5.3 (2023). DOI: 10.1038/s42254-023-00558-3 (cited on page 619).

[158] Valerio Scarani et al. The security of practical quantum key distribution. *Reviews of Modern Physics* 81 (2009). DOI: 10.1103/RevModPhys.81.1301 (cited on page 619).

[159] Nicolas Brunner et al. Bell nonlocality. *Reviews of Modern Physics* 86 (2014). DOI: 10.1103/RevModPhys.86.419 (cited on page 619).

[160] Stephanie Wehner, David Elkouss, and Ronald Hanson. Quantum internet: A vision for the road ahead. *Science* 362.6412 (2018). DOI: 10.1126/science.aam9288 (cited on page 619).

[161] William Kretschmer et al. Demonstrating an unconditional separation between quantum and classical information resources. *Preprint* arXiv: 2509.07255. (2025) (cited on pages 619, 620).

[162] Daniel Litinski. Magic State Distillation: Not as Costly as You Think. *Quantum* 3 (Dec. 2019). ISSN: 2521-327X. DOI: 10.22331/q-2019-12-02-205 (cited on pages 636, 659).

[163] Avimita Chatterjee, Koustubh Phalak, and Swaroop Ghosh. Quantum Error Correction For Dummies. *Preprint* arXiv: 2304.08678. (2023) (cited on page 658).

[164] César Benito et al. Comparative study of quantum error correction strategies for the heavy-hexagonal lattice. *Quantum* 9 (2025). DOI: 10.22331/q-2025-02-06-1623 (cited on page 659).

[165] Chetan Nayak et al. Non-Abelian anyons and topological quantum computation. *Rev. Mod. Phys.* 80 (3 Sept. 2008). DOI: 10.1103/RevModPhys.80.1083 (cited on page 659).

[166] Ady Stern and Netanel H. Lindner. Topological Quantum Computation—From Basic Concepts to First Experiments. *Science* 339.6124 (2013). DOI: 10.1126/science.1231473 (cited on page 659).

[167] Sergey Bravyi et al. High-threshold and low-overhead fault-tolerant quantum memory. *Preprint* arXiv: 2308.07915. (2023) (cited on page 659).

[168] Andrew Wack et al. Quality, Speed, and Scale: three key attributes to measure the performance of near-term quantum computers. *Preprint* arXiv: 2110.14108. (2021) (cited on page 659).

[169] D.-S. Wang. State-adaptive quantum error correction and fault-tolerant quantum computing. *Preprint* arXiv: 2508.06011. (2025) (cited on page 659).

[170] Ashley Montanaro. Quantum algorithms: An overview. *NPJ Quantum Information* 2 (2016). DOI: 10.1038/npjqi.2015.23 (cited on page 663).

[171] John Preskill. Quantum computing in the NISQ era and beyond. *Quantum* 2 (2018). DOI: 10.22331/q-2018-08-06-79 (cited on page 663).

[172] I. M. Georgescu, S. Ashhab, and Franco Nori. Quantum simulation. *Rev. Mod. Phys.* 86 (2014). DOI: 10.1103/RevModPhys.86.153 (cited on page 663).

[173] Alexander M. Dalzell et al. Quantum algorithms: A survey of applications and end-to-end complexities. *Preprint* arXiv: 2310.03011. (2023) (cited on page 663).

List of Figures

List of Tables

Journey Forward

A Journey Well Traveled

Congratulations on completing this book. If you engaged with the exercises and problem sets, you have not only read about quantum algorithms but also practiced essential habits: stating assumptions clearly, tracking resources, and reasoning carefully about error, precision, and success probability.

Across the chapters, a unifying framework emerges. Landmark algorithms rely on recurring primitives—Fourier-based estimation, amplitude amplification and estimation, Hamiltonian simulation, and polynomial-approximation methods via block encodings—while applications arise by combining these primitives with problem structure, input/output models, and resource constraints. With these patterns in mind, you can move from a problem statement to an algorithmic workflow, identify dominant costs, and assess when a speedup can translate into practical gains.

This book prioritizes concepts and principles over any particular programming stack. Tools will evolve, but clear understanding carries a long way. If you can articulate an algorithm's assumptions, access model, and resource drivers, you can adapt it across frameworks and evaluate new proposals as the field advances.

From Algorithms to Applications

Quantum computing is a field in flux. Noise, calibration drift, connectivity limits, and measurement overhead constrain what is practical today, while fault tolerance drives new architectures and sharper cost analyses. The key insight is that quantum advantage is end-to-end: it hinges on data access, encoding choices, precision requirements, and verification methods, not only on asymptotic scaling. The application chapters therefore emphasize transferable templates—from physics and chemistry simulation to optimization, quantum machine learning, and solvers for differential equations and linear systems—where familiar primitives recur under different bottlenecks and output criteria.

A Closing Note

The narrative of quantum computing continues to unfold, and you are equipped to contribute. The most meaningful advances will come from bridging theory and practice: translating applications into precise assumptions, choosing appropriate primitives, and assessing feasibility with candor. Read widely, test ideas against real constraints, and keep returning to fundamentals.

Thank you for embarking on this path. May curiosity, rigor, and the confidence to build guide your next steps.